National Audubon Society®
Field Guide to
North American Butterflies

A Chanticleer Press Edition

National Audubon Society®
Field Guide to
North American Butterflies

Robert Michael Pyle,
Consulting Lepidopterist,
International Union for Conservation
of Nature and Natural Resources

Visual Key by
Carol Nehring and Jane Opper

Alfred A. Knopf, New York

This is a Borzoi Book.
Published by Alfred A. Knopf, Inc.

Copyright © 1981 by Chanticleer
Press, Inc. All rights reserved under
International and Pan-American
Copyright Conventions. Published in
the United States by Alfred A. Knopf,
Inc., New York, and simultaneously in
Canada by Random House of Canada,
Limited, Toronto. Distributed by
Random House, Inc., New York.

www.randomhouse.com

Knopf, Borzoi Books, and the
colophon are registered trademarks
of Random House, Inc.

Prepared and produced by
Chanticleer Press, Inc., New York.

Color reproductions by Nievergelt Repro
AG, Zurich, Switzerland. Typeset in
Garamond by Dix Type, Inc., Syracuse,
New York. Printed and bound by Toppan
Printing Co., Ltd., Tokyo, Japan.

Published July 30, 1981
Seventeenth printing, August 2004

Library of Congress Cataloging-in-
Publication Number: 80-84240
ISBN: 0-394-51914-0

CONTENTS

11 Introduction
31 How to Use This Guide

Part I Color Plates
35 Key to the Color Plates
36 Thumb Tab Guide
45 *Color Plates*
Eggs, Caterpillars,
and Chrysalises 1–51
White Butterflies 52–84
Sulphurs 85–132
Folded-winged Skippers 133–240
Spread-winged Skippers 241–315
Swallowtails 316–360
Angled-winged Butterflies 361–381
Hairstreaks and Elfins 382–462
Blue Butterflies 463–510
Copper Butterflies 511–540
Metalmarks 541–555
Checkered Butterflies 556–591
Fritillaries and Orange
Patterned Butterflies 592–642
Boldly Patterned Butterflies 643–687
Eyespot Patterned Butterflies 688–759

Part II Text
321 Swallowtails and Parnassians
350 Whites and Sulphurs
400 Gossamer Wings
518 Metalmarks
534 Snout Butterflies
537 Brush-footed Butterflies
662 Satyrs or Browns

711 Milkweed Butterflies
716 True Skippers
846 Giant Skippers

Part III Appendices
857 Glossary
865 Butterfly-Watching Tips
869 How to Photograph Butterflies
873 Picture Credits
881 Host Plant Index
901 Butterfly Index

NATIONAL AUDUBON SOCIETY

The mission of NATIONAL AUDUBON SOCIETY, *founded in 1905, is to conserve and restore natural ecosystems, focusing on birds, other wildlife, and their habitats for the benefit of humanity and the earth's biological diversity.*

One of the largest, most effective environmental organizations, AUDUBON has nearly 550,000 members, numerous state offices and nature centers, and 500 + chapters in the United States and Latin America, plus a professional staff of scientists, educators, and policy analysts. Through its nationwide sanctuary system AUDUBON manages 160,000 acres of critical wildlife habitat and unique natural areas for birds, wild animals, and rare plant life.

The award-winning *Audubon* magazine, which is sent to all members, carries outstanding articles and color photography on wildlife, nature, environmental issues, and conservation news. AUDUBON also publishes *Audubon Adventures,* a children's newspaper reaching 450,000 students. Through its ecology camps and workshops in Maine, Connecticut, and Wyoming, AUDUBON offers nature education for teachers, families, and children; through *Audubon Expedition Institute* in Belfast, Maine, AUDUBON offers unique, traveling undergraduate and graduate degree programs in Environmental Education.

AUDUBON sponsors books and on-line nature activities, plus travel programs to exotic places like Antarctica, Africa, Baja California, the Galápagos Islands, and Patagonia. For information about how to become an AUDUBON member, subscribe to *Audubon Adventures,* or to learn more about any of its programs, please contact:

NATIONAL AUDUBON SOCIETY
Membership Dept.
700 Broadway
New York, NY 10003
(800) 274-4201
(212) 979-3000
http://www.audubon.org/

THE AUTHOR

Robert Michael Pyle is the author of
*The Audubon Society Handbook for
Butterfly Watchers, Watching Washington
Butterflies,* and many papers and articles
on butterflies and conservation. He
founded the Xerces Society, an
international organization for beneficial
and rare insect habitat protection, and
has served as Northwest Land Steward
for The Nature Conservancy. A
consulting lepidopterist, Pyle has
worked for the government of Papua
New Guinea on giant birdwing
butterfly conservation. Currently he is
consultant for the International Union
for Conservation of Nature and Natural
Resources and chairman of The
Monarch Project of the Xerces Society.

ACKNOWLEDGEMENTS

This book would not have been possible without the encouragement of many people. The authority of the text owes a great deal to Paul A. Opler, Office of Endangered Species, U.S. Fish and Wildlife Service, and Lee D. Miller, Allyn Museum of Entomology, who read the butterfly manuscript and made many valuable suggestions on the original edition as well as the revision; and to C. Don MacNeill, Department of Natural Sciences, Oakland Museum, and Ray E. Stanford, Research Associate, Denver Museum of Natural History, who did the same for the entire skipper text. John M. Burns, U.S. National Museum of Natural History, Smithsonian Institution, and Paul A. Opler and Lee D. Miller kindly furnished material for illustration as well as time-consuming identifications of difficult species. Charles L. Remington, The Yale Peabody Museum of Natural History, is also to be thanked for lending specimens. Significant aid with difficult groups was given me by John Lane, Santa Cruz Museum, and Larry Gall, Yale University, on gossamer wings, and by James A. Scott, University of Colorado at Boulder, on checkerspots and fritillaries. Donald Harvey, Allyn Museum of Entomology, helped with the life cycle accounts. Among other colleagues, I would especially like to thank Clifford D. Ferris, Jonathan P. Pelham, John Hinchliff, Rudy Mattoni, Karolis Bagdonas, Frank R. Hedges, Michael G. Morris, John Heath, Jeremy A. Thomas, Charles V. Covell, Jr., Jo Brewer, Lincoln P. Brower, Thomas C. Emmel, and F. Wayne King. Additionally, for the revised edition, I would like to thank Tim Friedlander for his help with the *Asterocampa* group, Dale W. Jenkins for the *Hamadryas* and *Doxocopa* sections; and Kenelm W. Philip for arctic butterflies. I would also like to recognize the contributions of the following authors: Alexander B. Klots; William H. Howe and his coauthors; Thomas C. and John F. Emmel, F. Martin Brown with Donald Eff and Bernard Rotger, Paul R. and Anne F. Ehrlich, William D. Field, Arthur Shapiro, Harrison M. Tietz, W. J. Holland, Ernst J. Dornfeld, John M. Burns, C. Don MacNeill, Lucien Harris, Jr., Charles P. Kimball, James A. Ebner, Larry J. Orsak, Cyril F. Dos Passos, and Roy O. Kendall. To my literary agent, Barbara Williams, I owe a special debt of gratitude. The most unselfish encouragement was given me by Marvyne Betsch, a true friend of butterflies. My wife, Sarah Anne Hughes, made the entire project possible with her active collaboration, suggestions, and especially her patience. I am very thankful to those at Chanticleer Press. Paul Steiner, Milton Rugoff, and Gudrun Buettner provided constant encouragement; Susan Rayfield developed the idea for the book. I am grateful to Jane Opper and Olivia Buehl, who with the assistance of Ann Whitman edited the text, and to Carol Nehring, who supervised the art and layouts. My thanks also goes to Helga Lose, who saw the book through production.

INTRODUCTION

Depicted in symbols and art since the Bronze Age, butterflies are among the most fascinating and beautiful animals. They live nearly everywhere—from gardens and mountains to acid bogs and arctic tundra. About 10,000 to 20,000 species occur worldwide; of these, almost 700, including occasional strays, are found in North America north of Mexico.

If approached carefully, butterflies are easier to observe than birds, and their variety is much less complex than the bewildering array of wildflowers. Yet butterflies display a range of behavior as fascinating as that of birds, and all the colors and brilliance of flowers. The life cycle of butterflies is one of the fundamental miracles of nature. And the presence of various species can tell us much about our environment.

This book is designed for everyone who wants to know how to identify butterflies in backyards, parks, and gardens, as well as in woods and fields. The combination of vivid color photographs and clear, nontechnical descriptions should make butterfly identification easy and enjoyable for all.

Geographic Scope: Most guides to butterflies have been regional in scope or have limited their coverage to selected groups of

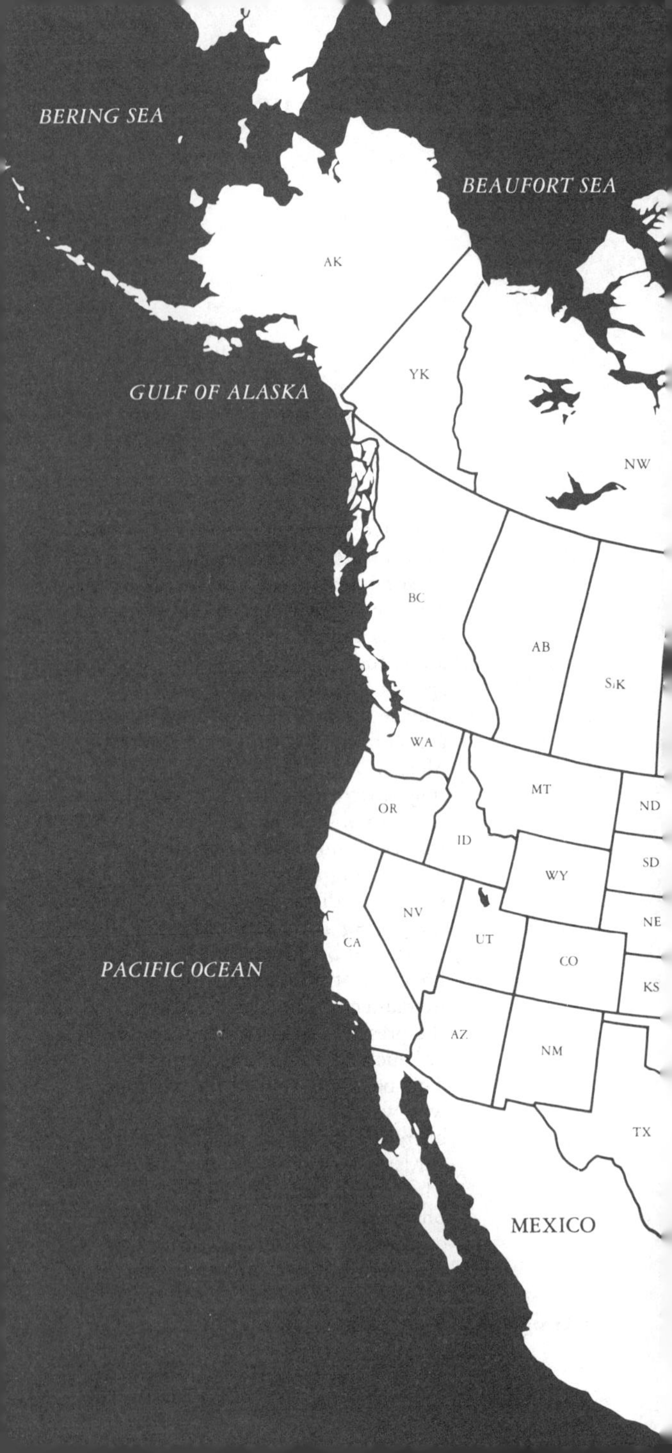

BERING SEA

BEAUFORT SEA

GULF OF ALASKA

PACIFIC OCEAN

AK

YK

NW

BC

AB

S₁K

WA

MT

ND

OR

ID

WY

SD

NV

UT

CO

NE

CA

KS

AZ

NM

TX

MEXICO

GREENLAND

BAFFIN BAY

HUDSON BAY

MB

ON

QU

NF

MN

WI

MI

NB

ME

NS

VT

NH

IA

IL

IN

OH

NY

MA

CT

RI

PA

MO

KY

WV

VA

MD

NJ

DE

ATLANTIC OCEAN

K

AR

TN

NC

MS

AL

GA

SC

LA

FL

GULF OF MEXICO

CARIBBEAN SEA

butterflies. This is the first field guide to cover, in one volume, all the native and introduced butterflies found in North America north of Mexico, as well as many emigrants, strays, and the native butterfly fauna of the Hawaiian Islands.

What Is a
Butterfly?

Butterflies are insects that belong to the large animal phylum Arthropoda, which includes all the jointed-leg invertebrates. In common with all insects (Class Insecta), butterflies have 6 jointed legs, 3 body segments, and 2 antennae. Along with the more numerous moths, butterflies make up the order Lepidoptera, whose members differ from all other insects in having scales over all or most of their wings, and often on the body as well. Within the order, butterflies and moths are separated from each other by their wing venation, body structure, and habits. Butterflies fly during the day, while most moths are nocturnal. Butterflies at rest tend to hold the wings vertically over the back; in contrast, moths may either fold the wings tentlike over the back, or wrap them around the body, or extend them to the sides. Virtually all butterflies have knoblike clubs at the tips of the antennae; moths lack antennal clubs.

Butterflies are divided into two superfamilies: the true butterflies (Papilionoidea) and the skippers (Hesperioidea). True butterflies tend to have narrow bodies, long antennae, and brightly colored, full wings. Skippers are stocky, compact, and hairy: their short triangular wings are most often a shade of tawny-orange, brown, black, or gray. Butterflies are further broken down into families, genera, and species. The families comprise large groups of butterflies with a number of common characteristics. For example, both hairstreaks and elfins belong to the family of Gossamer Wings. In North

America, there are 10 families. Genera are smaller groups of structurally related butterflies; the cloudywing skippers thus all belong to the genus *Thorybes*. Within each genus are various species, such as the Tiger Swallowtail or the Mourning Cloak. Members of a species breed with one another to produce similar offspring: there is relatively little hybridizing between species in nature. The species are further broken down into numerous subspecies, which are more or less distinct geographic populations. For example, the coloring of the Callippe Fritillary varies considerably among populations: the typical butterfly is yellow-brown, but the population in northern California and southern Oregon is orange-brown, while that found in the California Coast Ranges is dark brown with pale yellow spots. In this guide, we have included only those subspecies that are exceptionally distinctive.

Parts of a Butterfly:
An adult butterfly has 3 main body segments—the head, thorax, and abdomen—as well as 4 wings.

Head:

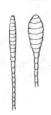

The head bears 2 clubbed antennae, 2 large, compound eyes, 2 furry palpi, and a long, coiled proboscis. The antennal clubs may be quite rounded, barely thickened, or, in skippers, hooked at the end like a shepherd's crook. Antennae are probably used for smelling as well as for touching and orientation. The globular, compound eyes are made up of thousands of facets, or ommatidia; each is a tiny lens linked to the optic nerve. Butterflies are thought to receive a single, integrated image in color, and they are sensitive to the ultraviolet range of light as well as to visible light. In front of and below the eyes protrude 2 furry palpi which receive sensations and protect the coiled, tonguelike proboscis. The

Parts of a Butterfly

fore wing

fore wing

antennae

hind wing

thorax

abdomen

fore leg

hind leg

tarsus

middle leg

Tiger Swallowtail

proboscis is constructed of 2 parallel, linked tubes, which work like a pair of drinking straws; the proboscis can be coiled up tightly against the face or extended fully to drink nectar or other liquids.

Body: Behind the head is a thick, muscular thorax. Its 3 segments bear the 6 legs and 4 wings. The abdomen is a long tube made of 11 segments behind the thorax; it contains the digestive and sexual apparatus as well as a row of spiracles along each side for the intake of air. The genitalia are located at the end of the abdomen; because they vary little within a species but considerably between species, specialists use the genitalia for classification. Studying these organs in detail, however, often requires dissection and magnification. The males possess a pair of grasping organs known as claspers which may be obvious, especially in swallowtails, or very hairy, as in the brush-footed butterflies. Females have slitlike genitalia, which in most species are hard to see. The abdomen of the female tends to be thicker and heavier due to a load of eggs.

Legs: Butterfly legs, like those of all arthropods, are jointed. The 5 sections end in clawed tarsi that serve as feet. Brush-footed butterflies and the males of some other families have the fore legs reduced to tiny paws that are useless for walking. Butterfly tarsi possess a sense similar to taste: tarsal contact with sweet liquids such as nectar causes the proboscis to uncoil, and females often scratch plants with the tarsi to find the proper host plant on which to lay eggs.

Wings: The 4 wings arise from the thorax above the legs. Fashioned of 2 taut membranes with rigid veins, the wings are used for flight and as color guides for purposes of sex recognition, camouflage, and mimicry. The fore wings are larger and longer than the hind wings, which are more rounded.

Parts of Butterflies

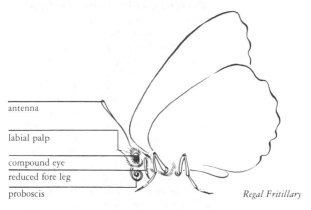

antenna

labial palp

compound eye

reduced fore leg

proboscis

Regal Fritillary

Fore wing (FW)

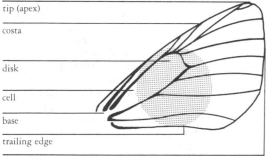

tip (apex)

costa

disk

cell

base

trailing edge

outer margin

Generalized Pierid

Hind wing (HW)

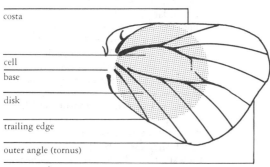

costa

cell

base

disk

trailing edge

outer angle (tornus)

outer margin

Generalized Pierid

Wing shape varies from triangular to nearly round, squarish, hooked, clipped, or long and drawn out. Millions of shinglelike, overlapping scales give the wings their colors and patterns. Solid colors derive from pigmented scales, whereas metallic, iridescent hues come from faceted scales that, like a rainbow, refract light. Among the many scales are sex scales, called androconia, that lie in thick velvety patches or in sharp black bars, called stigmata. The sex scales produce scent hormones, or pheromones, which generate odors that attract mates and aid in sex recognition.

Metamorphosis: After mating, female butterflies lay eggs either singly, or in rows, chains, or clusters of a few to several hundred eggs. The egg shape and texture varies greatly among species, ranging from spherical, flattened, or conical to smooth, ribbed, or ornamented with raised designs. Some eggs do not hatch until the following spring, while others hatch before winter. The caterpillar, or larva, has simple eyes, chewing jaws, and 3 pairs of jointed legs near the front as well as 5 pairs of grasping prolegs near the rear. Like the adult, the caterpillar has a row of spiracles along each side of the body. The caterpillar spends its life feeding; the more it consumes, the larger it grows. But because its skin cannot stretch, the caterpillar grows by molting or shedding its skin several times—each stage, called an instar, is larger than the previous one. The final molt produces the chrysalis, or pupa; it is a resting stage and does not feed. The chrysalises of most butterflies are naked, unlike those of moths, which are protected by a silken cocoon. Butterfly caterpillars can produce silk, which they use to bind leaves together for a shelter, which may be used either by the caterpillar or later by the

Egg

Caterpillar

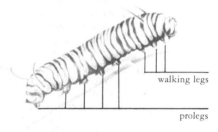

walking legs

prolegs

Chrysalis

cremaster

Emerging adult

chrysalis. The chrysalises of many butterflies hang by the tail end from a silken pad called the cremaster; others hang upright, supported by a silken girdle. Some chrysalises are green or brown, resembling leaves, stems, thorns, or bits of wood. Others are brightly variegated and covered with thornlike bumps or tubercles. As the adult butterfly begins to form inside the chrysalis, the shape of the compacted wings as well as the features of the head and body are visible in the surface of the chrysalis. When the adult is fully formed, the chrysalis' skin splits open, permitting the butterfly to crawl out. It soon begins to pump fluids from its swollen body into its shrunken wings. This transformation from egg to adult butterfly is known as complete metamorphosis. It is the life cycle of a butterfly.

As soon as the adult can fly, courtship begins, sometimes involving elaborate dances, prenuptial flights, or mutual wing strokings. Mating usually lasts several hours, and often occurs while the pair are flying. The adult's lifespan ranges from a week to 6 or 8 months, depending upon the species, with most averaging 2 weeks. In warm regions, several broods of a species are produced each year, but in mountains or the Arctic, it may require 2 years for a single brood to mature.

Survival: A combination of many factors affects the distribution of butterflies—the plants they eat, their tolerance of temperature and moisture, their ability to reach new areas, and changes in the landscape. Many species cannot withstand northern winters yet they emigrate annually to the North, where they produce one or more broods before dying. Other species emigrate in vast swarms because of population or resource pressures, or other unknown reasons, and do not return. Only the

Monarch has an annual, back-and-forth true migration.

Because butterflies are subject to many kinds of predators and parasites at all stages of their lives, they have evolved an impressive array of protective adaptations. Eggs, caterpillars, chrysalises, and even adult butterflies may blend superbly with their environment or resemble inedible objects such as bird droppings. Caterpillars may have fierce-looking horns and eyespots that deter small predators. Many adult butterflies bear small eyespots near the wing edges or conspicuous tails; these may help to deter the attack of birds and lizards from vital body parts. Birds learn to recognize the color of toxic species, like the Monarch, and to avoid them. Some butterflies have even evolved color patterns which fool predators. For example, edible butterflies such as the Red-spotted Purple gain protection from birds because they mimic distasteful species such as the Pipevine Swallowtail. Anglewings and some other butterflies, when seen from below, resemble leaves or bark but they flash brilliant and startling upperside colors when a predator comes too near. Perhaps no other group of animals demonstrates more vividly the evolution of defenses for survival.

Environment: All of the adaptive strategies may be useless if the habitats of butterflies are ruined. Both the caterpillars and adult butterflies require specific kinds of plants and habitats. For this reason, butterflies are especially vulnerable to land development. Widespread use of pesticides further reduces butterfly populations. Many North American species are declining in range and in numbers. At least one species, the Xerces Blue, has already become extinct through alteration of its habitat, while a number of others have been listed by the U.S. Fish and Wildlife Service as Threatened or Endangered species. Prairie grasslands, wetlands such as bogs and marshes, sand dunes, and virgin forests are among the most imperiled butterfly habitats. The loss of butterfly species may eventually affect our lives. Butterflies are significant plant pollinators as well as indicators of ecological quality in our environment. They fill important roles in the food chain, and very few species compete for plants that people value.

Observing Butterflies: In the past, most butterfly enthusiasts maintained collections of butterflies as part of their hobby. Today, however, many readers will prefer to know butterflies by learning how they live, and by watching, photographing, and rearing them. Butterfly behavior in some species is still little known; here is an area where observers can make genuine contributions. In many cases, the photographs in this guide represent the first photographs ever taken of a species. The general rules for butterfly watching and photography are essentially the same: haunt sunny, flowery places, be observant, and move very quietly and slowly. Patience is required but it is sure to be richly rewarded.

Another way to enjoy butterflies is to rear them from eggs or caterpillars. Few nature activities are more satisfying. Caterpillars or eggs may be easily brought in from their natural habitats. But after the new butterflies emerge and dry their wings, they should be returned to their preferred environment. By cultivating nectar flowers and host plants, it is also possible to attract many butterflies. A number of native species may quickly become established, and others will visit flowers for nectar, if not to breed. A good start can be made by letting some nettles and carrot plants grow in a chosen area, and then by planting buddleia and phlox in between. This arrangement will surely attract Red Admirals and swallowtails to the garden. Knowing the caterpillar's host plant as well as the adult butterfly's preferred nectar sources will help you determine which plants to cultivate. You will find all of this information in the Life Cycle section and comments of the species accounts.

Collecting butterflies is still a very important part of lepidoptera research. Nevertheless, those who collect for study should follow laws concerning endangered species and rules of common sense and consideration for the resource. Without these guidelines, we may lose some of our most treasured butterfly species.

How to Identify Butterflies:

With almost 700 species in North America, identifying the different butterflies may seem bewildering. The new photographic approach in this guide enables you to compare the butterflies you encounter with vivid color photographs of living butterflies in their natural habitats. Most other field guides to butterflies are illustrated with drawings or paintings of dead specimens, and are arranged by scientific order. In this guide you can quickly find the butterfly you have seen because the photographs are arranged by shape and color. Moreover, the photographs show the actual perching position and true coloring of the butterfly as it appears in life. For species with many variations or with differences between the sexes, more than one photograph has often been used. Many color plates also show both the wings from above and the wings from below. The photographs have been arranged so that similar-looking butterflies are close together. Nevertheless, distinguishing certain groups of look-alikes, such as some checkerspots, green hairstreaks, and skippers, may confound even the experts. Identifying the genus of these butterflies should satisfy most observers. Further identification to species often requires dissection of specimens by experts. Our arrangement of full color photographs by shape and color will enable you to identify most species and allow you to place all the butterflies in a given area at least among their close relatives.

Organization of the Color Plates: We have arranged the photographs of butterflies according to the features you see in the field—shape and color. For example, the group called Swallowtails contains, in addition to members of the swallowtail family, butterflies whose coloring mimics that of swallowtails. Similarly, the group of Checkered Butterflies is organized according to variations in their checkered patterns. The color plates are arranged in the following order:

Eggs, Caterpillars, and Chrysalises
White Butterflies
Sulphurs
Folded-winged Skippers
Spread-winged Skippers
Swallowtails
Angled-winged Butterflies
Hairstreaks and Elfins
Blue Butterflies
Copper Butterflies
Metalmarks
Checkered Butterflies
Fritillaries and Orange Patterned Butterflies
Boldly Patterned Butterflies
Eyespot Patterned Butterflies

Thumb Tab Guide: The grouping of the color plates is explained in a table preceding that section. A silhouette of a typical member of each group appears on the left. Silhouettes of butterflies within that group are shown on the right. For example, the silhouette of a hairstreak represents the group Hairstreaks and Elfins. This representative silhouette is inset on a thumb tab at the left edge of each double page of color plates devoted to that group of butterflies.

Captions: The caption under each photograph gives the plate number, common name, measurement, and page number of the text description. The measurement in inches indicates wingspan for butterflies and maximum body length for caterpillars (*ct*) and chrysalises (*ch*). Because butterfly eggs are extremely small, often far smaller than the fractions on rulers, we have used millimeters in the captions to help you compare relative sizes. (Fraction equivalents in inches are given in the text.) In some cases, male (δ), female ($\female$), wings above, or dorsal, (*d*), and wings below, or ventral, (*v*), are also indicated. The color plate number is repeated at the beginning of each text description.

Organization of the Text: The families, genera, and species are arranged in phylogenetic sequence, beginning with the most primitive and ending with the most advanced. However, because true butterflies are more familiar to most people than skippers, the text for true butterflies precedes that for the skippers even though skippers precede true butterflies in the phylogenetic order. The sequence of genera and species within true butterflies and skippers follows the phylogenetic order.

Each species account begins with the number of the color plate or plates, followed by the common name. Beneath this is the scientific name, always italicized, with the genus name capitalized and the species name in lower case. For example, the scientific name of the Monarch is *Danaus plexippus*.

Common Names: The common names of butterflies have never been standardized; a joint committee of the Lepidopterists' and Xerces societies is now working on an official list. The common names used here reflect prevailing usage. In a number of cases, where there is no suitable common name, a new name has been introduced.

Scientific Names: In this guide, the scientific names of species and genera conform with "A Catalogue/Checklist of the Rhopalocera of America North of Mexico," by Lee D. Miller and F. Martin Brown, published in the *Memoir of the Lepidopterists' Society,* No. 2, 1981. The classification of the hackberry butterflies (*Asterocampa* spp.) is based on the work of Tim Friedlander.

Description: Every description begins with the approximate wingspan measurement (both in inches and millimeters) from fore wing tip to fore wing tip of the fully spread adult butterfly. Females are usually larger than males but greater variation may occur than these average measurements suggest. For species that are comparatively small or large within a group of related butterflies, the measurement is followed by *tiny, small,* or *large.* Next comes the wing description, usually starting with the upper surfaces of the fore wings (FW) and hind wings (HW), then moving to the lower surfaces of the wings. Wing shape and any readily visible body structures are mentioned whenever these are important for identification. Prominent features are italicized for quick reference. The description refers to an average, fresh butterfly; however, much regional variation occurs, and often as the butterfly ages, it fades, or loses scales, changing the intensity and shade of colors. If the sexes are different, both are described. We have avoided technical details of structure which cannot be readily detected in the field.

Wherever possible, familiar terms are used; technical terms are defined in the glossary or illustrated with labeled drawings in the introduction. In a few cases, where no outstanding photograph was available, line drawings appear in the margins of the text description.

Similar Species: To help you distinguish between similar-looking species occurring in the same range, we describe the differences between the two species.

Life Cycle: This section describes briefly the egg, caterpillar, and chrysalis, and indicates the caterpillar's host plants. Sometimes, however, the immature stages have never been described, or the host plant may not be known in North America. In these cases, the text states that the life cycle is "undescribed," "unknown," or "unreported." Unpublished information generally has not been included.

Flight: The flight period of adult butterflies varies dramatically with altitude, latitude, and seasonal weather conditions. The months given are those in which the butterfly probably flies in most of its range or in a specific area. Whenever known, the number of broods flying each year precedes the flight period. If flight periods are not broken into separate groups, the sequence represents a collective time for all broods and is not necessarily a continuous period for a single brood.

Habitat: Many butterfly species frequent only certain kinds of environments. Habitats are therefore described as specifically as possible. If separate populations within the same species dwell in different habitats, these distinctions are also indicated. The habitats of wandering butterflies are obviously less well-defined than those of sedentary species.

Range: The range descriptions generally begin in the Northwest and then move eastward and southward to the southwestern and southeastern limits. Ranges outside North America are

summarized briefly. For emigrating
species, both breeding (resident) and
wandering ranges are given.

Comments: The species accounts conclude with
comments on the butterfly's behavior,
courtship patterns, nectar preferences,
and conservation status. Frequently,
the history of a butterfly's discovery
and the origin of its name, as well
as good places to view species are
also included. Very closely related
species which do not appear in separate
accounts are briefly described here,
including their range and the
caterpillar's host plant.

HOW TO USE THIS GUIDE

Example 1
*Butterfly along a
shoreline in the
South*

You see a brown butterfly with large purplish eyespots on the upperside of its wings. One eyespot on each fore wing and one eyespot on each hind wing are noticeably larger than the others.

1. Turn to the Thumb Tab Guide preceding the color plates and look for the silhouette that most resembles the butterfly you have seen. In the group called Eyespot Patterned Butterflies, you find the silhouette for buckeyes, calicoes, and the White Peacock, color plates 688–693, 758, 759.

2. Check the color plates. Two butterflies have the right coloring and eyespot patterns, the Buckeye and the West Indian Buckeye, color plates 688 and 689. The captions indicate the sizes and the text pages 630 and 632.

3. Reading the text, you find that only the Buckeye has large purplish eyespots. The West Indian Buckeye has smaller eyespots of nearly equal size, and they have little or no purplish shading. Your butterfly is a Buckeye.

Example 2
*Butterfly in a
vacant lot in the
Northwest*

You observe a boldly patterned orange butterfly resting on a flower. The tip of each fore wing is black, crossed by a prominent orange bar; the hind wings bear bluish spots. When the butterfly closes its wings, it reveals a marbled underside of pink, olive, tan, and white.

1. In the Thumb Tab Guide you find the group of Boldly Patterned Butterflies and you select the silhouette for hackberries, painted ladies, and the Red Admiral, color plates 661–666, 668–672.

2. Turning to the color plates, you narrow your choice to 3 photographs—the West Coast Lady, the American Painted Lady, and the Painted Lady, color plates 668–670. The captions refer you to text pages 623, 625, and 626.

3. Reading the text, you eliminate the American Painted Lady because it has 2 large blue eyspots on its underside, which your butterfly lacks. Of the other two, only the West Coast Lady has the orange bar across the black fore wing tip. You have seen a West Coast Lady.

Example 3
Butterfly in a meadow in the Midwest

You photograph an orange butterfly with thick black veins. It looks like a Monarch but you want to make sure of the identification.

1. Among the silhouettes preceding the color plates, you find the group of Fritillaries and Orange Patterned Butterflies. You select a silhouette that includes the Monarch, color plates 592–598.

2. Turning to the color plates, you compare the Monarch and the Viceroy, color plates 596 and 597. Your butterfly has the same curved black line across the hind wing as the Viceroy.

3. Checking the text on pages 637 and 711, you confirm your identification. Only the Viceroy has the black line across its hind wings.

SWALLOWTAILS AND PARNASSIANS
(Papilionidae)

More than 600 species worldwide;
fewer than 30 in North America.
Members of the family Papilionidae
vary greatly in appearance, from the
brilliant Australasian birdwings (largest
of all butterflies) to the pale, northern-
dwelling parnassians. The true
swallowtails, the most typical North
American members of this family, are
large, brightly colored butterflies, with
tailed hind wings. They have
wingspans of 2⅛–5½" (54–140 mm).
Many tropical swallowtails lack tails,
but one tropical group, the kites, has
the longest tails of all. In North
America, except for the 2 kites—both
zebra swallowtails—most swallowtails
fall into 4 general subgroups. Black
swallowtails (*Papilio* spp.) are black
with yellow spots that may become
broad bands; their caterpillars usually
feed on plants in the carrot family.
Giant swallowtails (*Heraclides* spp.) are
brown and yellow; their caterpillars
favor citrus plants. Tiger swallowtails
(*Pterourus* spp.) are generally yellow
with black stripes; their host plants are
mostly deciduous trees. Pipevine
swallowtails (*Battus* spp.) are blackish;
their caterpillars consume aristolochias,
woody vines whose pungent roots make
these butterflies distasteful to predators.
A few species, however, fit into none of
these groups.
All North American swallowtails have
somewhat spherical eggs. Most of the
caterpillars have prominent eyespots,
but members of a few genera lack
eyespots completely. All have an
osmeterium, an orange, red, or yellow
forked organ behind the head on the
back. This foul-smelling organ can be
turned inside out, and, along with the
eyespots, is thought to deter predators.
Swallowtail chrysalises resemble green

or brown bits of leaf or wood; they
hang upright through winter, slung by
a silken girdle around the middle.
The very different-looking parnassian
butterflies have wingspans of 1⅞–3″
(48–76 mm). Parnassians dwell in the
alpine and maritime West (as well as in
Europe and Asia), where the black and
yellow or black and orange caterpillars
feed on stonecrops, bleeding hearts, and
saxifrages, and pupate in primitive
cocoons. All female parnassians have a
sphragis, or waxy pouch, which forms
after the first mating and prevents
further coupling.
Swallowtails in the genera *Heraclides*
and *Pterourus* were formerly classified in
the genus *Papilio*.

Eversmann's Parnassian
(*Parnassius eversmanni*)

Description: 1⅞–2¼″ (48–57 mm). *Male yellow*
with 2 crimson spots on HW above and
below. *Female pale yellow or white above
with* gray scaling and transparent
margins on FW, and usually *3 red spots
on HW, the innermost spot extending to
inner HW edge.* Antennae black. *Female
has waxy white pouch (sphragis) at tip of
abdomen.*

Similar Species: Male Phoebus and Clodius parnassians
white. Female Phoebus has red spots on
FW and HW; HW spots of female
Clodius usually separate, rarely in band.

Life Cycle: Sluglike, downy caterpillar is black
with yellow spots; takes 2 years to reach
maturity, then pupates in a cocoon
among ground litter. Host plant is
corydalis (*Corydalis gigantea*).

Flight: 1 brood; June–July.

Habitat: Tundra and open mountain slopes.

Range: Alaska, Yukon Territory, NW. British
Columbia, and Northwest Territories.

The only yellow parnassian,
Eversmann's Parnassian lives in

northern Siberia, the northern islands of Japan, and Korea, as well as arctic North America. Only recently has it been found as far south as British Columbia.

70 Clodius Parnassian
(*Parnassius clodius*)

Description: 2⅜–3" (60–76 mm). Male *milk-white with* black checks, gray patches, and *red spots on HW above and below* varying in size according to geographical area. Female similar; often has large gray patches and transparent areas on outer FW and occasionally red spots near inner HW edge above and below; *waxy white pouch (sphragis) at tip of female's abdomen. Antennae of both sexes black.*

Similar Species: Female Eversmann's Parnassian has HW spots that extend into band. Phoebus Parnassian has red spots on both FW and HW, and checkered antennae; female Phoebus has gray or dark sphragis.

Life Cycle: Caterpillar usually black with rows of yellow or reddish spots; overwinters in decayed leaf litter and pupates in thin silken cocoon in spring. Bleeding heart (*Dicentra formosa*) serves as primary host plant in Pacific Northwest; other species of *Dicentra* are eaten elsewhere.

Flight: 1 brood; June–July in Washington, May–August in California.

Habitat: Forest edges at sea level in Northwest; moist, cool mountains and shaded canyons and ridges in drier parts of range.

Range: Coastal SE. Alaska, south in Cascade and Sierra Nevada mountains to N. California and Yosemite, and east to Idaho, Montana, Wyoming, and Utah.

The Clodius Parnassian is the only parnassian whose distribution is restricted to North America. It is sometimes found in the same mountains as the Phoebus Parnassian but Clodius usually flies at lower altitudes. Slow and deliberate, it often pauses to take nectar from hawkbit and other wild flowers. Once common in the Santa Cruz Mountains and the Snake River Canyon, the Clodius Parnassian is now no longer found in those areas—many conservationists blame logging and dam building, but others attribute the loss to drought.

28, 71 Phoebus Parnassian
(*Parnassius phoebus*)

Description: 2⅛–3″ (54–76 mm). *Male* cream- to snow-*white with* black and gray markings and *red spots on both FW and HW above and below,* varying in size and in hue from pale salmon to brilliant scarlet or, rarely, yellow to ocher; *black spots on FW outer edge.* Female dusky or largely transparent, with more black and gray markings and similar spots; *has waxy gray or black pouch (sphragis) at tip of abdomen. Antennae of both sexes banded in black and white.*

Similar Species: Clodius and Eversmann's parnassians have red spots only on HW. Clodius has black antennae; female has light sphragis.

Life Cycle: Chalk-white, button-shaped eggs hatch in summer. Caterpillar, to 1″ (25 mm), is black with yellow spots; overwinters and pupates in smooth tan chrysalis protected by loose cocoon in grass tussock or among debris. Life cycle may require 2 years. Host plants are stonecrops (*Sedum lanceolatum, S. obtusatum,* and perhaps other species of *Sedum*).

Flight: 1 brood; June–early September, emerging later at higher altitudes.

Habitat: Meadows, clearings, sage flats, and tundra; wherever host plant occurs.

Range: Sub-arctic Alaska south to central Sierra Nevada of California and down Rockies

to Utah and New Mexico; also
Eurasia.

A number of features probably reflect
the Phoebus Parnassian's adaptation to
life in the arctic-alpine zone with its
short, cool summers. The amount of
dark gray scaling seems to increase in
regions of high altitude and cold
climate, enabling the butterflies to
absorb the sun's warmth. Phoebus flies
at colder temperatures than many
butterflies, sometimes flying even in
snowstorms.

25, 318, 321   Pipevine Swallowtail
"Blue Swallowtail"
(*Battus philenor*)

Description: 2¾–3⅜" (70–86 mm). *Coal-black to
dark gray above with brilliant, metallic
blue,* especially toward HW margin
(male brighter than female); HW above
has row of cream to yellow spots around
rim. FW dull gray *below; HW has row of
big, bright orange spots curving through
blue patch along margin* and white
marginal spots.

Similar Species: Female Spicebush Swallowtail, female
Eastern Black Swallowtail, and dark
female Tiger Swallowtail all have 1 or
more orange spots on HW above.

Life Cycle: Clustered, rust-colored eggs. Mature
caterpillar, 1⅞–2⅛" (48–54 mm),
rust-black with black or red projections,
longest on head. Chrysalis, to 1⅛"
(28 mm), lavender to greenish-yellow
or pale brown; has sculptured curves,
angles, and horns. Host plants are
chiefly pipevines, Dutchman's pipe
(*Aristolochia macrophylla*) or Virginia
snakeroot (*A. serpentaria*) in East and 2
other species (*A. californica* and *A.
longiflora*) in West.

Flight: 2 broods in North, 3 in South; January–
October depending on latitude, late
April–early autumn in New England.

Habitat: Open woodlands, canyons, meadows, fields, gardens, streamsides, orchards, and roadsides.

Range: S. Ontario and New England south throughout East to Florida, west through Nebraska and Texas to Arizona and California, north to Oregon; also south into Mexico.

Horticulture has caused the spread of pipevines, and thereby extended the range of this butterfly. The adult favors honeysuckle, swamp milkweed, orchids, buddleia, azalea, lilac, and thistle. But the distasteful host plants of its caterpillars give this swallowtail an unpleasant flavor, causing birds to avoid it. Several butterflies—female Eastern Black Swallowtail, dark female Tiger Swallowtail, female Spicebush Swallowtail, female Diana Fritillary, and Red Spotted Purple—have evolved so that they resemble the Pipevine Swallowtail. This kind of similarity, known as Batesian mimicry, may protect the mimics from predators.

329, 332 Polydamas Swallowtail
"Gold Rim"
(*Battus polydamas*)

Description: 3–4″ (76–102 mm). *Tailless; long, scalloped wings. Black above with band of yellowish spots* within FW margin and yellow-green band within HW margin. *Below,* brown-black with yellow-green spot band on FW; *red chevrons rim HW.* Body has orange-red spots underneath.

Life Cycle: Egg nearly spherical, rust-colored, with warty surface. Mature caterpillar black, tan, or rust-colored, 2–2⅜″ (51–60 mm); has dark red tubercles. Chrysalis green or brown. Host plants are various species of pipevine (*Aristolochia*).

Flight: March–December in Texas; April–November in Florida.

Habitat: Disturbed habitats, such as cleared

areas among new growth, gardens, and fields.

Range: East and west coasts of Florida, rarely Georgia and Gulf States but common in Texas; also south to Argentina and in West Indies.

The Polydamas Swallowtail varies greatly within its range, both in the amount of yellow and red marking and in population size. Nearly absent during some years, it reappears later in fair numbers, but generally is more common in the southernmost parts of the United States. Like the Pipevine Swallowtail caterpillar, the Polydamas Swallowtail caterpillar feeds on pipevine and is distasteful to birds. Adult butterflies take nectar from lantana and other flowers.

324, 335, 338 **Eastern Black Swallowtail**
(*Papilio polyxenes*)

Description: 2⅜–3½″ (67–89 mm). *Black to blue-black above with blue* cloud on outer HW (more blue on female). Small *cream-yellow spots and chevrons rim wings* above and below (larger on male). *Bright orange eyespot with round, black-centered pupil* at corner of HW toward body. Sometimes band of yellow spots across outer third of wings inside row of blue patches, more commonly present or enlarged on male. *No yellow stripes on thorax.*

Similar Species: Western Black Swallowtail has yellow bars along thorax and irregular off-center pupil in eyespot. Short-tailed Swallowtail and Short-tailed Black Swallowtail have stubby tails; Short-tailed Swallowtail also has orange in bands of dorsal spots. Spicebush Swallowtail has 2 orange spots on HW above. Pipevine Swallowtail has no orange spots on HW above.

Life Cycle: Egg yellow. Mature caterpillar, to 2″

(51 mm), white to leaf-green with black bands on each segment broken by yellow or red-orange spots. Chrysalis woodlike brown or leaf-green; overwinters. Host plants are Queen Anne's lace (*Daucus carota*) and other members of carrot family (Apiaceae), as well as some members of the citrus family (Rutaceae), including rue (*Ruta graveolens*) and Texas turpentine broom (*Thamnosma texana*).

Flight: 2 or 3 broods; February–November, depending on latitude. Late spring, midsummer, and early autumn flights in mid-continent.

Habitat: Open spaces including gardens, farmland, meadows, and banks of watercourses; seldom in woodlands.

Range: S. Canada along E. Rockies into Arizona and Mexico, and east to Atlantic.

Eastern Black Swallowtails may be attracted to gardens by parsley or carrot plants, and nectar sources such as phlox and milkweed. Less blue and less rapid than the Pipevine Swallowtail, the Eastern Black likes to flit among the vegetation, drifting and stalling again and again, until disturbed; then it takes off in a direct line, making it most difficult to pursue. A similar-looking species of the Missouri Ozarks, known as the Ozark Swallowtail (*Papilio joanae*), has dramatically different habits. It dwells in forests, where caterpillars feed on meadow parsnip (*Heracleum*). Another related black swallowtail, the Kahli Swallowtail (*Papilio kahli*), flies in Manitoba and Saskatchewan and may be a hybrid of the Old World Swallowtail. It frequents bare hilltops, where females lay their eggs on cow parsnip (*Heracleum lanatum*).

325, 359 Desert Swallowtail
(*Papilio rudkini*)

Description: 2⅛–2¾″ (67–70 mm). *3 forms:* black
with yellow spots; black with yellow
bands of larger spots; *typical form mostly
yellow* since broad yellow bands of spots
cover up most of black. Tailed; with
blue patches and orange spot with pupil
in from tail on HW margin. *Outer edge
of yellow spots tends to be scalloped outward
toward margin.*

Similar Species: Anise Swallowtail has straight-edged
yellow spots. Black form of Desert
Swallowtail resembles Eastern Black
Swallowtail but former is smaller,
paler, and has less orange overall.

Life Cycle: Highly variable. Egg light cream-
colored. Caterpillar ringed with green,
ivory, and black in varying proportions;
feeds on turpentine broom (*Thamnosma
montana*), as well as on Queen Anne's
lace (*Daucus carota*) and other members
of carrot family (Apiaceae). Chrysalis
light tan to greenish- or grayish-tan;
can withstand years of drought.

Flight: Year-round; most common early spring
and fall in 2 broods following rainy
periods.

Habitat: Dry washes and mountain canyons in
desert.

Range: S. Nevada, SE. Utah, S. California, and
W. Arizona.

Since it occurs in several different
forms, the Desert Swallowtail could be
extremely hard to identify if not for its
clear-cut habitat preference. The Anise
Swallowtail and both Eastern and
Western Black swallowtails enter its
range on the edges; in these areas
confusion is possible. The Desert
Swallowtail may represent a desert
subspecies of the Eastern Black
Swallowtail. It is treated here as a
distinct sibling species with few
physical distinctions.

326 Short-tailed Swallowtail
"Maritime Swallowtail"
(*Papilio brevicauda*)

Description: 2¾–3⅜" (70–86 mm). Rounded wings
with *just a hint of tails.* Inner halves of
wings black, followed by *band of large
yellow spots usually shot with orange,* a row
of bright blue patches (most evident on
HW), and a series of small yellow and
orange spots around margin. Red-
orange eyespot with black pupil appears
along inner HW edge. Below similar,
with more orange, yellow, and blue.

Similar Species: Eastern Black Swallowtail has longer
tails, more blue above, and less or no
orange among yellow bands above.

Life Cycle: Caterpillar, to 1½" (38 mm), is light
green with yellow-spotted black bands;
feeds on Scotch lovage (*Ligusticum
scothicum*), cow parsnip (*Heracleum*),
angelica (*Angelica*), and various other
members of the carrot family.
Caterpillar pupates in late summer;
chrysalis overwinters near ground.

Flight: 1 brood; June–July.

Habitat: Edges and glades in evergreen forests,
grassy sea cliffs, and old cottage
gardens in villages; mountainsides.

Range: Canadian Maritime Provinces,
S. Labrador, Newfoundland, and Gaspé
Peninsula, along shores of Gulf of St.
Lawrence.

Unlike other black swallowtails, Short-
tailed Swallowtails fly near the ground.
When alarmed they flee over cliffs,
toward the sea, or into dense forests. In
the woods, they take nectar from the
blossoms of blueberry and Labrador tea,
and can also be found on irises, orchids,
honeysuckle, and garden flowers.

326, 334, 337,   **Western Black Swallowtail**
358   "Baird's Swallowtail"
  (*Papilio bairdii*)

Description: 3–3½" (76–89 mm). *Large. Male black above with row of yellow dots near margin;* blue band inside *and band of larger yellow patches* across midwing. *Female very black* with yellow marginal dots but *lacks yellow patches;* metallic blue patches usually outstanding. Both sexes have *1 orange eyespot on each HW with pupil off-center* and often joining margin. *Dark form has yellow side stripes* on body. Some populations, known as "brucei" and "hollandi," have yellow bands covering most of wings.

Similar Species: Eastern Black Swallowtail has centered, round pupil in orange eyespot, and lacks yellow bars on thorax. Short-tailed Black Swallowtail has stubbier tails and paler yellow bands on wings. Dark form Desert Swallowtail is paler beneath, with centered pupil in orange eyespots. Anise Swallowtail resembles yellow ("brucei") Western Black, but Anise has black abdomen with yellow side stripes.

Life Cycle: Eggs green. Green caterpillar has black bands dotted with orange. Green or brown chrysalis resembles stick or rolled leaf. Host plant is the composite dragon wormwood (*Artemisia dracunculus*).

Flight: 2 broods; May–September. Spring brood often smaller than late summer brood, but varies with rainfall.

Habitat: Semi-arid mountains and Upper Sonoran zone up to 9000' (2745 m) for breeding.

Range: S. Great Basin northeast to South Dakota, south to New Mexico, Arizona, and San Bernardino Mountains of California.

The ranges of the Western and Eastern Black swallowtails sometimes overlap in the Rocky Mountains. But the Western Black cannot tolerate suburban

habitats, which are perfectly acceptable
to the Eastern Black. The western
butterfly uses dragon wormwood,
unlike relatives that favor host plants in
the carrot family. Another black
swallowtail of the West, the so-called
Papilio nitra, which may or may not be
a distinct species, dwells along the
eastern flank of the Rocky Mountains
from Alberta into Colorado. Its
midwing yellow band is quite broad,
and it can be distinguished from all
forms of the Western Black by the
centered pupil in its orange hind wing
eyespot.

352, 355   Oregon Swallowtail
(*Papilio oregonius*)

Description: 2⅝–3⅜″ (67–86 mm). Wings full,
drawn out toward FW tip; tails long.
Pure yellow above with yellow to white
dots on FW and HW margins; blue
patches inside HW and broad, yellow
midwing spot band, the rest black.
*Patches in spot band may be scalloped away
from margin; orange HW eyespot has
comma-shaped, off-center pupil.* Similar
but paler below, with row of orange to
rust-colored spots between midwing
band and blue patches. *Abdomen yellow
with black side stripes below* and broad
black midline above.

Similar Species: Anise Swallowtail is lemon-yellow with
yellow spots that are straight-edged
toward margin; abdomen black with
yellow side stripes. Old World
Swallowtail has orange HW eyespot
without pupil.

Life Cycle: Pale green caterpillar black-banded and
yellow-spotted; feeds on dragon
wormwood (*Artemisia dracunculus*);
pupates on dry sticks or beneath stones.

Flight: 2 distinct broods; April–May and July–
September. May be delayed, reduced,
or overlapping.

Habitat: Basalt canyonlands, river shores and

terraces, sagelands, ravines, and gullies.

Range: S. central British Columbia, E. Washington, E. Oregon, Idaho, and W. Montana; possibly Alberta and North Dakota badlands.

The Oregon Swallowtail is one of the few butterflies restricted to the Pacific Northwest. Much of its riverside habitat has disappeared under dams on the Columbia and Snake rivers. This butterfly is the official state insect of Oregon, and it may be spotted visiting thistles along riverbanks. Because of its beauty and strictly regional range, the Oregon Swallowtail was one of the first 4 butterflies to be depicted on U.S. postage stamps in 1976.

354, 357 Old World Swallowtail
(*Papilio machaon*)

Description: 2⅝–2¾" (67–70 mm). *Wings fairly rounded,* tails rather short. Black background above with yellow spots along outer margins, band of shiny blue spots across HW; *broad midwing band of dull yellow spots* fills much of wings. *Orange HW eyespot has no pupil,* just black rim, which extends partly into orange on eastern individuals. Below similar but paler, with more orange.

Similar Species: Old World Swallowtail rarely overlaps with other yellow and black swallowtails; its longer tails, longer wings, and pupilless eyespots are distinctive.

Life Cycle: Undescribed for North American populations. Suspected host plant in Alaska is arctic sagebrush (*Artemisia arctica*).

Flight: 1 brood; June–July.

Habitat: Forest edges in mountain country, wooded and shrubby flats, shores, and arctic and alpine tundra.

Range: Arctic Ocean in Alaska south into N. British Columbia, east across Yukon and Northwest Territories, and southeast across boreal Canada to northern shores of Lake Superior and Hudson Bay. Absent from Maritimes, Labrador, and northeast arctic Canada. Also throughout much of Eurasia.

In Britain, this species is known simply as "The Swallowtail" because no other members of this family appear there. The Old World Swallowtail is the only true swallowtail with a circumpolar range, although there are 2 parnassians whose range does stretch around the Northern Hemisphere. Despite its broad range, the Old World Swallowtail is relatively inaccessible and its biology in North America is comparatively unknown.

36, 353, 356 Anise Swallowtail
(*Papilio zelicaon*)

Description: 2⅜–3″ (67–76 mm). Yellow-banded; more yellow than black. Lemon-yellow *spots in midwing band above are straight-edged* and fill spaces between veins of wing; HW flushed with blue between broad, yellow band and medium-long tail; *HW eyespot orange with large, round, centered pupil.* Similar below. *Black abdomen has yellow side stripes.*

Similar Species: Desert Swallowtail has outwardly convex, yellow spots on wings. Oregon Swallowtail is darker yellow, has yellow abdomen with black lines, and concave spots that are scalloped inward, away from margin. Yellow ("brucei") form of Western Black Swallowtail has off-center pupils, and yellow abdomen with black stripes.

Life Cycle: Immature stages variable. Generally, yellow eggs hatch green, black-banded and orange-spotted caterpillar, to 2″ (51 mm), which forms green or brown twiglike chrysalis, to 1¼″ (32 mm). Both young leaves and flowers or buds

of many species are consumed,
including fennel (*Foeniculum vulgare*),
seaside angelica (*Angelica lucida*),
and cow parsnip (*Heracleum lanatum*), as
well as carrots and parsley (Apiaceae),
and citrus trees (Rutaceae).

Flight: Varies greatly: year-round in S.
California; distinct spring and fall
broods on Northwest Coast, 1 brood
midsummer at high altitudes in
Rockies.

Habitat: Everywhere from sea-level tidelines to
14,000' (4,270 m) mountaintops,
vacant city lots to sage deserts,
canyons, and parks. Absent from dense
forests except clearings or roadsides.

Range: Pacific Coast from British Columbia to
Baja California, east to Black Hills and
eastern edge of Rockies; scarce or absent
in most of SE. California deserts.

The Anise Swallowtail is one of the
most adaptable native butterflies, along
with the Eastern Black Swallowtail.
Males are especially ardent hilltop
fliers, and seek out any eminence on the
horizon as a courtship rendezvous.

330, 333 **Short-tailed Black Swallowtail**
"Indra Swallowtail"
(*Papilio indra*)

Description: 2⅛–3⅜" (54–86 mm). *Small with
negligible to medium-sized tails. Brownish-
black with pale cream-colored* (not yellow)
spots and bands; HW has diffuse blue
patches, *orange eyespot with centered,
round, black pupil. Great variation among
subspecies:* some have inner ⅔ of wing all
black, no midwing band, and very
extensive blue; others have narrow
midwing band; still others have broad
midwing band of pale cream cut off
sharply by black on inner edge, so that
basal third or half of wing is all black.
2 narrow gold lines on thorax; abdomen
chiefly black.

Similar Species: Eastern Black Swallowtail has longer
tails, bands of yellow spots, more blue
above. Other black swallowtails in same
range have longer tails and brighter
yellow spot bands.

Life Cycle: Early stages extremely variable.
Usually, cream-colored egg hatches
caterpillar which becomes black-banded
with white, pink, or another light
shade, and spotted with yellow or
orange. Chrysalis gray, tan, or olive,
more blunt than that of other black
swallowtails; overwinters. Host plants
include members of carrot or parsley
family (Apiaceae).

Flight: Most of range, 1 brood in spring and
early summer; also 2nd, late summer
broods on Colorado Plateau and
California deserts.

Habitat: Various arid lands, including canyons,
pinyon-juniper slopes, desert ranges,
cliffs, boulder-strewn foothills, and
serpentine barrens; also moister high
mountain slopes.

Range: Mountain ranges, deserts, and
canyonlands of all western states.

The Short-tailed Black Swallowtail
differs from other black swallowtails in
its array of well-developed subspecies.
Exceptionally powerful in flight, Short-
tails can be difficult to approach; but
males often congregate at damp earth to
drink, and the retiring, seldom-seen
females sometimes come down from the
canyon rims to take nectar from mint
flowers beside streams. Males of some
races put on dramatic aerial displays,
battering one another in territorial
battles or courtship flights.

327 Thoas Swallowtail
(*Heraclides thoas*)

Description: 4–5½" (102–140 mm). *Very large.*
Above, blackish-brown crossed by broad
bands of pale yellow and smaller yellow

spots around margins. Long, dark, spoon-shaped tails have yellow or white center. Predominantly yellow *below; HW crossed by row of blue crescents with a rust-colored middle spot.* Orange crescent on inner HW angle above and below.

Similar Species: Giant Swallowtail very similar but generally darker yellow; 4th yellow spot from bottom on FW margin usually smaller; noticeable notch on top of abdomen near rear, which Thoas lacks.

Life Cycle: Caterpillar olive and off-white, blotched; resembles bird dropping; feeds on peppers (Piperaceae) in Central America.

Flight: Year-round in Tropics.

Habitat: Forest edges, glades, gardens, open areas—anywhere with abundant sunlight and flowers.

Range: Much of American Tropics; S. Texas, rarely, north to Colorado and Kansas.

Outwardly difficult to distinguish from the Giant Swallowtail, the Thoas Swallowtail differs in both genitalic characteristics and chromosome numbers. Both the caterpillar, in its unappealing camouflage, and the adult are unwary. Adult butterflies visit lantana and viburnum.

27, 328, 331 **Giant Swallowtail**
"Orange Dog"
(*Heraclides cresphontes*)

Description: 3⅜–5½" (86–140 mm). *Very large.* Long, dark, spoon-shaped tails have yellow center. Dark brownish-black *above with 2 broad bands of yellow spots* converging at tip of FW. Orange spot at corner of HW flanked by blue spot above; both recur below, but blue continuing in chevrons across underwing, which also has orange patch. Otherwise, yellow below with black veins and borders. Abdomen yellow with broad black midline tapering at tip; *notch on top of abdomen near rear.* Thorax has yellow lengthwise spots or stripes.

Similar Species: Thoas Swallowtail smaller and paler, 4th yellow spot from bottom on FW margin larger, no notch on top of abdomen. Schaus' Swallowtail smaller, with longer, thinner tails.

Life Cycle: Mature caterpillar, 2⅛–2⅜" (54–60 mm), brown or olive, resembling large bird dropping, with dirty buff patches and saddles and usually red scent horns; feeds on various citrus trees (Rutaceae), including common prickly-ash (*Zanthoxylum americanum*), hoptree (*Ptelea trifoliata*), and rue (*Ruta graveolens*). Chrysalis, to 1⅝" (41 mm), mottled gray-brown.

Flight: Year-round in South in multiple broods, but scarce in mid-winter. May–June and August–September broods farther north produce summer and autumn individuals unpredictably throughout range.

Habitat: Sunny, open areas, forest edges, glades, roads, rivers, and citrus groves.

Range: S. Canada south through U.S., east of Rockies through Mexico, and west along border into Arizona and California. More common in South and south-central states than farther north. Local despite broad tastes.

Known as the "Orange Dog" by citrus growers, the Giant Swallowtail is sometimes considered a citrus pest and is subjected to massive spraying. It is capable of flying long distances and often strays into northern and midwestern districts. Adults take nectar from lantana, orange blossoms, and other flowers. The Giant, the Thoas, and the female Tiger swallowtails are the largest North American butterflies.

Schaus' Swallowtail
(*Heraclides aristodemus*)

Description:	3⅜–3¾" (86–95 mm). *Rather rounded. Above, both wings brown with dark yellow spots* within brown border *and bold yellow band across middle; long, brown, yellow-edged tail;* 1 orange and 1 blue spot on HW toward abdomen. Below, dirty yellow; darker patches and outlines on FW; *blue in broad crescents across wings lined by chestnut-brown patches on HW below.*
Similar Species:	Giant Swallowtail larger, has yellow tails with dark edges, and much less pronounced chestnut patch on HW below. Palamedes Swallowtail larger, more rounded, and darker below.
Life Cycle:	Egg green. Caterpillar rich maroon with cream-colored patches and blue spots. Tubular, tapered, and horned chrysalis overwinters. Host plants are torchwood (*Amyris elemifera*) and wild lime (*Zanthoxylum fagara*).
Flight:	1 brood; April–early June. Partial 2nd brood in late summer in some years.
Habitat:	Restricted to shady hardwood hammocks.
Range:	Extreme S. Florida, only on certain keys; separate subspecies in West Indies.

The effects of population growth and development have almost eliminated this butterfly because much of its undisturbed hardwood hammocks habitat has been ruined. The Schaus' Swallowtail is officially listed as a Threatened Species. Efforts are underway to conserve remaining colonies in Key Biscayne National Monument and on Key Largo. Another, rather similar butterfly also placed on the Threatened Species list is the related Bahaman Swallowtail (*H. andraemon*). Although not a regular resident of southern Florida, it can be distinguished by the duller brown wings, the broader, straighter yellow

band, and the black-rimmed yellow
tail, opposite to that of the Schaus'.

Ruby-spotted Swallowtail
(*Heraclides anchisiades*)

Description: 3⅜–3¾" (86–95 mm). *Long, drawn-out
FW, coal-black above* with raylike effect
of lighter scales and white bar running
upward from lower edge (usually absent
on males). *HW long, scalloped, no tails;*
black above with *large ruby or rose-red
patch* consisting of close spots and bars.
Below similar with less intense red,
more white in streaks.

Life Cycle: Caterpillar green and brown, streaked
and mottled with white; has 2 rows of
short, fleshy horns on back; feeds on
citrus (Rutaceae) and probably prickly
ash relatives (*Zanthoxylum*) in Tropics.

Flight: Early spring–late fall; 2 or more
broods.

Habitat: Citrus groves, gardens, hedges.

Range: Lower Rio Grande Valley in Texas,
south to Bolivia.

An immigrant naturalized in gardens
and groves, the Ruby-spotted
Swallowtail is only marginally part of
the North American fauna. Like many
tropical and subtropical butterflies,
it may become established under
normal domestic conditions, and then
die back during hard winters. Formerly
Papilio anchisiades.

6, 49, 322,   **Tiger Swallowtail**
348, 351   (*Pterourus glaucus*)

Description: 3⅛–5½" (79–140 mm). Males and
some females above and below are *yellow
with black tiger-stripes across wings* and
black borders spotted with yellow.
Long, black tail on each HW. HW
above and below usually has row of blue

patches inside margin, with orange spot above and sometimes much orange below, running through yellow. Dark form females are black above with border-spotting of yellow, blue, and orange (blue sometimes becomes cloud on HW), below brown-black with shadowy "tiger" pattern. *Yellow spots along outer edge of FW below are separate in all but northernmost populations. Most have orange uppermost spot on outer margin of HW above and below* and orange spot on trailing edge.

Similar Species: Western Tiger Swallowtail has spots on outer margin of lower FW that run together into band; uppermost spot on border of HW is yellow; blue spots are more violet-tinted. Spicebush Swallowtail is distinguished from dark female Tigers by bluish-green spotting around margins above and orange spot on costa of HW (not outer border).

Life Cycle: Yellow-green, globular egg, $\frac{1}{32}$″ h × $\frac{3}{64}$″ w (0.8 × 1.2 mm), very large for a butterfly. Young caterpillar brown and white, resembling bird droppings; mature caterpillar, to 2″ (51 mm), is green, swollen in front, with big, false, orange and black eyespots and band between 3rd and 4th segments. Mottled green or brown sticklike chrysalis, to 1¼″ (32 mm), overwinters. Great variety of host plants, mostly broadleaf trees and shrubs; favorites include willows and cottonwoods (Salicaceae), birches (Betulaceae), ashes (*Fraxinus*), many cherries (*Prunus*), and tulip-poplars (*Liriodendron tulipifera*).

Flight: 1–3 broods; spring–autumn, actual dates vary with latitude.

Habitat: Broadleaf woodland glades, gardens, parks, orchards, and roads and rivers.

Range: Central Alaska and Canada to Atlantic; southeast of Rockies to Gulf. Rarer at northern and southern edges of range.

This species is the most widely distributed tiger swallowtail, and one of the most common and conspicuous butterflies of the East. Alaskan, Canadian, and northeastern butterflies are smaller and paler than those of the eastern states. Feeding in groups, adults take nectar from a wide range of flowers. The black female form has evolved to mimic the distasteful Pipevine Swallowtail; its presence in the population reflects the abundance of the species it mimics.

347, 350   **Western Tiger Swallowtail**
(*Pterourus rutulus*)

Description: 2¾–3⅞″ (70–98 mm). Above and below, *lemon-yellow with black tiger-stripes across wings* and black yellow-spotted margins. 1 or 2 orange spots and several blue spots near *black tail* on HW; blue continuous all around outer margin of HW below. *Yellow spots along outer black margin of FW below run together into band; uppermost spot on border of HW above and below is yellow.*

Similar Species: Tiger Swallowtail usually has distinct separate yellow spots on margin of FW below; HW uppermost spot is normally orange. Two-tailed Tiger Swallowtail has narrower stripes, more tails. Pale Tiger Swallowtail is paler, with darker veins below.

Life Cycle: Egg deep green, shiny, spherical. Caterpillar, to 2″ (51 mm), deep to light green, swollen in front, accentuating large yellow eyespots with black and blue pupils. Dark brown, woodlike chrysalis overwinters slung from a twig or tree trunk. Hosts more limited than those of eastern Tiger, include willows, poplars, and aspens (Salicaceae), several alders (*Alnus*) and sycamores (Platanaceae).

Flight: February in S. California, May in Washington, normally June–July in mountain areas. Up to 3 broods in low altitudes and latitudes, 1 in cooler places with shorter seasons. Present most of summer.

Habitat: Widespread, but normally near moisture—canyons, watersides, trails, roadsides, parks and gardens; sagelands and mesas with creeks.

Range: British Columbia south to Baja California, east through Rockies to Black Hills, and High Plains of Colorado and New Mexico. Rare east of Rockies.

The Western Tiger may be the most conspicuous butterfly in the West. The eastern and western species essentially replace each other along a diagonal line, northwest to southeast, although there may be some slight hybridizing along the dividing line. Such east-west species-pairs are not unusual among butterflies. In Western canyons, males of several species of swallowtails gather in spectacular numbers around mud puddles or beside streams, with the Western Tiger usually predominating.

346, 349, 360 Two-tailed Tiger Swallowtail
(*Pterourus multicaudatus*)

Description: 3⅜–5⅛" (86–130 mm). *Large; 2 tails on each wing. Yellow with narrow black stripes.* HW edge above and below lined with bright blue patches, black border spotted with yellow and orange; between border-spots each wing points outward in a series of partial tails, and 1 long and 1 medium tail. Female has broader stripes, more blue on HW, and orange cast.

Similar Species: Western Tiger and Pale Tiger swallowtails have only 1 tail and broader stripes; Pale Tiger creamier overall.

Life Cycle: Yellow-green egg. Apple-green caterpillar has yellow, black-rimmed eyespots on black-banded hump; becomes reddish before pupating. Feeds on various cherries (*Prunus*) and ashes

(*Fraxinus*), and on common hoptree
(*Ptelea trifoliata*) in Texas; folds leaves
into protective tent while feeding.

Flight: February–November in Texas; spring
and fall broods in California; single
June–July brood in Rockies and
Cascades.

Habitat: Semiarid canyonland, mid-level
mountains, and gardens; also moister
parts of dry areas, especially canyon
bottoms.

Range: E. British Columbia and E.
Washington along Canadian borders to
Dakotas, south to Oklahoma, Texas
and through California to Baja; also
Mexico and Guatemala.

The largest western tiger, this species is
equally at home in the wild canyons of
the Colorado Desert or in a Denver
suburban garden. The smaller Three-
tailed Tiger Swallowtail (*P. pilumnus*), a
Mexican species, appears rarely in the
lower Rio Grande Valley of Texas and
in southern Arizona. It is yellow
with broad black stripes; the corner
projection of the hind wing is
developed into a third tail. Its
caterpillars feed on laurels (Lauraceae).

342, 345   Pale Tiger Swallowtail
(*Pterourus eurymedon*)

Description: 3–3¾" (76–95 mm). *Pale cream with
bold, broad black stripes* and black
borders above; HW borders spotted
with cream, blue, and pale to bright
orange. Wings less brightly colored
below, with distinctive dark veins.
1 long black tail on each HW.

Similar Species: Western Tiger Swallowtail lemon-
yellow, has paler veins below, and
narrower black bands. Two-tailed
Tiger Swallowtail has more tails,
narrower black bands, and is deeper
yellow.

Life Cycle: Egg chartreuse. Caterpillar, to 1¾" (44

mm), soft green with yellow and black eyespot patterns. Chrysalis barklike, dark brown streaked with black. Preferred host plants mostly in buckthorn family, including mountain lilac and mountain balm (*Ceanothus*), also holly-leaf cherry (*Prunus ilicifolia*) and coffeeberry (*Rhamnus californicus*) in California, and alders (*Alnus*).

Flight: 1 brood; late spring—midsummer, varying with latitude and altitude.

Habitat: Mountainous or hilly country with broadleaf scrub or chaparral, dry slopes, canyons, and roads; ocean bluffs in Northwest, sea level to timberline.

Range: British Columbia and Montana south to New Mexico and Baja California; and in Cascades, Sierra Nevada, Coast Ranges, and Rockies. Absent from drier, hotter parts of basins and deserts.

Less numerous than other tigers, the Pale Tiger Swallowtail occurs widely throughout the West. It tends to fly rapidly in complex patterns between buckthorn bushes. Unlike its yellow relatives, it does not course up and down canyons all day but rather seeks hilltops. It takes nectar from mints, thistles, and buckeyes. Males settle more often on damp earth, sometimes in great congregations.

51, 316, 319 **Spicebush Swallowtail**
"Green-clouded Swallowtail"
(*Pterourus troilus*)

Description: 3½–4½" (89–114 mm). Both sexes black-brown above with cream-white to cream-yellow spots around FW outer margin; HW above black at base, with *1 bright orange spot on costa, 1 on trailing edge. Male HW above is broadly clouded with diffuse but brilliant blue-green* (rarely blue) between spots. Series of sharply defined, pale blue-green lunules runs around outer curve of wing. *Female bluer*

above than male; HW has broad cloud of
dark green and blue scales and lime-green
lunules. Below, HW of both sexes has
2 curved rows of bright orange-red spots
(1 around margin, 1 parallel but
midwing), *enclosing area heavily spotted
with blue.* 1 rounded black tail on HW.

Similar Species: Eastern Black, dark female Western
Tiger, and Pipevine swallowtails all
lack orange spot on HW costa above.

Life Cycle: Pale green egg. Caterpillar, to 1⅝"
(41 mm), is dark green with 2 pairs of
yellowish eyespots on front and rear of
hump. Winter chrysalis, to 1¼" (32
mm), smooth, bark-colored, swollen
about wing cases; summer chrysalis
may be green. Chief host plants include
spicebush (*Lindera benzoin*), sassafras
(*Sassafras albidum*), and various bays
(*Persea*).

Flight: Spring—early autumn; dates and
number of broods vary, depending on
latitude.

Habitat: Woods, forest edges, and pine barrens;
also meadows, fields, rights-of-way,
along streams, and in gardens.

Range: E. North America below S. Canada,
becoming more rare westerly; absent
from prairies except where wooded, and
only rarely reaching High Plains or
Rockies.

A grand and beautiful butterfly, the
Spicebush Swallowtail takes nectar from
Joe-Pye weed, jewelweed, and
honeysuckle. Like black female Tigers,
it mimics the Pipevine Swallowtail.

50, 336, 339 **Palamedes Swallowtail**
(*Pterourus palamedes*)

Description: 3⅛–5½" (79–140 mm). *Very large
with rounded wings. Blackish-brown
above, rimmed by yellow spots and crossed
midwing by yellow band* which is broken
on FW but is entire on HW, where it
ends in bright blue spot. Below, FW

has 2 rows of yellow border-spots; HW
has orange spots enclosing *row of blue
and gold to olive clouds and long, straight
yellow bar parallel to abdomen down inner
third of HW below.*

Similar Species: Other yellow and black swallowtails are
smaller, have more pointed fore wings.
Schaus' and Giant swallowtails are
browner, have larger wings, and more
extensive yellow beneath.

Life Cycle: Yellowish-green egg. Grass-green
caterpillar, to 2" (51 mm), has double
set of rimmed, orange, false eyespots
with black pupils. Possibly overwinters
as a caterpillar as well as a chrysalis,
unusual for group. Chrysalis, to 1⅝"
(41 mm), slightly mottled, greenish.
Red bay (*Persea borbonia*), sassafras
(*Sassafras albidum*), and sweet bay
(*Magnolia virginiana*) are host plants.

Flight: April–August in mid-range; February–
December in up to 3 broods in Florida.

Habitat: Subtropical wetlands, coastal swamps,
and humid woods with standing water.

Range: Resident from S. Maryland to S.
Florida, throughout Southwest, around
Gulf to S. Texas and N. Mexico. North
in Mississippi Valley to Missouri.

This butterfly is the signature
swallowtail of the great swamps—the
Everglades, the Great Dismal,
Okefenokee, Okeechobee, and Big
Cypress. In common with many swamp
skippers, the adults love to take nectar
from pickerelweed, and they are
reported to roost communally in oaks
and palmettos.

48, 340, 343 **Zebra Swallowtail**
(*Eurytides marcellus*)

Description: 2⅜–3½" (60–89 mm). *Long triangular
wings with swordlike tail. Above, chalk-
white* to hint of blue-green *with black
stripes* and bands; HW above has 2 deep
blue spots at base and bright scarlet

spot closer to body. *Below*, black-
bordered *scarlet stripe runs through middle
of HW*. Width of black stripes and
length of tails vary with season and
brood: spring zebras are paler, smaller,
shorter tailed; summer individuals are
larger and darker with very long white-
edged tails, exceeding 1″ (25 mm).
Antennae rust-colored.

Similar Species: Dark Zebra has straight, parallel black
and white markings near margins; more
heavily black with black antennae.

Life Cycle: Green egg. Caterpillar, to 2⅛″ (54
mm), banded with black and yellow,
band on hump broader than others.
Overwintering chrysalis, to 1″ (25
mm), is green or brown, stockier and
more compact than that of other U.S.
swallowtails. Host plant is pawpaw
(*Asimina triloba*) in North and other
Asimina species in South.

Flight: March–December in 4 broods on Gulf,
April–October in 2 broods in Midwest;
1st brood most numerous.

Habitat: Waterside woodland passageways,
shrubby borders, meadows, riversides,
lakeshores and marshes; absent from
mountains.

Range: Lake States and Ontario, east to S. New
England, south along Atlantic to
central Florida and Gulf, and west to
E. Great Plains.

The aptly named Zebra is the most
abundant regular North American
representative of the kite swallowtails,
named for their triangular wings and
long sharp tails. Despite a large range,
the Zebra occurs only near pawpaw or
its relatives—it usually fails to adapt to
suburban growth and development of
the countryside. Yet the Zebra is very
common along the banks of the
Potomac near Washington, D.C., and
by small rivers in Virginia. Formerly
called *Graphium marcellus*.

341, 344 **Dark Zebra Swallowtail**
(*Eurytides philolaus*)

Description: 2½–3½" (64–89 mm). *Long triangular wings with long tails. Milk-white with* submarginal, parallel, *broad black bands; crimson spots* at HW inner angle. *Alternate female form all black* with small white crescents around HW margin ending with crimson spots. Similar below, with long scarlet HW streak. Antennae black.

Similar Species: Zebra Swallowtail has submarginal black band curving out to meet black margin; midwing white band also curved. Zebra antennae rust-colored.

Life Cycle: Unreported. Host plants are pawpaws (Annonaceae).

Flight: Early spring and midsummer broods in Mexico; more common at beginning of rainy season.

Habitat: Lowland dry forests, roads, clearings, pastures, and forest openings.

Range: Reported once from Texas (Padre Island, Cameron County); Mexico south through Central America.

The Dark Zebra should be watched for in Texas. In Yucatan, it gathers by the hundreds around damp spots. The Maya knew this butterfly and named it "X-Chail." Formerly called *Graphium philolaus*.

WHITES AND SULPHURS
(Pieridae)

Approximately 1000 species worldwide; 55–60 species in North America. Most pierids are some shade of white, yellow, or yellowish-green; a few have bright orange tips and greenish-yellow marbling. There is often dramatic color variation seasonally and between sexes. North American species are usually medium-sized with wingspans of 1¼– 2″ (32–51 mm), although there are also very small and giant sulphurs.

Pierid eggs are spindle- or vase-shaped, and somewhat resemble the eggs of milkweed butterflies. The caterpillars tend to be green and cylindrical. Host plants are mostly crucifers for whites and legumes for sulphurs, although some pierids feed on conifers, mistletoes, heaths, willows, spurges, and plants in the composite family.

A few species compete with humans, consuming cabbage and other food or fiber crops. The cone-headed chrysalis hangs from a support by a silken girdle and usually overwinters. Adult butterflies have 6 fully developed legs, which they use to crawl over damp earth and flowers during frequent stops to drink. Some hardy, northern pierids fly earlier, later, and farther north than most other, non-overwintering species. Other tropical or southern pierids emigrate northward in vast numbers some springs, but unable to withstand the northern winter, and unable to return south, they die out annually. Whites classified in the genera *Artogeia* and *Pontia* were formerly grouped under *Pieris*. Nomenclature for sulphurs remains unchanged.

61 Pine White
(*Neophasia menapia*)

Description: 1¾–2" (44–51 mm). Male and female above *white with heavy black markings around FW tips and along FW costa; female also shows network of black veins around HW margin* above and below; on male this pattern is only below. FW below black at rear margin with pattern of large white spots; *female has bright orange or crimson outlining to HW pattern below.*

Similar Species: Veined White has more delicate dark markings on veins below and lacks heavy black markings above.

Life Cycle: Green egg tiny, pear-shaped with bright white beads around neck; deposited in rows, unlike singly-laid eggs of most pierids. Caterpillar dark green with white back and side stripes. Chrysalis slender, also green striped with white; lodges at bases of host pines (*Pinus*), true firs (*Abies*), and Douglas-fir (*Pseudotsuga*).

Flight: July–September, most abundant late summer.

Habitat: Pine and fir forests from sea level to mid-montane.

Range: S. British Columbia and Alberta south through mountains and forests into S. California, N. Arizona, central New Mexico, and Black Hills of South Dakota and Nebraska. Absent from northwest coastal forests.

A close relative of the South American whites in the genus *Catasticta*, the Pine White bears little similarity to most other North American pierids aside from its color. Instead of having a rapid flight and a taste for mustards, this strange white normally flutters weakly, high among conifers. It is difficult to see close up, although males and occasional females can be observed as they take nectar on the forest floor in early morning and late afternoon. Except in southern Arizona or Mexico, it is the only white butterfly that frequents pines. Some consider the Pine White a forest pest because the caterpillars can, during rare outbreaks, completely defoliate conifers.

66, 598 **Chiricahua Pine White**
(*Neophasia terlootii*)

Description: 1⅞–2¼" (48–57 mm). *Male white with heavy black markings around FW tips, completely filling FW cell and along veins of HW, which may be fringed with scarlet. Female deep orange or brick-red with same* pattern of black and yellowish spots along FW margin.

Life Cycle: Caterpillar green, white-lined. Feeds on conifers; ponderosa pine (*Pinus ponderosa*) a common host plant.

Flight: October–November.

Habitat: Medium elevation mountain pine and fir forests.

Range: W. Mexico and SE. Arizona, perhaps into SW. New Mexico.

The sexual dimorphism of the Chiricahua Pine White is one of the most dramatic in North America. Neither the white male nor orange female is commonly encountered due to this butterfly's very late flight period and isolated range. Visitors to Arizona' Chiricahua and Huachuca mountains in late fall may see this remarkable insect around flowers and mud puddles on warm mornings.

---

69 **Florida White**
(*Appias drusilla*)

Description: 1⅝–2⅜" (41–60 mm). *Male all white on both surfaces except for black along FW costal margin. Female variable,* may be off-white or have dark-bordered FW and pale orange HW, with FW below orange at base and HW below pale orange. *Both sexes have drawn out, rather pointed FW* (especially male) concave on outer margin, *and silky sheen to inner portion of wings above and HW below.*

Similar Species: Great Southern White has more rounded wings, dark tips. Giant White is black-spotted.

Life Cycle: White eggs turn yellow before hatching; deposited in groups of 2 or 3. Caterpillar, to 1¼" (33 mm), dark green, gray and white-lined on sides, stippled with fine yellow and black hair. Chrysalis pale gray-green with reddish spots and some black and white markings. Host plants include capers (*Capparis*) and Guiana plum (*Drypetes lateriflora*).

Flight: Virtually year-round in southern part of range.

Habitat: Shaded hardwood hammocks, evergreen river forests, and open areas with flowers.

Range: Neotropics: resident in S. Texas and S. Florida, rarely emigrating northward as far as Nebraska and New York.

The Florida White is one of the few North American butterflies that prefers shaded thicket interiors to glades or open sunny spaces. However, it will go into the sun to take nectar from yellow composite flowers. Other whites prefer sunshine and lack this species' narrow, pointed wings.

79 Becker's White
"Sagebrush White"
(*Pontia beckerii*)

Description: 1⅜–1⅞" (35–48 mm). White above with *open, squarish black spot in FW cell* and bold black spotting around FW tip and margin. Female may also have black spots on HW above. *Below, gray-green or moss-green scaling broadly outlines veins of HW, broken by clear white band across wing;* FW tips below have some green scaling and black spot in FW cell. Brightness of green scaling variable.

Similar Species: Western White has more but lighter spotting, duller olive veins, and lacks white band. Spring White much smaller, with grayer-green veins. Marblewings are smaller with less distinct green patterns.

Life Cycle: Egg spindle-shaped, ridged. Caterpillar pale green, black-dotted, yellow cross-banded. Chrysalis smooth gray with white wing case; resembles bird dropping. Host plants include bladder-pod (*Isomeris arborea*), golden prince's plume (*Stanleya pinnata*), black mustard (*Brassica nigra*), and probably other crucifers.

Flight: 2 distinct broods; May–June, August–September. More broods in some southerly locales.

Habitat: Arid lands such as sage flats, dry coulees, foothill canyons, and lower mountains.

Range: Drier, intermontane W. North America from interior British Columbia through Montana to Black Hills of South Dakota, SE. Colorado, and across Great Basin to Baja California.

A hardy butterfly of harsh environments, Becker's White is at home on a hot wind in a dusty canyon. Its association with sagebrush arises from a common adaptation to the desert. Look for this bright green-marked butterfly on rabbit brush, where it takes nectar, or find it roosting among sparse vegetation at dawn or dusk.

64 **Spring White**
"California White"
(*Pontia sisymbrii*)

Description: 1¼–1½" (32–38 mm). Male milk-white or sometimes pale yellow above, female cream-colored; both sexes have *closed charcoal FW cell spot and black spots around FW tip and margin. Below, veins black or gray,* lined with gray-green scales on FW tip and all of HW.

Similar Species: Becker's and Western whites both much larger, bulkier, with greener veins below. Dark form Checkered White larger, markings browner.

Life Cycle: Egg yellow, ribbed, long, conical. Caterpillar boldly black-banded against yellow. Chrysalis dark, grainy brown, rounded. Host plants are crucifers, including desert candle (*Caulanthus*), jewel flower (*Streptanthus*), and rock cress (*Arabis*).

Flight: 1 brood; February–May in lowlands, June–July in higher mountains.

Habitat: Rocky deserts, hills, arroyos, steppes; and alpine slopes in Olympics and Sierra Nevada.

Range: All western provinces and states from Yukon Territory to Baja California, east through Great Basin and Rockies to Black Hills and E. New Mexico.

The species name of *sisymbrii* refers to this butterfly's preference for mustards as host plants, a common trait among Northern Hemisphere whites. The name Spring White accurately describes this butterfly's precocious appearance on the cold flats and coulees weeks before most butterflies emerge.

60, 63, 81 **Checkered White**
"Common White"
(*Pontia protodice*)

Description: 1¼–1¾" (32–44 mm). Generally *white above with charcoal or brown markings;*

HW veins below lined with brown- or olive-green. Female more heavily marked than male; spring brood more marked than summer—heavily checkered with charcoal above and lined with olive below. *Summer male nearly immaculate white* except for black cell spot above and tan tracery below. All gradations in between can occur.

Similar Species: Western White has similar variations but summer male greenish-veined beneath and summer female more darkly checkered. In most of range best separated by habitat; at eastern base of Cascade and Sierra Nevada mountains only an expert can distinguish species.

Life Cycle: Yellow, spindle-shaped eggs. Caterpillar becomes blue-green, black-speckled, and downy with 4 lengthwise yellow stripes. Blue-gray chrysalis has black speckles; overwinters in California. Host plants include many kinds of native and alien crucifers as well as some capers, such as bee plant (*Cleome*).

Flight: Early spring–late fall in several broods; year-round in parts of California.

Habitat: Mostly lowland open spaces, especially disturbed areas, fields, vacant lots, railroad yards, and other weedy plots.

Range: North America and N. Mexico from lower margins of Canada southward; absent from most of Pacific Northwest. Scarcer in East.

The distributions of native white butterflies have probably changed considerably since pre-colonial times, possibly due in part to competition from the introduced Cabbage White. However, these changes may be caused to a greater extent by the expansion and contraction of various habitats due to use or disuse by people. The Checkered White is thought to be less common now in its native East and more abundant in disturbed parts of the

West, into which it has spread on the heels of agriculture. Although it can reach enormous numbers, it seldom approaches the abundance of the Cabbage White.

62 Western White
(*Pontia occidentalis*)

Description: 1¼–1¾" (32–44 mm). Male *white above with network of black marginal and submarginal spots on FW tip* and black FW cell spot. Male has *olive-green scaling on veins and toward tips on HW below.* More marking in female, with black network extending onto HW above. Spring brood more heavily marked than summer; all have greenish vein-scales below.

Similar Species: Becker's White has brighter green veins, less but bolder black marking. Checkered White summer male lacks greenish veins; summer female lacks heavy black, but male and female spring butterflies are very similar.

Life Cycle: Caterpillar dull green, light- and dark-banded; feeds on various crucifers (Brassicaceae) and spider plants or bee plants (*Cleome,* family Capparaceae).

Flight: April–September, depending on altitude.

Habitat: All sunny mountain habitats including arctic-alpine, but also lowlands in clearings, fields, and roadsides.

Range: Alaska and British Columbia to Manitoba, south between Pacific Coast ranges and eastern fringes of Rockies to central California and New Mexico.

The Western and Checkered whites are very closely related, but they have definite ecological differences. The Western White is strictly adapted to mountains and northern habitats and the Checkered White to lowlands. The ranges of the 2 species overlap between the highlands and the plains, where these butterflies can be extremely confusing, even for practiced lepidopterists. Western Whites often fly high into the arctic-alpine zone, where they are one of the most common butterflies.

54, 65 Veined White
"Mustard White"
(*Artogeia napi*)

Description: 1½–1⅝" (32–41 mm). *Rounded wings.* Above, delicate white. *Below, cream-colored to yellowish; HW veins have light to heavy gray-olive or brown scaling;* extreme forms also have scales along veins of other wing surfaces. Some summer broods lack dark vein-scaling. Sex spots (1 for male, 2 for female) occasionally appear on FW above, often blurred. If FW tips gray, vein-scaling is heavy.

Similar Species: Cabbage White virtually always has gray FW tips and black sex spots on FW and lacks dark-dusted veins below. West Virginia White has smokier, diffused vein-scaling, rounder FW tips, and lacks yellowish tint to HW below. Pine White has black markings above and heavier dark markings below.

Life Cycle: Pale, vase-shaped egg. Caterpillar forest-green with darker or yellowish back and side stripes. Host plants include wide array of crucifers, such as cresses (*Thlaspi, Arabis,* and *Barbarea*) and toothworts (*Dentaria*).

Flight: April–August in 2 or 3 broods.

Habitat: Deciduous and coniferous woodlands, forest clearings and edges, roadsides, and in cool, moist places.

Range: Cool, temperate portions of Northern Hemisphere. In North America, Alaska to Labrador south to Arizona, Montana, Lake States, and New York. Absent from S. California and entire Southeast.

Many lepidopterists have speculated about the Veined White's changed habitat following the introduction of the Cabbage White in 1860. The Cabbage White is reputed to have taken over as the common white of open landscapes, while the Veined White appears to have retreated to the woodlands, where it is better able to compete. But recent studies also suggest that these habitat changes may be the effect of human land use and not just interspecies competition. Both species have been found to coexist in many marginal places. The Veined White may well consist of more than one biological species, but until more distinctions are made it will be considered one variable, circumpolar species.

67 West Virginia White
(*Artogeia virginiensis*)

Description: 1⅜–1⅝" (35–41 mm). *Dusky white with smoky gray-brown scaling suffused along veins of HW below.* Sometimes very pale gray markings above on FW tip or HW costa. *HW below* may be cream-colored but *has no yellowish cast.*

Similar Species: Veined White clearer white, usually yellowish below, more discretely scaled along veins.

Life Cycle: Caterpillar chartreuse with lighter yellow-green stripes on side and back; feeds on toothwort (*Dentaria diphylla*) and pupates well before winter. Slim chrysalis has long, curved "beak."

Flight: 1 brood; April–May.

Habitat: Rich, moist woodlands.

Range: Margins of Great Lakes to Quebec and New England, south in Appalachians to N. Georgia.

Formerly considered the same species as the Veined White, the West Virginia White is in fact a completely distinct butterfly, occupying moister, shadier, generally more southern habitats. From

Maryland to North Carolina in the
Appalachians it is rather widespread
and common. But populations farther
north are becoming fewer and sparser as
the butterfly's native woodlands are
altered. In Ontario, the West Virginia
White is an Endangered Species.

46, 56   **Cabbage White**
"Small White"
"European Cabbage White"
(*Artogeia rapae*)

Description: 1¼–1⅞" (32–48 mm). Milk-white
above with *charcoal FW tips, black
submarginal sex spots on FW* (1 on male,
2 on female) and on HW costa. *Below,
FW tip and HW pale to bright mustard-
yellow,* specked with grayish spots and
black FW spots.

Similar Species: All whites of similar size have dark-
scaled veins on HW below, or lack
yellow there or gray on FW tips.

Life Cycle: Egg yellowish, vase-shaped.
Caterpillar, to ¾" (19 mm), bright
green with yellowish back stripes and
side stripes; covered with short pile.
Chrysalis, to ¾" (19 mm), speckled,
green or tan. Many crucifers
(Brassicaceae) acceptable as host plants,
as well as nasturtiums (Tropaeolaceae)
and sometimes other plants.

Flight: 3 or more broods; from last to first hard
frost.

Habitat: Gardens, agricultural and abandoned
fields, cities, plains, foothills;
wandering virtually everywhere except
where the most extreme climatic
conditions exist.

Range: All of North America south of
Canadian Taiga, including Hawaii.
Native to Eurasia.

Exceedingly well known to most
gardeners, the Cabbage White has
spread throughout North America after
its unintentional introduction to

Quebec in 1860. No other butterfly is so successful over such a great variety and expanse of landscape. Some of the pierid emigrants may temporarily exceed this species in numbers, but they are neither as persistent nor as widespread. Its spectacular success has been blamed—probably erroneously—for the decline or retreat of some of its indigenous relatives. Farmers and gardeners consider the Cabbage White a pest, so great is its appetite for cabbages, radishes, and nasturtiums.

52, 53, 84 Great Southern White
(*Ascia monuste*)

Description: 1¾–2¼″ (44–57 mm). *Large. Male white with pointed, full wings. Charcoal scales dust tips and surrounding veins of FW above;* below, FW tips and entire HW creamy, unmarked yellowish. *Female dimorphic, either like male but with darker margins, or entirely suffused with smoky brown or gray scales;* dark form mostly migratory phase.

Similar Species: Florida White lacks dark tips and has narrower more pointed wings. Giant White has black cell spot on FW.

Life Cycle: Caterpillar pale or bright yellow, striped greenish-maroon; feeds on wide array of cultivated and wild crucifers, capers, and saltworts, including pepper grass (*Lepidium virginicum*), saltwort (*Batis maritima*), and spider flower (*Cleome spinosa*).

Flight: Year-round in southernmost part of range, warm months farther north.

Habitat: Beaches, salt marshes, coastal plains, offshore islands, and sandy flats.

Range: Neotropics: S. Texas, Gulf Coast, and Florida, emigrating up Mississippi Valley to Kansas and up Atlantic to Virginia.

Because the striking dimorphism of the female depends upon length of day, a higher proportion of the summer population is of darker color. Generally common, the Great Southern White builds up to enormous numbers prior to northerly emigrations, which occur irregularly and are probably due to a local scarcity of food. The mass movements are very impressive—for days, the large butterflies pass in small clusters, flying in a rapid and very directed manner 15–20′ (5–6 m) in the air.

55 Giant White
(*Ganyra josephina*)

Description: 2½–2⅞″ (64–73 mm). *Very large. Clear white above and below, with drawn-out wings and black spot at end of FW cell;* female may have black along FW outer veins and additional black smudges; sometimes female has salmon shading above.

Similar Species: Other big whites lack black cell spot.

Life Cycle: Unreported.

Flight: Probably most months.

Habitat: Scrubby openings in Rio Grande Valley.

Range: Resident in S. Texas, wandering into Arizona; south to Central America.

The name Giant White is appropriate to the largest of all North American whites. It is not well known in spite of its conspicuous dimensions because it is resident in this country only in the lower Rio Grande Valley of Texas.

40, 77 Creamy Marblewing
(*Euchloe ausonides*)

Description: 1½–1¾″ (38–44 mm). *Creamy, off-white above with* black network in upper FW tip and *narrow, rectangular black spot across end of FW cell.* Yellow veins and rich, *yellow-green marbling across entire HW below in fairly separate bars* and on FW tips below. Shiny blue-gray body hair.

Similar Species: Pearly Marblewing is smaller, whiter, with square spot off FW cell. Northern Marblewing is more densely marbled, other markings grayer. Olympia Marblewing has more open marbling on HW below and rosy flush above and below.

Life Cycle: Long, jug-shaped egg. Caterpillar, to ¾" (19 mm), dark green with bluish back stripe and yellow side lines as well as many tiny black projections. Chrysalis, to ¾" (19 mm), purple at first, growing gray. Caterpillars feed on soft buds, flowers, and, rarely, seed pods of host mustards, including several rock cress species (*Arabis*), mountain tansy mustard (*Descurainia richardsonii*), and other mustards.

Flight: February to mid-August, depending upon altitude and latitude; 2 broods in coastal California, 1 brood elsewhere.

Habitat: Moister mountain areas, often among pine and aspen forests, trails, clearings, meadows, creek sides, and weedy lowlands in California.

Range: Western mountains from Alaska and Alberta to central California and New Mexico; also in Great Lakes region to N. Michigan and Ontario.

A freshly emerged marblewing is a lovely but ephemeral sight since marblewings fly only briefly during springtime in any one locality. Marblewings appear later higher in the mountains, where spring comes in midsummer and all seasons except winter are compressed. Flying low, fast, and erratically, the Creamy Marblewing seldom pauses except to take nectar or to roost.

Northern Marblewing
(*Euchloe creusa*)

Description: 1¼–1⅜″ (32–35 mm). *Small. Pure white above* with narrow black FW cell spot; *FW tips vaguely patterned with gray scaling. HW below heavily but loosely marbled with yellow-green (not in distinct bars);* also green below on FW tips and margins.

Similar Species: Creamy and Pearly marblewings have green marbling in separate bars, blacker wing tips.

Life Cycle: Host plants are drabas (*Draba*) and other crucifers.

Flight: 1 brood; May–July.

Habitat: Subalpine moraines, forest clearings, and glades.

Range: Alaska and W. Northwest Territories, south in Rockies of British Columbia and Alberta to Waterton National Park on Montana border.

Until relatively recently, the Northern and Pearly marblewings were considered to be the same species, perhaps not even distinct from the Creamy Marblewing. All of these have now been established as separate species, although they have many features in common and can be difficult to tell apart. However, each has its own ecological niche in the American West, both within regions where they occur together and on a wider basis.

78, 80 Olympia Marblewing
(*Euchloe olympia*)

Description: 1½–1¾″ (38–44 mm). *Chalk-white, with narrow, angular to sharply rounded wings;* FW above has gray-marked tip, black-checked cell. *Yellow-green marbling in discrete pattern on HW* below, showing through to upperside. Most fresh individuals on prairies and fewer farther east show *bright rosy wash on base of HW above and below.*

Similar Species: Creamy Marblewing has denser marbling on HW below, no rosy flush, and much heavier black marking at tip of FW above.

Life Cycle: Caterpillar bright green, striped lengthwise with gray and yellow; feeds on buds and flowers of several kinds of rock cress (*Arabis*), hedge mustard (*Sisymbrium officinale*), and other mustards. Rosy purple chrysalis changes to gray-brown; overwinters.

Flight: 1 brood; March–June.

Habitat: Open woods and meadows in eastern portion of range; watercourses and nearby fields and bluffs, foothill ridges and open grasslands farther west, sandy flats and dunes in Great Lakes area.

Range: SE. Alberta to S. Ontario southward to West Virginia, N. Texas east of Rockies.

The easternmost marblewing in America is also the most singular looking, especially with its rosy color when newly emerged. It seems to remain rather strictly within its range and is considered uncommon, although it occasionally wanders and at times can be fairly numerous locally. The number of adults on the wing fluctuates dramatically. This butterfly's anonymity is partly accounted for by its brief flight period early in the season and the excellent camouflage of its folded wings against yellow-green cresses.

58 Pearly Marblewing
(*Euchloe hyantis*)

Description: 1¼–1⅜" (32–35 mm). *Small. Pure white with pearly sheen,* especially between marbling below. Above, FW has charcoal to black checked tips and *broad, nearly square black bar at end of FW cell. Below, HW marbling is grass-*

green in solid, separate bars; FW tips
below also green.

Similar Species: Northern Marblewing's marbling is
fuzzy and ill-defined. Creamy
Marblewing has yellower marbling with
some white showing through and
narrow cell spot.

Life Cycle: Long, flask-shaped, yellow egg. Mature
caterpillar apple-green, with dark
specks and broad white side stripes.
Chrysalis beaked, yellow to wood-
brown. Caterpillar feeds only on seed
pods of western tansy mustard
(*Descurainia pinnata*), desert candle and
wild cabbage (*Caulanthus*), as well as
Arizona jewel flower, mountain jewel
flower, and other *Streptanthus* species.

Flight: 1 brood; April–May, rarely March–
June.

Habitat: Sagelands, pinyon-juniper forests, rocky
ravines, dry watercourses, serpentine
outcrops, granite moraines, and canyon
walls in California.

Range: Basin and foothill country between
E. Coast Ranges and W. Rockies from
S. British Columbia and W. Montana
south to Baja California, SE. Arizona
and N. New Mexico.

While the Creamy Marblewing prefers
moist environments in mountain
forests, the Pearly Marblewing has
claimed the arid lands, although in
some places both species fly together.
Like the Spring White, the Creamy
Marblewing can be found flying over
sage flats when, despite the chilly April
wind, the big desert violets are in
bloom.

76 **Desert Orangetip**
(*Anthocharis cethura*)

Description: 1⅛–1½″ (28–38 mm). Upperside
*white or sometimes pale yellow with pale
yellow-orange patch near FW tip*
contained within black and white

barred tip and margin. (Female sometimes lacks orange tip.) *Below HW* yellow-washed, with heavy *yellow-green marbling fusing into fairly separate swirls or bands.*

Similar Species: Sara Orangetip has deep red-orange patch, sometimes lacks barred border outside patch; marbling is loose, not banded.

Life Cycle: Egg orange, vase-shaped, ribbed. Caterpillar has orange and green crossbands and white patches, meeting at shiny black spots on underside. Host plants include long-beaked twist flower (*Streptanthella longirostris*) and tansy mustards (*Descurainia*). Striated brown to gray chrysalis is pointed at both drawn-out ends, and swollen.

Flight: 1 brood; late February into May.

Habitat: Desert chaparral, juniper hills, canyons, ridgeline meadows, and open deserts.

Range: S. California, N. Nevada and W. Arizona into Baja California, especially in Mojave and Colorado deserts; also Santa Catalina Island.

The Desert Orangetip is rarer and more limited in range than the superficially similar Sara Orangetip. Good spring rainfall brings the Desert Orangetip out on the arid desert stretches. One race limited to Santa Catalina Island off Los Angeles was feared extinct, but recent field work shows that it still flies there in good numbers.

Pima Orangetip
(*Anthocharis pima*)

Description: 1¼–1½″ (32–38 mm). Both sexes *lemon-yellow above and below; FW* above and below *broadly tipped with bright orange.* Checked, black and yellow border surrounds orange tip; black bar at end of FW cell. *HW below* whitish and *covered by olive-green marbling.*

Similar Species: Yellow forms of other orangetips paler.

Life Cycle: Unreported.
Flight: 1 brood; February–April.
Habitat: Foothill canyons, watercourses leading into desert, and low desert hills.
Range: Extreme SE. California, S. Nevada, SE. Utah, much of S. Arizona, and SW. New Mexico.

This brilliant butterfly flits among the saguaros, chollas, and wildflowers of the Sonoran Desert after winter rains. The males fly to desert promontories, while the females favor lower spots.

75 Sara Orangetip
(*Anthocharis sara*)

Description: 1¼–1⅛" (32–41 mm). *Male white above, female white or yellow. Orange patch near FW tip normally brilliant red-orange,* sometimes paler; some populations have black and white barred tip and black margin outside orange patch; paler patch repeated on FW below. *Marbling on HW below grass-green, loose and granular rather than in separate bars.* Often some green on FW tip below and black at ends of HW veins above.
Similar Species: Desert Orangetip has banded marbling and paler orange tips with barred margins.
Life Cycle: Egg yellow becoming deep orange, spindle-shaped. Caterpillar moss-green, stippled along side with white and darker green. Chrysalis resembles silvery-gray or brownish-green thorn. Many kinds of crucifers (Brassicaceae) acceptable as host plants.
Flight: Occasionally 2 broods in central California; February–April and May–July. 1 brood elsewhere; March–July, depending upon latitude, altitude, temperature, and moisture.
Habitat: Aspen woods and meadows in Rockies; elsewhere desert canyons and arid slopes, mountain roads, ridges, streamsides, alpine seeps, sea-level

pastures, and hay fields; many kinds of sunny places.

Range: Coastal SE. Alaska to Baja California, north and east to E. Rockies at mid-elevations.

The variable Sara Orangetip is nearly as successful as the introduced Cabbage White in its ability to exploit a wide diversity of habitats across a broad area. However, it cannot carry on summer and fall generations, the caterpillars being limited to flowers and seed pods for their food. In the far West, Sara flies over the deserts, near the seaside, and up to the alpine, but the Rocky Mountain race is restricted to cool mountain places where it lives in tight colonies.

5, 37, 57, 74 **Falcate Orangetip**
(*Anthocharis midea*)

Description: 1⅜–1½" (35–38 mm). *Male white above with bright orange patch near tip of FW* and black FW cell spot. *Female white.* with black FW cell spot, *lacks orange patch on FW* but occasionally has pale orange flush at FW tip. *Both sexes* checked around the edges with black and *hooked at tip of FW. Below. marbled loosely over HW with bright chartreuse or greenish-brown.*

Similar Species: Female could be confused with Olympia Marblewing, but latter lacks hooked tips and marbling is in tight bands.

Life Cycle: Egg, ⁵⁄₁₂₈" h × ¹⁄₆₄" w (1.0 × 0.4 mm), greenish-yellow to pale orange, elongated. Grown caterpillar, to ⅞" (22 mm), moss-green, striated with blue, green, and yellow, with orange back stripes and white side stripes. Chrysalis looks like green thorn, thickened in middle. Host plants are various cresses (*Arabis. Cardamine.* and *Barbarea*), hedge mustard (*Sisymbrium*) and shepherd's purse (*Capsella bursa-pastoris*).

Flight: April—May; 2 weeks in any locality.
Habitat: Eastern deciduous forests or mixed oak-pine woods or pine barrens; especially glades, roads, rocky sites, and low, moist woods near streams.
Range: S. Wisconsin and central Massachusetts south to Kansas, central Texas and N. Louisiana along coast to Georgia.

The unmistakable Falcate Orangetip is a familiar spring butterfly in the eastern United States. The name "falcate" refers to the hooked tip of the fore wing. Population levels of this butterfly fluctuate from year to year, and can easily drop to near zero when woodlands are developed. For many years, butterfly seekers have visited West Rock, a protected state park outside New Haven, Connecticut, in the early spring to find the Falcate Orangetip. It appears even on cool and semicloudy days, perhaps an adaptation to the brevity and unpredictability of its early season.

83 Gray Marble
(*Anthocharis lanceolata*)

Description: 1⅝–1¾" (41–44 mm). White; *FW above has gray, slightly hooked tip* and black cell spot. *Below,* FW tip and *entire HW smoked and marbled with gray or brownish-gray.*
Similar Species: Male Cabbage White is yellowish and HW below mustard-yellow.
Life Cycle: Caterpillar green with white specks and white side stripes. Chrysalis gray-brown with long, curved hook on head. Rock cresses (*Arabis*) and some other cresses serve as host plants.
Flight: February—May, according to location.
Habitat: South slopes of wooded, walled canyons, desert washes and edges, and mountain ravines.

Range: Siskiyou and S. Cascade mountains of Oregon south through California mountains and Carson Mountains of Nevada into Baja California.

An uncommon, drab, and frail inhabitant of hotter mountains, the Gray Marble is not known to many people. Its liking for rock walls, where cresses cling, makes it difficult to observe closely.

85, 111, 114 **Common Sulphur**
"Clouded Sulphur"
(*Colias philodice*)

Description: 1⅜–2″ (35–51 mm). Above, male light yellow with sharp black borders; female yellow or white with yellow-spotted black border. *Below*, both sexes greenish yellow in spring and fall broods, clear yellow in midsummer; *double red-rimmed silvery spot* near end of cell *and row of brown spots slightly inward from edges* of both wings. (Hybrids with Orange Sulphur may have orange patches over about half of wing area.) Albino female has pink fringes.

Similar Species: Other sulphurs usually lack row of brown spots around HW below except the Orange Sulphur, which is larger, orange above, and yellow rather than greenish below. Albino female Orange Sulphur is larger and greener beneath.

Life Cycle: Chartreuse egg laid singly on leaves of various legumes, especially clovers (*Trifolium*). Caterpillar bright green with darker back stripe and light side stripes. Green chrysalis overwinters.

Flight: Several broods; March–December, weather permitting.

Habitat: Almost any open country; especially numerous in clover meadows, parks, and pastures; absent from dense forests and extreme deserts.

Range: Transcontinental except for most of Florida.

Because it feeds on clover, alfalfa, vetches, and many other pervasive legumes, the Common Sulphur has spread dramatically. It was probably originally a northern and eastern species, while the Orange Sulphur occupied the West and the South. Now, due to the spread of agriculture, their territories have largely merged.

<table>
<tr><td>86, 97, 100,
130</td><td>**Orange Sulphur**
"Alfalfa Butterfly"
(*Colias eurytheme*)</td></tr>
</table>

Description: 1⅝–2⅜" (41–60 mm). Male and female *above bright gold-orange with pink blush and black borders* (broken by yellow spots of female); black spot in FW cell, red-orange spot in HW cell. *Below, orange, yellow, or greenish-yellow with single or double red-rimmed silvery spot in HW cell and brown submarginal row of spots.* Both wings have pink fringe. Spring individuals may be yellow above with orange cell flush.

Similar Species: Common Sulphur is normally yellow. Orange form Queen Alexandra's Sulphur and Arctic Sulphur are smaller, paler, and lack prominent row of brown spots below.

Life Cycle: Egg whitish, long and pitcher-shaped; laid singly on top or bottom of leaves. Caterpillar dark grass-green with pink stripes low on sides, white stripes higher, and covered with tiny white hair. Chrysalis green, dashed with yellow and black; overwinters. Host plants include great array of herbaceous legumes; alfalfa (*Medicago sativa*) and white clover (*Trifolium repens*) are preferred.

Flight: Overlapping broods; March–December, shorter period farther north.

Habitat: Nearly any open space, particularly alfalfa fields.

Range: Almost ubiquitous in North America; rarer north of Canadian border, in subtropical Florida, and in coastal Northwest.

Like the Common Sulphur, this species breeds with enormous success on native and naturalized legumes. These food preferences, along with strong and probing flight habits, allow the Orange and Common sulphurs to colonize disturbed and cultivated areas as well as natural habitats. When these 2 sulphur species are superabundant in an alfalfa field, they often hybridize and produce many part-orange, part-yellow butterflies that are difficult to classify. However, the species have not merged; they breed true where density is lower.

103, 110, 113 Queen Alexandra's Sulphur
(*Colias alexandra*)

Description: 1½–1⅞″ (38–48 mm). *Male clear, lemon-yellow above. FW pointed and yellow-fringed* with small black spot near end of FW cell; brighter yellow spot in HW cell; *narrow black FW and HW border. Female yellow or white with* fainter cell spots and suggestion of dusky, *interrupted border on FW above. Below, FW yellow; HW granular, lime- to olive-green with silvery, unrimmed cell spot.* Some northern populations are orange.

Similar Species: Western Sulphur is dusky yellowish below with pink-rimmed cell spot. Pink-edged Sulphur has rounder wings with bright pink edge. Common Sulphur has row of brown spots below. Scudder's, Blueberry, and Giant sulphurs have pink fringes. Orange form Greenland Sulphur is rounder, smaller, and darker at base of wings above. Orange Sulphur has row of brown spots below.

Life Cycle: Eggs pale yellowish, pitcher-shaped; laid singly on golden banner (*Thermopsis*) and many other legumes, including some species of milkvetch (*Astragalus*),

wild pea (*Lathyrus*), and locoweed
(*Oxytropis*); lupine (*Lupinus*) may also be
used. Caterpillar green with lighter and
darker lengthwise stripes; overwinters
in 3rd instar after aestivating through
drier part of summer.

Flight: 1 brood; June–July.

Habitat: Clearings, meadows, roadsides, and
other openings, and edges in
mountainous ponderosa pine and true
fir forests; also sagelands.

Range: British Columbia and Alberta south to
Nevada and New Mexico in mountains
and Great Basin (yellow populations);
Northwest Territories, Manitoba, and
South Dakota (orange populations),
with mixtures in between.

When these butterflies congregate by
the score at damp earth following a
Colorado mountain shower, the
stationary green triangles and frenetic
yellow blurs create a striking spectacle.
This species is named for Queen
Alexandra, wife of Edward VII of
England, who also lent her name to the
world's largest butterfly, Queen
Alexandra's Birdwing (*Ornithoptera
alexandrae*) of Papua New Guinea.

131 Western Sulphur
(*Colias occidentalis*)

Description: 1½–2" (38–51 mm). *Male above bright
lemon-yellow with black borders,* orange
spot on mid-HW, and usually black
spot near end of cell. *Female paler yellow
or whitish above* with vestiges of black
border suffused with yellow. Both sexes
dusky burnt-orange to olive-yellow beneath
with a pearly, pink-rimmed spot in
HW cell; minor dark spot near front
tip of FW cell.

Similar Species: Queen Alexandra's Sulphur is greenish
below; its HW cell spot below lacks
pink rim. Pink-edged Sulphur is
smaller, more rounded, with more

prominent pink fringe. Common
Sulphur has a row of brown spots
below.

Life Cycle: Unreported. Host plants include a
variety of native and introduced
legumes, such as vetch (*Vicia
angustifolia*) and white sweet clover
(*Melilotis alba*).

Flight: Mid-June to mid-August.

Habitat: Open ocean bluffs, mountain slopes,
and subalpine meadows; also brush and
sunny clearings in forests.

Range: SE. Alaska to California on protected
coasts, Coast Ranges, and W. Cascade
and Sierra Nevada mountains.

Although its caterpillars feed upon an
array of legumes, the Western Sulphur
has not expanded across the countryside
as the Common and Orange sulphurs
have. Overall, the Western Sulphur is
confined to a northwestern coastal and
mountain belt, and to a limited set of
habitats within that region. In
Washington, one population flies at sea
level, on coastal bluffs and among
manzanita heaths; others dwell in the
Olympic, Coast, and Cascade ranges,
up to rather high altitudes. A dusky,
broad-bordered California race prefers
north and east slopes of open Douglas-
fir forests.

96 Mead's Sulphur
(*Colias meadii*)

Description: 1¼–1½" (32–38 mm). *Deep burnt-
orange above, bright grainy olive-green
beneath* with red spot and silver spot
respectively on HW cell above and
below. *Male's black margin above entire,
female's broken by yellow spots. Hair,
antennae, and wing fringes pink.*

Similar Species: Greenland Sulphur paler; golden orange
above, dusky beneath.

Life Cycle: Slim caterpillar green with black
speckles and hair; overwinters in 3rd

instar. Feeds upon alpine clovers, such as Parry's clover (*Trifolium parryi*) and whiproot clover (*T. dasyphyllum*).

Flight: 1 brood; July–August.

Habitat: Arctic-alpine tundra and rocky slopes; sometimes extending along streams and trails into forests below treeline late in summer and northward in its range.

Range: Restricted to Rocky Mountains: British Columbia south to N. New Mexico.

This species is named for Theodore Mead, a 19th-century Colorado collector who discovered many butterflies of the Rockies. The butterfly's dusky green undersides serve as solar collectors, warming its body when the sulphur lies flat against wind-buffeted tundra. Stubby wings carry it high above the next ridge if it should catch the wind. One of the few places where Mead's Sulphur may be seen with relative ease is Mesa Seco, a vast, flat tundra tableland more than 12,000′ (3660 m) in altitude, between Slumgullion Pass and Lake City in SW. Colorado.

89 Greenland Sulphur
(*Colias hecla*)

Description: 1⅜–1⅞″ (35–48 mm). *Male dull orange above with dark borders and dusky scaling near body.* Female dull orange or yellowish above, often nearly completely covered with dusky scaling. Both sexes have pink fringe and red-orange HW cell spot above. *Below,* green- to olive-yellow with black FW cell spot and *elongated HW cell spot, silver- and pink-edged.*

Similar Species: Mead's Sulphur is deeper, brighter orange above, greener below. Orange form Queen Alexandra's Sulphur has larger, more pointed wings and is pale near wing base above. Orange Sulphur has brown spots on HW below.

Life Cycle: Caterpillar green and laterally striped; it or chrysalis may overwinter; possibly takes 2 years to mature. Alpine milkvetch (*Astragalus alpinus*) is recorded host plant.

Flight: Late June–early August.

Habitat: Arctic tundra and grassy slopes and swales.

Range: Circumpolar. High Arctic Eurasia and North America including Baffin Island and Greenland, south along Hudson Bay to Churchill; in mountains to mid-Alberta and N. British Columbia.

Truly holarctic, this butterfly skirts the Arctic Circle around the globe. It is one of only about half a dozen butterflies resident in Greenland. Living on the margins of human habitability, it flits quickly and low over the tussocks and lichen-encrusted stones of the Northland.

94 Labrador Sulphur
(*Colias nastes*)

Description: 1⅛–1⅝" (28–41 mm). *Male above dull green sometimes shot with yellow; black margin heavily broken by yellow spots. Female above pale greenish-yellow or white* with border similar to male's but more obscure. Below, olive; both sexes have smeared red edge to pearly cell spot on HW.

Similar Species: Thula Sulphur yellower, less greenish; fringes and rims of cell spots redder and more prominent.

Life Cycle: Moss-green caterpillar has lighter head and side stripes that are red-edged below; feeds on alpine milkvetch (*Astragalus alpinus*) and other arctic legumes; overwinters.

Flight: Normally July–August.

Habitat: Higher altitudes, more barren slopes than other arctic *Colias* species. Stony, windswept ridges, screes, and summits.

Range: Alaska to British Columbia, south to

borders of Washington and Montana, east to Labrador on cold mountain ranges, and north to Arctic Archipelago; also Old World Arctic.

One of the few butterflies in Labrador, this sulphur can only be found in the lower 48 states in Glacier National Park, Montana, and the Pasayten Wilderness Area, Washington. Like most other northern sulphurs, the Labrador Sulphur can be either a powerful long-distance flier or a local stay-at-home, depending upon wind, sexual condition, age, and local environmental factors.

88 Thula Sulphur
(*Colias thula*)

Description: 1⅛–1⅝" (28–41 mm). *Male pale yellow with slight dusky scaling* near body above and across HW below. *Narrow charcoal border above, faced within by band of yellow spots blending into disk.* Pink fringes. *Below,* dusky yellow with *silvery HW cell spot pink-rimmed.* Female similar but paler.

Similar Species: Labrador Sulphur greener with lighter pink fringes and cell spot rims.

Life Cycle: Unreported. Lupine (*Lupinus arcticus*) is host plant.

Habitat: Rocky areas and tundra.

Range: Shores of Arctic Ocean and Seward Peninsula in Alaska, inland across far North Slope.

Some consider the Thula a race of the Labrador Sulphur, while others think it may be a subspecies of another extreme arctic sulphur, Booth's Sulphur (*C. boothii*). Yet since the Labrador Sulphur can be found in the range of the Thula, it is more likely that these are 2 distinct species.

90 Palaeno Sulphur
(*Colias palaeno*)

Description: 1⅜–1¾" (35–44 mm). *Male clear yellow above with black border; female yellow or white with spotted dusky border.* Both have pink fringes and *small, light cell spots* on FW above and HW above and below. *HW below olive suffused with duskiness,* sometimes very dark.

Similar Species: Blueberry Sulphur paler yellow with thick pink rim around HW cell spot below. Common Sulphur lighter with brown spots on HW below. Pink-edged Sulphur pale below with more prominent, pink-rimmed cell spot.

Life Cycle: Unreported. Arctic bilberry (*Vaccinium uliginosum*) and dwarf bilberry (*V. caespitosum*) considered host plants.

Flight: 1 brood during hotter part of short Arctic summer.

Habitat: Arctic heaths.

Range: Bering Sea south through Alaska and Yukon to S. Baffin Island, Labrador, and subarctic Canada; also N. Europe.

These blueberry-feeding sulphurs sometimes reach the Canadian subarctic taiga as well as the High Arctic. Churchill, Manitoba, a reasonably accessible railhead and port on Hudson Bay, is one place arctic butterfly enthusiasts have visited for many years to find this northern sulphur.

91, 92 Behr's Sulphur
"Sierra Sulphur"
(*Colias behrii*)

Description: 1¼–1⅝" (32–41 mm). Narrow wings. Rather *dusky; male cool, olive-green above with dark border; female greenish or dirty-white above with vestiges of dark border.* Both have faint pink fringe on FW and HW and whitish cell spot above and below on HW. Wings below lack dark border.

Life Cycle: Green, light-lined caterpillar feeds on dwarf bilberry (*Vaccinium caespitosum*).

Flight: 1 brood; late July–early September.

Habitat: High altitude subalpine meadows and streamsides.

Range: Restricted to Sierra Nevada of California at high altitudes between Tuolumne and Tulare counties.

Greenest of all the sulphurs, Behr's Sulphur is a Pleistocene riddle— isolated in California, it has no stepping-stones to relatives for miles to the north. This range pattern is known as a disjunction. An old legend relates that an early California collector lost his life to Indians for taking this presumably sacred creature. Modern butterfly watchers may observe it with impunity at Tuolumne Meadows and Tioga Pass in Yosemite National Park.

95 Pink-edged Sulphur
(*Colias interior*)

Description: 1⅜–1¾" (35–44 mm). Rounded wing with *bright pink fringes* and body hair. *Above, male yellow with black borders; female yellow or white with vestiges of borders on FW*. Both sexes have yellow to orange HW cell spot above and silver, pink-rimmed cell spot on HW beneath. Below, FW dull yellow, HW slightly olive.

Similar Species: Common Sulphur has row of brown spots on HW below. Palaeno Sulphur darker beneath; its cell spot lacks pink rim. Western Sulphur dusky below, wings longer with very light pink fringe. Blueberry Sulphur dusky below paler above, with broader black margins. Great Northern Sulphur larger, pointed wings and brighter red rims on HW cell spot below. Queen Alexandra's Sulphur lacks pink fringes.

Life Cycle: Pale, pitcher-shaped egg laid singly on

velvet-leaf blueberry (*Vaccinium myrtilloides* or *V. canadense*) and other species of blueberries. Caterpillar bright yellow-green with lighter back stripes and red-edged side stripes surrounded by bluish; overwinters.

Flight: 1 brood; early June–August, depending on latitude and altitude.

Habitat: Transitional open spaces such as burns, clearings, meadow edges, woodland roadsides; also marshes and bogs.

Range: British Columbia to Newfoundland, south in mountains and cool woodlands to Oregon, Lake States, New York, Pennsylvania, and Virginia.

For a century the Pink-edged Sulphur has been better known than some of the other northern sulphurs may ever be. Certain other *Colias* have pink fringes, but one or more other marks usually distinguish them; none is edged with such a rich rose when fresh. Like other sulphurs, the males of the slow-flying Pink-edged Sulphur gather at mud puddles and take nectar often; bristly sarsaparilla is visited in Maine, various asters in Washington.

87, 132 Blueberry Sulphur
(*Colias pelidne*)

Description: 1¼–1½" (32–38 mm). *Pale yellow above* (female sometimes white), *black margin entire in male and merely hinted in female.* HW cell spot faint, both above and below. *Pink fringes. Below,* dark cloudy olive; *HW black-dusted* quite heavily.

Similar Species: Pink-edged Sulphur darker above with narrower margins and lighter below with larger cell spot. Palaeno Sulphur lacks dark or pink rim on cell spot. Common Sulphur lighter below with brown spots on HW below. Scudder's Willow Sulphur much brighter below with larger, more silvery cell spot.

Life Cycle: Unreported. Host plants are blueberry
(*Vaccinium*), and, in Idaho,
creeping wintergreen (*Gaultheria hemifusa*

Flight: Mid- to late summer.

Habitat: Tundra in North and East; subalpine
mountains in West, at 6,000–11,500′
(1830–3508 m).

Range: E. British Columbia and W. Alberta
in Rockies; N.E. Oregon, Idaho,
Montana, and mountainous Wyoming;
E. Hudson Bay, James Bay, Labrador,
and Newfoundland.

Huckleberry heaths in the higher parts
of Yellowstone National Park are good
places to seek clear examples of this
sulphur.

Great Northern Sulphur
(*Colias gigantea*)

Description: 1½–2″ (38–51 mm). *Large. Bright
yellow* (some females white); *narrow
charcoal border on male, absent except for
smudge on female.* Pink fringes, yellow
HW cell spot above, *silvery spot below
with bright pink edging* (often double.)
Wings below yellow with trace of olive-
gray on HW.

Similar Species: Other sulphurs are much smaller and
usually have less bright pink fringe and
cell spot rim. Common and Orange
sulphurs have brown spots below.
Queen Alexandra's Sulphur lacks pink
cell spot rim.

Life Cycle: Caterpillar overwinters in 3rd instar;
feeds on netvein dwarf willow
(*Salix reticulata*) and other willows.

Flight: 1 brood; mid- to late summer.

Habitat: Hudsonian Zone of Canadian taiga,
southward in relict acid bogs.

Range: Alaska and Yukon to Hudson Bay,
south to British Columbia, Wyoming,
and Manitoba.

Like Scudder's Willow Sulphur, the
Great Northern Sulphur employs small,

tough bog willows rather than legumes or heaths as host plants. While individuals from the extensive Hudson Bay bogland are larger and fly farther, the Great Northern Sulphurs of the Rockies tend to be smaller and to remain close to their limited willow bogs. The Great Northern Sulphur resembles the giant sulphurs of the genus *Phoebis*, but they are larger and occur further south.

93 Scudder's Willow Sulphur
(*Colias scudderii*)

Description: 1½–1¾″ (38–44 mm). *Male clear yellow above, slightly greenish below; female usually white, sometimes pale yellowish. Wing margin above black on male, virtually absent on female.* Cell spot black on FW above and below, yellow on HW above, silver on HW below. Pink fringe; *no submarginal row of brown spots.*

Similar Species: Common Sulphur has brown spots beneath. Queen Alexandra's Sulphur greener below, lacks pink fringes. Blueberry Sulphur dusky below with smaller cell spot.

Life Cycle: Unreported. Hosts are willows (*Salix*).

Flight: 1 brood; late June at southern edge of range into late August at high altitudes.

Habitat: Willow bogs and willow-fringed meadows from Douglas-fir zone up to timberline.

Range: NE. Utah and SW. Wyoming through Colorado into N. New Mexico in higher mountain ranges.

Scudder's Willow Sulphur bears the name of a major 19th-century New England naturalist, Samuel Hubbard Scudder, whose monograph of New England butterflies is still an unparalleled regional classic. However, it is doubtful that Scudder ever saw a living example of this sulphur, as it is restricted to the middle Rockies.

Generally, Scudder's Willow Sulphur remains in wet or dry meadows near its host plants, even though it is a stronger flier than most sulphurs.

98, 101 **Dogface Butterfly**
(*Zerene cesonia*)

Description: 1⅞–2½" (48–64 mm). Both sexes *yellow above and below, with FW curved and pointed; black margin above scalloped in poodle-head design;* black FW cell spot suggests dog's eye. *Autumn–winter individuals may have* much black marking above and intense *rosy scaling beneath,* especially on HW and FW tips. Below, varies greatly, mostly yellow with reddish-pink mottling. Female sometimes white.

Similar Species: California Dogface male has purple flush on FW above and more orange all over; both sexes less heavily marked with black. Mexican Yellow much smaller, paler, with FW less pointed, HW more pointed.

Life Cycle: Caterpillar green, black-stippled; may be plain, striped lengthwise, or crossbanded with yellow and black. Host plants include false indigo (*Amorpha californica*), lead plant (*A. fruticosa*), and clovers (*Trifolium*). Chrysalis and/or adult may overwinter.

Flight: Mid- to late summer in North; most of year in South.

Habitat: Open woodlands; dry, sandy oak scrub; deserts in S. California, straying over many kinds of habitat.

Range: S. California to Florida, irregularly north through Midwest to Canada and Northeast, south to Argentina.

Although individuals and seasonal broods exhibit much variation, the Dogface Butterfly is unique—only the California Dogface resembles it. The Dogface is not uncommon throughout much of the South, and to be expected here and there in northern states; it is well worth singling out from other sulphurs for its great beauty. The magenta-flushed winter form (called "rosa") is especially striking. Also called *Colias cesonia*.

99, 102 California Dogface
"Flying Pansy"
(*Zerene eurydice*)

Description: 1⅝–2½" (41–64 mm). *Male saffron-orange above with curved and pointed FW tips* and thick *black FW margin indented in poodle-head pattern.* (Some populations also have black HW margin.) Male may have *brilliant plum-red or violet-purple sheen to dog-face. Female pale yellow* with pointed wing tips; barely suggesting or *lacking dark FW margin.* Below, deep yellow or greenish with black FW cell ring, silvery spots in HW cell.

Similar Species: Dogface lacks orange coloring and purple on male FW above; female more heavily marked with black. Mexican Yellow is smaller, paler, with less pointed FW and more pointed HW.

Life Cycle: Egg yellow-green. Caterpillar, to 1" (25 mm), dull green bearing either white, orange-edged side lines or thin, light crossbands. Chrysalis light green. Host plants include false indigo (*Amorpha californica*), clovers (*Trifolium*), and some members of the pea family (*Dalea*).

Flight: 2 broods; spring–autumn, midsummer at higher elevations.

Habitat: Mountains and foothills, open oak slopes, and Douglas-fir clearings; usually absent from deserts.

Range: California coast ranges and lower W. Sierra Nevada from Lake and Napa counties in north to Baja California; occasionally in W. Arizona.

California's spectacular state insect appeared on a U.S. postage stamp in 1976. It flies together with the Dogface

Butterfly in the San Bernardino Mountains where the 2 species may hybridize. Also called *Colias eurydice*.

White Angled Sulphur
(*Anteos clorinde*)

Description: 2¾–3½" (70–89 mm). *Large with FW angled into point,* short tail on each HW. *White above; red or blackish cell spot on each wing; bright yellow bar across FW* cell (pale or absent on female). Below, pale jade-green.

Life Cycle: Host plants are senna (*Senna spectabilis*) and some members of the mimosa family (*Pithecellobium*).

Flight: Late summer–autumn; year-round in Tropics.

Habitat: Openings and edges in tropical thorn scrub, riverside evergreen forests, and open spaces to north.

Range: Resident in S. Texas, occasionally straying into Arizona, New Mexico, and Colorado.

These very large angled sulphurs resemble no other butterflies. Although the shape and robustness of these sulphurs is reminiscent of the Old World Brimstones (*Gonepteryx*), the 2 genera are not closely related.

105 Yellow Angled Sulphur
(*Anteos maerula*)

Description: 2⅞–3½" (73–89 mm). *Large. Clear yellow with curved and bluntly hooked FW* black spot at end of FW cell. *HW bluntly tailed.* Below, jade-green to yellow.

Similar Species: Cloudless Giant Sulphur has smooth outline, lacking FW hook.

Life Cycle: Host plants are legumes, such as senna (*Senna*).

Flight: Any month possible, usually autumn.

Habitat: Disturbed, weedy, open areas, tropical dry forests, and along rivers.

Range: Mexico and Caribbean; wandering north to Dallas, Omaha, and S. Florida; perhaps resident in S. Texas.

Like the White Angled Sulphur, this high-flying species engages in northern emigrations that occasionally send some into the United States. Vacant lots with Spanish needles around Miami and weedy canals in southern Texas are good places to search for this rare stray.

32, 115, 118 Cloudless Giant Sulphur
(*Phoebis sennae*)

Description: 2⅛–2¾″ (54–70 mm). *Large with fairly elongated but not angled wings. Male clear yellow above* and yellow or mottled with reddish brown below. *Female* lemon-yellow to golden or white on both surfaces, with varying amounts of *black spotting along margin and black open square or star at end of FW cell.*

Similar Species: Yellow Angled Sulphur has angled wings. Other sulphurs much smaller.

Life Cycle: Egg pitcher-shaped; white at first turning pale orange. Mature caterpillar, 1⅝–1¾″ (41–44 mm), yellow to greenish, striped on sides, black-dotted in rows across back; hides in day tent formed with silk and leaves of host plant, which may be partridge pea (*Chamaecrista cinerea*), sennas (*Cassia*), clovers (*Trifolium*), or other legumes (Fabaceae). Chrysalis very drawn out, pointed at both ends and humped in middle; pink or green striped with yellow and green.

Flight: Multiple broods in South, 2 farther north. Year-round where warm enough, midsummer to fall where emigrant rather than resident.

Habitat: Many kinds of open spaces, gardens, glades, seashores, and watercourses.

Range: S. California, Gulf States, and Mexico

to Florida, emigrant through Midwest and all of East to Canada; rarer northward.

Summer movements bring this sulphur to states far north of its winter range, and autumn emigrations greatly reinforce its northern numbers, sometimes introducing millions to relatively small areas. This butterfly's appearance in the Rockies or New York is a real event. Yet all of these northern emigrants die without returning south; the function of the emigration is not really known. The Tailed Giant Sulphur (*P. neocypris*) somewhat resembles the Cloudless, but it is larger and bears stout, turned-in tails on the hind wings. It emigrates seaward in great numbers, yet is rarely encountered north of Mexico.

116, 119 Orange-barred Giant Sulphur (*Phoebis philea*)

Description: 2¾–3¼″ (70–83 mm). Large. *Male* yellow above and below, *with broad orange bar nearly crossing FW above* and broadly bordered with orange on HW above. *Female* golden-yellow or whitish with traces of orange areas of male and *black square-spotting in FW cell, FW tip, and along margins of both wings.* Wings below variable, blend of pink, lavender and orange with black cell spots on FW and HW.

Similar Species: Orange Giant Sulphur resembles female but has much lighter marginal markings.

Life Cycle: Egg spindle-shaped, pale. Caterpillar chartreuse with rust-colored to black side band, yellow band below and grainy black dots and hairs; wrinkled; tapered at both ends. Chrysalis strongly bowed, variable in color. Host plants are sennas (*Cassia*).

Flight: 2 or more broods; any month.

Habitat: Gardens and parks in Florida; tropical
scrub.

Range: Much of New World Tropics; S.
Florida, and S. Texas, infrequently
straying north as far as Nebraska and
New York.

The Orange-barred Giant Sulphur has
been established in southern Florida
since the 1930's. It is a powerful
speedster in flight, but pauses often to
probe for nectar from many kinds of
flowers.

108 Argante Giant Sulphur
(*Phoebis argante*)

Description: 2¼–2½″ (57–64 mm). *Male above
clear, bright orange; female* bright orange
to very pale whitish pink *with brown
markings above on border* and across both
wings below. Both sexes have *zigzag,
broken line on FW below.*

Similar Species: Orange Giant Sulphur has straight,
unbroken line on FW cell below.
Orange-barred Giant Sulphur much
larger; female has darker marginal
markings.

Life Cycle: Unreported. Host plant in Costa Rica is
a member of the legume family
(*Pentaclethra macroloba*).

Flight: Year-round in Tropics.

Habitat: Wet tropical lowlands and forests.

Range: Texas south into Mexico, Central
America, and probably S. America.

For many years this species was
confused with the very similar Orange
Giant Sulphur. Although still a rarity,
the Argante is now believed to occur
more frequently in North America than
had previously been thought. The very
pale female form is more common than
the orange form in the American
Tropics.

121 Orange Giant Sulphur
(*Phoebis agarithe*)

Description: 2¼–2½″ (57–64 mm). *Male bright golden-orange all over; female* ranges from golden-orange to pale pinkish *with delicate brown markings* along upper edges and across wings below. *Both sexes have straight, continuous cell line on FW beneath.*

Similar Species: Orange-barred Giant Sulphur female has darker marginal markings. Argante Giant Sulphur has zigzag, broken line on FW below.

Life Cycle: Silky yellowish-white egg turns orange-red. Caterpillar green with thin yellow side stripe edged with dark green. Chrysalis green with purplish mottling and purple and yellow stripes. Host plants are ram's horn (*Pithecellobium guadelupense*) and perhaps senna (*Cassia*).

Flight: March–December where resident in U.S.; year-round in Tropics.

Habitat: Clearings and edges of subtropical scrub.

Range: S. Florida and Mexico; wandering to Arizona, Texas, Kansas, and, very rarely, up Atlantic Seaboard.

The Orange Giant Sulphur is a strong flier, and fond of visiting puddles and flowers. It is more common in our area than its look-alike, the Argante Sulphur, but is still rarely seen.

127 Statira
(*Aphrissa statira*)

Description: 2¼–2½″ (57–64 mm). *Male bright yellow* on inner ⅔ of wings above, *with broad pale yellow satiny or mealy border;* below, pale yellow. Female creamy white above and below; lacks border. Both sexes have dark brown edge—just at FW tip on male (or absent), narrowly edging both wings of female.

Life Cycle: Florida host plants include powder

puff (*Calliandra*) and black mangrove (*Dalbergia*).

Flight: 2 broods; June–September, November–February.

Habitat: Generally coastal: islands, keys, shores, and flowery patches some distance inland.

Range: Tropical America extending into S. Florida and S. Texas.

Statira, along with some species of *Phoebis,* takes part in immense flights out to sea from South America. Unlike the Cloudless Giant Sulphur, Statira exhibits no such mass movements in North America, and in fact is considered rather an uncommon resident species.

104 Lyside
(*Kricogonia lyside*)

Description: 1½–2″ (38–51 mm). Variable, with *pointed FW tips.* Above, male usually whitish, female cream-colored or yellow (but some females white, some males yellow). Both have *golden patches at base of FW. Male often has black mark near outer edge of HW above* parallel to abdomen. Below, both sexes lime-green with yellow patch at FW base.

Similar Species: Cabbage White has black FW tip; Lyside's yellow flush at base of FW above is distinctive.

Life Cycle: Caterpillar, to ¾″ (19 mm), dull green with silvery or gray back stripes edged in brown. Chrysalis bluish-gray with white bloom. Host plant is Texas lignumvitae (*Porliere angustifolia*) in Texas; related species elsewhere.

Flight: Summer months.

Habitat: Arid plains and open places where host plants occur; apparently unrestricted in emigration.

Range: Tropical America, resident in S. Arizona and Texas, straying into Florida and Midwest.

Bursting out of Texas, Lyside often invades the Arkansas Valley and other watersheds of the western Great Plains.

125, 126, 128   **Tailed Orange**
(*Eurema proterpia*)

Description: 1⅜–1¾" (35–44 mm). *Male bright orange with black on FW costa; female more ocherous with broadly black FW tip.* Below, male yellow, female dark orange, both with rust-colored mottling. *Summer brood barely pointed* at tips of FW and HW; male has black veins. *Winter butterflies truly tailed* on HW, FW quite pointed; male's veins not blackened.

Similar Species: Little Yellow lacks HW tails; its black FW margin covers only half costa, HW margin above is black.

Life Cycle: Unreported.

Flight: July–October in 2 or 3 broods; throughout year in Tropics.

Habitat: Openings among brush and watercourse margins.

Range: Mexico into S. Arizona and Texas; also Central and South America.

The tailed winter form and angled summer form of this butterfly were once considered 2 separate species. It is not uncommon for members of this genus to exhibit 2 or more different seasonal appearances, as well as distinct sexual dimorphism. In any season, this butterfly's deep, rich orange and black costa make it absolutely distinctive.

107, 129   **Little Yellow**
(*Eurema lisa*)

Description: 1–1½" (25–38 mm). *Small. Bright yellow above with black FW tips and margins* on male, black often reduced to spot on female or series of connected spots at ends of veins. *Female sometimes white* with black markings. *Below,* yellow-green with dark smudges, including *rust-colored spot in upper HW margin.*

Similar Species: Mimosa Yellow is smaller with reduced areas of black on edges and orange spot on HW tip.

Life Cycle: Egg minute. Caterpillar, to ¾" (19 mm), green with fine pile and white side stripes; feeds on legumes such as senna (*Cassia*) and partridge pea (*Chamaecrista fasciculata*), clovers (*Trifolium*), as well as hog peanut (*Amphicarpa*).

Flight: Year-round in far South; May–October farther north.

Habitat: Many disturbed and natural open areas, especially roadsides and fields.

Range: Lake States and New England south, between Mississippi Valley and Atlantic to Gulf, South America, and West Indies.

Although common fairly far north in late summer, the Little Yellow cannot survive temperate or northern winters. The species refills the Northeast and Midwest every year with fresh immigrants, which furnish 1 or 2 more broods before the autumn chill kills them. Vast numbers of Little Yellows emigrate to the Caribbean and Atlantic. Columbus is supposed to have witnessed from the decks of the Santa Maria one such mass movement, probably consisting of this species or the Cloudless Giant Sulphur.

122 Mimosa Yellow
(*Eurema nise*)

Description: 1–1¼" (25–32 mm). *Small. Male bright yellow,* female paler, both *with black FW tips* sometimes extending narrowly along margins; *female also has dark smudge on HW margin.* Below, some slight mottling and rust-colored spot on HW.

Similar Species: Little Yellow is larger, more heavily marked; best distinguished by behavior and range.

Life Cycle: Caterpillar green, light and dark lined; known to feed on sensitive plant (*Mimosa pudica*).

Flight: 3 broods; most of year.

Habitat: Forest borders.

Range: S. Texas and Florida to Brazil.

The Mimosa Yellow and the Little Yellow behave very differently, and their ranges barely overlap in North America. The latter usually frequents wood margins as well as open spaces, while the Mimosa Yellow sticks to forest borders, darting into the cover of woods when alarmed. Two additional small sulphurs stray occasionally into our southernmost area. The Dina Yellow (*E. dina*) is bright golden orange-yellow, with a narrow black border and spotting beneath. Like the Mimosa Yellow, it occurs in southern Florida and Texas, but far less frequently. Chamberlain's Yellow (*E. chamberlainii*) is a Bahaman species known to appear in Florida; it is golden and has rounded wings.

117, 120 Sleepy Orange
(*Eurema nicippe*)

Description: 1⅜–1⅞" (35–48 mm). Male *bright golden-orange above with broad, uneven black border and a black FW cell spot;* yellow beneath with small brown blotches. Female orange or yellow; black border breaks down halfway along HW; below, HW cocoa-brown with darker blotches; FW orange. In autumn individuals HW below is darker.

Similar Species: Tailed Orange has HW tails and black FW margin covers entire costa; it lacks black margin on HW above.

Life Cycle: Egg long, narrow. Caterpillar, to 1″ (25 mm), slender, green, downy, and side-striped with white, yellow, and black. Host plants include senna (*Cassia*), clovers (*Trifolium*), and other legumes. Chrysalis ash-green to brown-black.

Flight: March–November in South with 2–3 overlapping broods; shorter flight period northerly. Adults may be seen in all months in Deep South.

Habitat: Old fields, wood edges, desert scrub, open pine woods, mountain canyons, watersides, and wet meadows.

Range: Throughout South and Southwest, northward east of Rockies, rarely well into Northeast.

This sulphur cannot withstand cold winters yet annually penetrates the northerly latitudes—a characteristic pattern of many North American butterflies. The common name, Sleepy Orange, may have come from the butterfly's habit of hibernating through the cooler days of the southern winter. In summer it seems anything but sleepy with its rapid flight. This species may be extremely prolific in the South, flying over meadows or clustering at mud puddles. Northern individuals are usually more dispersed, and may turn up in surprising places—well up a Colorado canyon, for example, or along a Great Lakes shoreline.

68, 124 **Fairy Yellow**
"Barred Sulphur"
(*Eurema daira*)

Description: 1–1⅜″ (25–35 mm). *Small. Variable* according to season. Male and female *light yellow above with broad black wing tips and margins;* male also has black bar along lower margin of FW. *HW below dark rust-brown on winter individuals, white on summer brood.* A tropical race that sometimes appears in S. Florida has white HW above.

Similar Species: Little Yellow has yellow HW above. Dwarf Yellow much smaller with heavier black markings, olive below.

Life Cycle: Egg white, irregularly spindle-shaped. Caterpillar, to ¾″ (19 mm), 2-toned green with light side stripe in between; feeds on legumes, such as joint vetch (*Aeschymomene viscidula*) and pencil flower (*Stylosanthes biflora*).

Flight: Throughout the year.

Habitat: Dry disturbed sites, waysides, shores, and scrub.

Range: Arkansas and E. Texas through Florida and the Carolinas, straying slightly north and west.

The various forms of the Fairy Yellow were once thought to represent more than one species. The forms might be better described as wet and dry season forms rather than winter and summer forms. They seem to have behavioral differences as well: dry season individuals tend to remain in woods and thickets, while wet period butterflies fly in open spaces.

123 Boisduval's Yellow
(*Eurema boisduvaliana*)

Description: 1⅛–1⅝″ (28–41 mm). *Short, sharp-tailed. Yellow with highly angular black borders on male,* restricted to *black FW tip on female.* Underside yellow with rust-pink mottling, especially at tips; more so in winter butterflies, less in summer.

Similar Species: Mexican Yellow is paler, much less reddish below; its black border makes narrow dog's head pattern.

Life Cycle: Unreported.

Flight: Any month possible.

Habitat: Sonoran scrub and openings.

Range: Mostly Mexico, possibly resident in

S. Texas, straying to S. Arizona and
Florida.

This species is named for Jean Baptiste
Boisduval, a 19th-century French
entomologist who dispatched collectors
and named material from much of the
New World. Another tailed yellow, the
Salome Yellow (*E. salome*), is more
golden-yellow and quite differently
black-margined; it rarely crosses the
border from Mexico.

106 Mexican Yellow
(*Eurema mexicana*)

Description: 1⅜–1⅞" (35–48 mm). *Cream-white to
pale yellow above*, male deeper yellow
along upper margins of HW above and
FW below; HW below bright yellow to
yellow-green with light to heavy rust-
colored dusting. *Both male and female
have deeply indented black FW margin
creating shape of a long-snouted dog's head;*
on male, margin continues to
HW tail.

Similar Species: Dogface Butterfly larger with blunter
face pattern, more pointed FW, and no
HW tail. Boisduval's Yellow is
yellower, lacks pattern of face.

Life Cycle: Sennas (*Cassia*) are host plants.

Flight: Early spring–late fall; summer in
northern part of range.

Habitat: Open oakwoods, meadows among pine
forest, desert chaparral, and mountain
canyons.

Range: Resident populations from Arizona to
Texas, emigrants periodically to S.
California and north through Rockies,
Mississippi Valley, and Lake States to
Ontario. South to Central America.

Although the Mexican Yellow has been
considered a resident in Colorado,
probably none of the *Eurema* yellows is
more than an emigrant and summer
breeder in states with cold winter

climates. Spring arrivals showing up in the same places each year give the appearance of having permanent "colonies." The Painted Lady also gives this impression, but neither it nor the Mexican Yellow overwinters in the North. The related Caribbean Dainty White (*E. messalina*) is shiny white above and below with a narrow, inky margin; it visits southern Florida, where it sometimes forms colonies.

109, 112 Dwarf Yellow
"Dainty Sulphur"
(*Nathalis iole*)

Description: ¾–1⅛" (19–28 mm). *Smallest N. A. pierid. Variable, yellow above with broadly black FW tips and diffuse black bars along costal and inner margins* of HW and FW, respectively. *Black more extensive on female,* which often also has black borders; HW above of female may be orange. Below, orange area lies along FW costa, black along inner FW and in submarginal spots; yellow *undersides olive-scaled* lightly or heavily.

Similar Species: Fairy Yellow is larger, reddish or whitish, but not olive beneath.

Life Cycle: Caterpillar, to ⅝" (16 mm), deep green with purple back stripes and parallel black and yellow side stripes; uniquely among our pierids, feeds on weedy composites, such as sneezeweed (*Helenium autumnale*), bur marigold (*Bidens pilosa*), and garden marigold varieties (*Tagetes*), as well as many flowers in the pink family, such as common chickweed (*Stellaria media*). Chrysalis green, smooth, lacks drawn-out head horn possessed by virtually all other pierid chrysalises.

Flight: Any month in southern, year-round range; progressively later farther north, usually distributed in migratory range by June or July and lasting until frost.

Habitat: Disturbed areas, grazed drylands,

canyons; particularly corridors such as rivers, roadsides, railroads, and rights-of-way.

Range: Resident in S. California, Arizona, and Gulf States, south into Mexico. Emigrant most years throughout Midwest to Manitoba, rarely into Northwest and Northeast.

Each year, if conditions are favorable in the South, cadres of Dwarf Yellows advance northward from Mexico and the southwestern desert, reaching as far as interior Canada. Recently, a much less frequent or obvious emigration to the Northwest and Northeast has been detected, recorded many years and miles apart, but always along major river courses such as the Snake and the Shenandoah. The secret of the Dwarf Yellow's remarkable travels is its ability to follow such river corridors, colonize weedy spots, and breed very rapidly. Whether individuals travel great distances themselves, or leapfrog northward through successive broods is not yet clear. Unable to withstand the frosts of winter, the Dwarf reenacts its invasion of the North each year, only to perish with the coming of autumn.

GOSSAMER WINGS
(Lycaenidae)

Approximately 5000 species described of an estimated 7000 species worldwide; 100 in North America. These small butterflies, with wingspans of ⅞–2″ (22–51 mm), include 4 groups—blues, coppers, hairstreaks, and the Harvester. Most species are found in the South; but the coppers and blues are better represented in the North. The 4 groups differ in wing venation as well as other characteristics: blues tend to be blue (especially males); hairstreaks are usually tailed; and most coppers are copper-colored. However, there are some tailed blues and coppers, a few blue coppers and hairstreaks, and some copper-colored hairstreaks and blues. The Harvester is distinctive. The coloring of gossamer wings derives from 2 different types of scales: the browns, grays, and oranges originate in pigmented scales, while the blues, greens, purples, and fiery coppers are generated structurally by light-refracting scales.

Unlike their close relatives the metalmarks, gossamer wings hold the wings over the back when at rest. In most species, the males have reduced fore legs and only the 4 hindmost legs can be used for walking. Females have all 6 legs well-developed.

Eggs are flattened and turban-shaped. Caterpillars are flattened, oval, and sluglike. Most caterpillars feed on plants, usually consuming the buds, flowers, or fruit, but the Harvester's caterpillars prey upon aphid nymphs. Caterpillars of many blues and hairstreaks possess honeydew glands; ants milk these and, in return, protect the caterpillars from predators. The rounded, compact chrysalises are usually formed in ground litter; some are girdled with silk for support. While pupating, the chrysalis produces faint sounds by flexing the body and rubbing together the membranes between body segments. Lepidopterists believe this sound may help ward off parasites or small predators. Egg and chrysalis overwinter but the caterpillar does so only rarely.

All coppers were formerly grouped in the genus *Lycaena,* but recently they have been subdivided into several genera.

---

518 Harvester
(*Feniseca tarquinius*)

Description: 1⅛–1¼" (28–32 mm). *Above, orange-brown to orange-yellow (sometimes nearly white) with blackish-brown borders and splotches; FW outer margin bulges outwards. Below,* FW mostly brownish with yellowish and darker spots; *HW has pattern of dots and fine grayish circular markings.*

Life Cycle: Caterpillar carnivorous, feeds exclusively upon woolly aphids of genera *Schizoneura* and *Pemphigus.* When mature, greenish-brown caterpillar, about ½" (13 mm), buries itself among dead prey; becomes covered with whitish aphid secretions and debris caught on long body hair. Back of chrysalis has a "monkey's face" design.

Flight: Multiple broods in far South, 3 broods north to New York, 1–2 broods farther north; usually April–May and July–September.

Habitat: Damp areas, such as swampy glades, wooded riverbanks, and forest trails; often around alders, but host aphids also occur on witch hazel, wild currant, hawthorn, beech, ash, and others.

Range: Nova Scotia and the Maritime Provinces south to Florida, west to Ontario and central Texas.

The Harvester is the sole American representative of an Old World tropical

group of carnivorous gossamer wings.
The adults are rather sluggish and fly
slowly, never moving far from their
host aphids. They usually do not visit
flowers, but do alight on twigs and
leaves to take aphid honeydew. Males
perch on sunlit patches of foliage in the
late afternoon. No other North
American butterfly looks or behaves
like the Harvester.

422, 517, 539 Tailed Copper
(*Tharsalea arota*)

Description: ⅞–1¼" (22–32 mm). *Short spikelike
tail. Above, male dull copper-brown, often
with purplish sheen;* HW has few black
dots and some orange marks near
margin. *Female light orange with brown
spots, shading,* and wide FW margins;
HW margin marked with orange. Both
sexes orange to gray *below with dark spots
and thin wavy lines* over inner ⅔ of
wings, outer third *with alternating bands
of soft brown and white,* straight on FW
and jagged on HW.

Life Cycle: Egg off-white; overwinters. Caterpillar
green, darker toward head, with fine
double white line down back, yellowish
line down side, and covered with
minute yellowish-white hair. Chrysalis
brown or mottled yellow-brown. Host
plants are various wild currants and
gooseberries (*Ribes*).

Flight: 1 brood; mostly July–August, late
May–early September in California.

Habitat: Usually below 8000' (2440 m), often
along watercourses or in moist areas,
mountain meadows, and clearings.

Range: S. Oregon south to S. California, east
to edge of Great Plains, north to S.
Wyoming, and south to central New
Mexico.

The Tailed Copper can be found
perched on the host plant or at flowers,
such as rabbit brush, buckwheat, and

clovers. Although it can be very common, this butterfly is often rather local. Beneath, the ground color and the amount of white banding are quite variable, the darkest individuals coming from Virginia City, Nevada, the Great Basin, and Rocky Mountains, and some of the lightest from around Los Angeles, California. In spite of its tails, the Tailed Copper should not be confused with hairstreaks because of its distinctive coloring.

513 American Copper
"Flame Copper"
(*Lycaena phlaeas*)

Description: ⅞–1⅛" (22–28 mm). *Above. FW bright copper or brass-colored* with dark spots and margin; *HW dark brown with copper margin. Undersides mostly grayish* with black dots; FW has some orange, *HW has prominent submarginal orange band.*

Similar Species: Female Bronze Copper is similarly patterned but larger. No other copper, except the duskier Bog Copper, is as small as the American.

Life Cycle: Egg pale green with pronounced ribbing. Mature caterpillar downy, either green with rose side markings, or dull rose with yellowish side markings; overwinters as chrysalis. Host plants are sheep sorrel (*Rumex acetosella*) and curly dock (*R. crispus*) in eastern lowlands, and mountain sorrel (*Oxyria digyna*) in arctic and alpine habitats.

Flight: Lowland form has multiple broods: 2 broods in North and 4 or more in South from April–October. Arctic and alpine forms have 1 brood in August.

Habitat: Lowland form in waste places, pastures, yards, and old fields. Arctic and alpine forms found above treeline on barren ground, talus slopes, and fell-fields.

Range: Lowland form widespread over East from Nova Scotia and Gaspé south and east to North Dakota, NE. Kansas,

Arkansas, and N. Florida. Arctic and
alpine forms from Greenland west to
Alaska, south to Hudson Bay; also in
western mountains to Colorado and
central California. Holarctic, occurring
in Europe and Asia.

The eastern lowland form is very
common in the North but rare in the
South; eastern populations are more
common than those in the West. Often
considered pugnacious, the American
Copper displays much seasonal and
individual variation. The rare arctic and
alpine forms are sometimes quite dingy.

512 Lustrous Copper
(*Lycaena cupreus*)

Description: 1–1¼″ (25–32 mm). *Male above lustrous
bright copper to red-orange with black dots
and narrow black border.* Female darker,
less shiny, with darker markings.
Underside of FW has orange flush; *HW
below pale pink or dark pinkish gray-brown
with* black dots and *thin, red submarginal
line.*

Similar Species: American Copper lacks copper coloring
on HW above. Ruddy Copper has
clear, silvery, almost unspotted HW
below; male has fewer spots above,
especially on HW.

Life Cycle: Unknown. Caterpillars were once found
on mountain sorrel (*Oxyria digyna*) in
Colorado above timberline. Females lay
eggs on dock (*Rumex paucifolius*) below
timberline in California and NE.
Nevada. Apparently overwinters as egg.

Flight: 1 brood; July–August.

Habitat: High mountain altitudes in flowery
meadows and around rocky streambeds
and stream benches; also talus slopes
and fell-fields above timberline in
Rockies and Cascades.

Range: N. Washington, SE. Oregon, and
Sierra Nevada of California, east to
Rockies; N. British Columbia south to

New Mexico. Scattered and local
between 3 major mountain
ranges.

Lustrous Coppers tend to be highly
variable. Some populations that occur
below timberline are darker and often
quite numerous. Arctic-alpine
populations tend to be scarcer and to
have more brass coloring. The rockslide
form from Colorado, Snow's Copper
(*L. c. snowii*), was once considered
a distinct species. Its stony habitat
above 11,000' (3355 m) is certainly
different from the Lustrous Copper's
ponderosa pine forests; this is perhaps
an argument for the Snow's status as
a separate species.

511, 522 Ruddy Copper
(*Chalceria rubidus*)

Description: 1⅛–1¼" (28–32 mm). *Male above
bright copper,* sometimes with darker
cast; *female dull gold with black spots and
brownish shading. Below,* both sexes
nearly white to pale yellow; FW has
black spots, *HW immaculate or with
black dots (lacking any red markings).*

Similar Species: Lustrous Copper has darker, more
heavily spotted HW above and below.
Female Ruddy Copper may resemble
female Blue Copper or Common Blue
but latter 2 have heavier black spots
and lack an orange marginal zigzag on
HW above.

Life Cycle: Largely unknown. Host plants include
wild rhubarb (*Rumex hymenosepalus*) and
dock (*R. triangularis*).

Flight: 1 brood; July–August.

Habitat: Moderate to high elevations, up to
11,000' (3355 m) in open, dry areas;
sagebrush near meadows or streams;
sandy watercourses.

Range: Sierra-Cascade axis to E. Rockies and
adjacent prairies; Alberta south to N.
Arizona and New Mexico.

Brilliantly colored and rapid fliers, Ruddy Coppers can be observed taking nectar at flowers such as wild buckwheats, rabbit brush, and bush cinquefoil. Ferris' Copper (*C. ferrisi*) is a similar, very recently described species that differs in having an orange flush across the fore wing below; it is known only in the White Mountains, Arizona.

43, 466, 492, **Blue Copper**
516   (*Chalceria heteronea*)

Description: 1⅛–1¼″ (28–32 mm). *Male above bright sky-blue; female gray-blue or slate-gray* with dark spots and soft brown shading. Both sexes *below nearly white* with black spots on FW; *HW variable, with or without blurred darker spots (lacking any red* markings).

Similar Species: Large Copper is larger and more brilliant. Common Blue has more and better defined black spots on HW below. Female Blue Copper resembles female Ruddy Copper but lacks orange zigzag and highlights. Female Common Blue is darker, slate-gray.

Life Cycle: Egg overwinters. Caterpillar, to ⅝″ (16 mm), pale green with silver-white body hair. Feeds on various species of wild buckwheat (*Eriogonum fasciculatum, E. latifolium, E. umbellatum, E. nudum* complex, *E. microthecum*). Chrysalis, to ⅜″ (10 mm), dull green flecked with dark brown; darker gray-brown predominates at tip.

Flight: 1 brood; July–August.

Habitat: Low to middle elevation mountain canyons, sagelands, and flowery river flats and plateaus.

Range: Mainly east of Sierra-Cascade axis, across Great Basin and through Rockies to E. Wyoming; SW. Canada to N. Arizona and New Mexico. Also west of Sierra Nevada in N. California to coast; locally in Tehachapi and Tejon ranges, south of Bakersfield.

Despite its bright sky-blue color, the Blue Copper's venation and structure ally it with the coppers rather than with the blues. This common color demonstrates convergent evolution between the Blue Copper and blues— through the process of natural selection the 2 species have grown to resemble each other. Common over most of their range, both male and female Blue Coppers frequent flowers, especially the blossoms of wild buckwheat.

---

425 Hermes Copper
(*Hermelycaena hermes*)

Description: 1–1⅛" (25–28 mm). *Short. spikelike tail. Above dark brown; golden-yellow FW disk* with bold black spots and yellow or orange zigzag on HW near tail. Below, bright yellow; FW has dark spots, HW almost solid yellow.

Similar Species: Tailed Copper female has orange on both HW and FW above; male lacks orange on both HW and FW above, except for small submarginal band on HW. Tailed Copper also has more complex spotting below.

Life Cycle: Egg white; overwinters on stem of buckthorn (*Rhamnus crocea*). Mature caterpillar apple-green with darker green and yellow-green stripes on back and indistinct stripe on sides. Grass-green chrysalis has yellow stripes and bars; attaches itself to host plant.

Flight: 1 brood; late May–late July, peaking about June 20.

Habitat: Hillsides and canyon bottoms in chaparral and coastal sage scrub.

Range: Limited to vicinity of San Diego, California; found within 50–100 mile (80–160 km) radius.

The Hermes Copper takes nectar from California buckwheat, which commonly grows near buckthorn—the caterpillar's host plant. Because buckthorn covers

steep slopes and gully bottoms, this
butterfly survives close to suburban San
Diego. However, accelerating land
development has already destroyed
many colonies, threatening the future
of this species.

528, 536, 537   **Great Gray Copper**
(*Gaeides xanthoides*)

Description: 1¼–1¾″ (32–44 mm). *Large. Male
above uniform gray-brown, with orange*
scaling and small black dots *along HW
margin. Female above has few black spots*
and often *light orange* scaling, especially
on FW; orange margins on HW more
extensive than male. Both sexes *below
very light gray with* small black spots
and either *prominent orange HW margins
(East) or thin orange HW marginal band
(West)*.

Similar Species: Edith's Copper is smaller and darker
below with large gray as well as small
black spots and less prominent orange
zigzags.

Life Cycle: Egg white. Mature caterpillar green,
yellow-green, or magenta with dark
orange stripes. Chrysalis pink-buff;
found in loosely constructed cocoon of
silk and soil in debris at base of host
plants—several docks (*Rumex
hymenosepalus, R. conglomeratus, R.
crispus, R. obtusifolia*).

Flight: 1 brood; late May–early August.

Habitat: In West, dry slopes, sandy flats, and
dry riverbeds. In East, prairie swamps,
marshes, and meadows.

Range: Oregon south to Baja California, east
through Rockies and across to Great
Plains from Manitoba to Oklahoma.

The Great Gray Copper is the largest
American copper and one of the largest
gossamer wings. Its rapid and jerky
flight make this butterfly difficult to
watch, except when it stops to take
nectar from milkweed. In prairie

gullies, the Great Gray Copper may fly
side by side with Viceroys, and
sometimes the 2 species clash in flight.

521, 529 Edith's Copper
(*Gaeides editha*)

Description: 1⅛–1¼" (28–32 mm). *Above, male
dark gray* with faint dark spots; *HW
rear margin has faint orange scaling.
Female above brownish-gray with yellow
suffusion* and numerous and larger spots;
HW has extensive orange strip along outer
margin. Both sexes *below pale tan to
gray;* black spots on FW; *HW has
blotchy gray-brown spots over inner* ⅔,
often minutely ringed by lighter scales.

Similar Species: Great Gray Copper larger, lighter
below with smaller, more widely spaced
spots; its habitat is rarely the same.

Life Cycle: Poorly known. Host plants are horkelia
(*Horkelia fusca, H. tenuiloba*), dock
(*Rumex*), or cinquefoil (*Potentilla*).

Flight: 1 brood; June–August.

Habitat: Mountain meadows, watercourses,
forest openings, and roadsides.

Range: SW. Alberta south through Blue
Mountains of Washington, Cascades of
Oregon, and Sierra Nevada, Great
Basin, and Rockies to California,
Nevada, and Colorado.

Edith's Copper is sometimes
tremendously abundant, especially in
flowery glades and meadows at middle
and high elevations below timberline.
Adults frequently take nectar from
dogbane and yarrow. This butterfly was
named by Theodore Mead for W.H.
Edwards' daughter, whom Mead later
married.

520, 524 Gorgon Copper
(*Gaeides gorgon*)

Description: 1⅛–1¼" (28–32 mm). *Male above generally copper-brown with purple reflections* and a few indistinct dark spots; *HW margin slightly yellowish. Female above pale yellow with* brown spots and smudges and *orange marginal zigzags.* Both sexes *below pale whitish to tan-brown* with many small black spots; *HW near margin has row of separate, black-capped orange crescents.*

Similar Species: Male Purplish Copper usually darker tan below with orange zigzags rather than crescents. Female Great Gray Copper lacks yellow suffusion above.

Life Cycle: Egg cream-colored; overwinters on stem of host plant, wild buckwheat (*Eriogonum elongatum, E. nudum*). Mature caterpillar pale turquoise with long white hairs. Chrysalis is blue-green with some yellowish coloring.

Flight: 1 brood; late May–June, sometimes into July.

Habitat: Restricted to areas where wild buckwheat grows; commonly chaparral, oak, and pine woodlands, brushy hillsides, gulches, and roadcuts.

Range: Pacific border from S. Oregon south to N. Baja California, east to Warner Mountains of NE. California, Sierra Nevada, and desert edge in S. California.

Although the male and female Gorgons are similar to many other coppers, the 2 sexes are dramatically dissimilar to each other in color and pattern. This dissimilarity is known as sexual dimorphism. There have even been instances of what is known as bilateral gynandromorphism in the Gorgon— this is a condition in which a single butterfly exhibits female characteristics over one half of its body and male characteristics over the other half.

514, 527 Bronze Copper
(*Hyllolycaena hyllus*)

Description: 1¼–1⅜" (32–35 mm). *Large. Male above dark copper-brown with violet sheen on FW;* shiny gray with *orange margin on HW. Female above bright orange* to yellowish with dark spots and margin on FW; HW dark gray with dark spots and *prominent orange or yellowish HW margin. Below both sexes have* FW bright orange with black spots and often gray margins; *HW bluish-white to gray with small black spots and prominent fiery orange margin.*

Similar Species: Purplish and Dorcas coppers smaller, less prominent zigzag on HW below.

Life Cycle: Host plants are various docks, especially curly dock (*Rumex crispus*) and knotweeds (*Polygonum*). Mature caterpillar is bright yellowish-green with dark line down back.

Flight: Mostly 2 broods; June–July, August–October, possibly a 3rd brood in some areas. 1 brood in Colorado; June–July.

Habitat: Strongly restricted to vicinity of host plants; moist areas such as grassy and sedgy margins of wet meadows, swamps, small streams, roadside ditches, and even salt marsh edges.

Range: Northwest Territories east through Great Lakes of Canada to Maine, south to S. Alberta, E. Idaho, E. Colorado, Kansas, Ohio, Maryland, and N. New Jersey; also Mississippi.

The beautiful Bronze Copper is most common in the East, becoming more local to the West. Even there it can be common in its restricted habitat, perching on grasses and sedges, often remaining still until disturbed, and seldom visiting flowers. Previously known as *Lycaena thoe*.

519 Mariposa Copper
(*Epidemia mariposa*)

Description: 1–1⅛" (25–28 mm). *Male above copper-brown with purple sheen* and some faint darker spots, suggestion of orange zigzag marginal strip on HW. *Female above dark brown with darker spots and bright orange spotting.* especially on FW. *Below.* both sexes have yellowish or gray FW with dark spots and grayish margin; HW mottled soft gray or gray-brown, *HW darker on inner ⅔ with fine. mostly curved black markings.*

Similar Species: Nivalis, Purplish, and Dorcas coppers somewhat similar above; below, no other copper resembles Mariposa.

Life Cycle: Early stages not yet formally described. Caterpillar may feed on knotweed (*Polygonum douglasii*).

Flight: 1 brood; July–August.

Habitat: Mountains below timberline; near sea level in far Northwest; moist areas, such as bogs, meadows, seeps, and clearings.

Range: Southernmost Alaska and Queen Charlotte Islands, south in Cascades and Sierra Nevada to central California; in Rockies to central Wyoming.

The Mariposa Copper favors cool northern climates—glaciers probably determined its distribution in coastal and mountain pockets. This species occurs locally and sparingly in California, more abundantly in the Cascades of Washington; in the Queen Charlotte Islands it is the most common butterfly. "Mariposa" means butterfly in Spanish.

540 Nivalis Copper
"Lilac-bordered Copper"
(*Epidemia nivalis*)

Description: 1⅛–1¼" (28–32 mm). *Above. male pale reddish-copper with purple sheen* and a few

dark spots; *orange HW marginal band.
Above, female dark brown with yellow-
orange shading* and HW marginal band.
Below, both sexes light yellow-orange
with spots on FW, and *HW pinkish or
2-toned* with yellow near base and pale
to deep lilac-purple toward margin,
with or without fine black dots.

Similar Species: Purplish Copper does not have 2-toned
underside. Mariposa Copper has gray
border on FW below; HW gray, not
pinkish, with many fine black lines and
dots.

Life Cycle: Pale bluish egg overwinters. Mature
caterpillar pale green with 1 red and 2
white lines on back, covered with fine
brownish hair and raised white points.
Chrysalis straw-yellow with brown
spots and stubby hair. Host plant
knotweed (*Polygonum douglasii*).

Flight: 1 brood; July–August.

Habitat: Mountains, flowered slopes and
meadows, forest clearings, rocky
outcrops, stream benches, and
sagebrush flats.

Range: British Columbia south in mountains
to Sierra Nevada in California; east to
Wyoming and Colorado.

Few butterflies are lovelier than a
freshly emerged Nivalis Copper, due to
the beautiful contrast of mauve and
gold below. The scales responsible for
these delicate hues fall away quickly
under the rigors of adult life, making
these coppers rather dingy in a few
days. Butterfly distribution offers many
mysteries—for example, this butterfly
dwells in the Olympic Mountains and
the eastern Cascades of Washington
State but not in the intervening western
Cascades.

525, 530 Bog Copper
(Epidemia epixanthe)

Description: ⅞–1″ (22–25 mm). *Small. Above, male dark brown with purple gloss* and some dark spots; *female duller* and somewhat grayish, often with bright orange margins. *Below, both sexes yellow to white,* with some black spots (heavier on FW); *red marks form submarginal row on HW.*

Similar Species: Both Dorcas and Purplish copper males have purple reflections but are larger with more prominent orange zigzags, and their undersides are darker tan or brown.

Life Cycle: Largely unknown. Egg overwinters on host plant, wild cranberry (*Vaccinium macrocarpum*).

Flight: 1 brood; late June–late July or August.

Habitat: Closely restricted to acid bogs and boggy marshes where wild cranberry grows.

Range: Newfoundland west to the Riding Mountains of Manitoba (and possibly E. Saskatchewan), south to Minnesota, Michigan, Wisconsin, N. Ohio, NW. Indiana, Pennsylvania, and New Jersey.

The Bog Copper is found in relatively small numbers. Some colonies in New York are known to have only 50–100 individuals year after year. This species is more common in the East and more local and scarce farther west. There is considerable geographic variation, as well as variability within populations, especially in the prominence of the black spots beneath. Observers can find Bog Coppers perching on cranberries and other bog shrubs, or flying feebly among plants.

515, 526 Purplish Copper
(Epidemia helloides)

Description: 1–1¼″ (25–32 mm). *Above, male dull copper-brown with purplish reflections* and scattered black spots; HW with usually prominent submarginal orange crescents; margins of both wings narrowly darkened. *Above, female orange to tawny with brown shading, brown margins* and heavier dark spots; orange crescents on HW more prominent. *Below,* both sexes have ocherous FW with black spots; *HW dull pinkish-tan to grayish-tan with* fine black spots and *scalloped red submarginal band.*

Similar Species: Nivalis Copper has 2-toned HW beneath. Bog Copper smaller and darker above, white to yellowish beneath. Dorcas Copper usually darker above and with less prominent red line on HW beneath (Rocky Mountain populations indistinguishable from this species).

Life Cycle: Egg whitish. Mature caterpillar grass-green with yellow stripes longitudinally on back and sides and also obliquely on sides; feeds mainly on plants in the buckwheat family (Polygonaceae), such as docks, sorrels, knotweeds (*Rumex* and *Polygonum*), but some specialists have observed use of cinquefoil (*Potentilla*) as host plant. Stubby, greenish-gray chrysalis marked with gray and brown.

Flight: Multiple broods; most of year in S. California, May–September (mostly July–August) farther north.

Habitat: Sea level to over 10,000' (3050 m) in a wide variety of habitats: urban weed fields, tidal marshes, mountain canyons; often in conjunction with moist areas.

Range: Pacific Coast from S. Canada to Mexico east to Great Lakes region of S. Canada, Michigan, Illinois, W. Kansas, and New Mexico.

This commonest California copper becomes less prevalent farther east, often consorting with other coppers. In the Rocky Mountains, especially at higher elevations, the Purplish Copper is very similar in appearance to the Dorcas Copper, and in this area the 2 species cannot be distinguished by most experts unless their natural host plants are known. There is speculation that they may be forms of the same species. The Purplish Copper, along with the Mylitta Crescentspot and the introduced Cabbage White, flies in many parts of the Northwest otherwise nearly devoid of butterflies. Hardier than their size suggests, these butterflies persist late into the autumn.

523   Dorcas Copper
(*Epidemia dorcas*)

Description: 1–1¼" (25–32 mm). *Above, male dark copper-brown with bright deep purple reflections,* dark spots and borders and red-orange submarginal crescents absent to well developed on HW; *female mostly dark brown with orange or buff markings restricted mostly to outer third* of wings. Both sexes *below dull yellowish- to pinkish-brown* with variable dark spotting and a thin submarginal reddish line (often absent).

Similar Species: Bog Copper smaller and even darker above with a yellow to white underside. Purplish Copper usually brighter above and more ocherous below (Rocky Mountain populations indistinguishable from this species).

Life Cycle: Caterpillar feeds on various cinquefoils (*Potentilla*); in East associated with shrubby cinquefoil (*P. fruticosa*)

Flight: 1 brood; July–August.

Habitat: Moist meadows, forest clearings, bogs, and salt marshes.

Range: Alaska east around Hudson Bay to Newfoundland, south to Maine, Ohio, and Minnesota; in Rockies to N. New Mexico.

The Dorcas Copper more or less replaces the Purplish Copper in cooler environments of the North and East.

Dorcas Coppers from higher elevations
on the Rocky Mountains closely
resemble the Purplish Copper, and
where both occur, only an expert can
distinguish them. Dorcas Coppers are
very much alike throughout their
ranges with 2 exceptions: near Lincoln,
Maine, a small, dull, dark population
occurs, and near Bathurst, New
Brunswick, individuals are notably
darkly marked below. Adults of all
populations commonly visit flowers,
and seem to favor those of composites.

382, 498 Colorado Hairstreak
(*Hypaurotis crysalus*)

Description: 1⅜–1½" (35–38 mm). *Large. Above,
deep purple with* wide dark margins and
usually a few *marginal golden-orange spots
on both wings;* male has incomplete dark
FW band, female has complete band.
Both sexes *below light to brownish-gray
with white-edged darker line* and short
orange submarginal band on both
wings; blue patch and black-centered
orange spot on HW near tail.

Life Cycle: Adults strongly associated with Gambel
oak (*Quercus gambelii*); eggs are found on
bark and caterpillars have been reared
on leaves.

Flight: 1 brood; June–September (mostly
July–August).

Habitat: Oak canyons and mountain foothills,
at 5000–9000' (1525–2745 m).

Range: NE. Utah and SW. Wyoming south
through E. Nevada and Colorado to SE.
Arizona and New Mexico; probably also
Mexico.

The Colorado Hairstreak, along with
the Golden Hairstreak, bears a closer
relationship to Old World hairstreaks
than it does to our other North
American species. It is surely one of our
most brilliant and exotic looking
butterflies. Unlike many butterflies,

this hairstreak remains active past sunset and sometimes on cloudy or even rainy days.

426, 427 Golden Hairstreak
(*Habrodais grunus*)

Description: 1–1¼" (25–32 mm). *Short stubby tail. Sexes similar above, dull golden-brown with broad, diffuse, brownish borders;* male darker. *Below, golden-tan to silvery-yellow with metallic gold flecks* and several inconspicuous marks; brownish postmedian band over wings; HW has marginal row of blue or gray chevrons.

Similar Species: Behr's Hairstreak darker below with heavier pattern; tailless. Nelson's Hairstreak reddish rather than golden above and below.

Life Cycle: Egg overwinters on twigs of host trees: chinkapin (*Chrysolepis chrysophylla*), canyon oak (*Quercus chrysolepis*), huckleberry oak (*Q. vaccinifolia*), and tanbark oak (*Lithocarpus densiflorus*). Caterpillar pale bluish-green with light yellowish lines, covered with silver-white hair. Pale bluish-green chrysalis has brownish points and short white hair; attaches to leaf, producing adult in about 2 weeks.

Flight: 1 brood; late June–August.

Habitat: Oak-covered canyons, slopes, ridges; also forest roads in Pacific Northwest. Absent from desert mountains.

Range: S. Washington, SW. Idaho; Oregon south through California and W. Nevada to coastal Arizona.

Perched on its host tree, often on a protruding branchlet, this hairstreak looks like a dried leaf. It spends most o its time high up in the trees, only occasionally descending to take nectar from everlasting flowers. Noted for its crepuscular habits, the Golden Hairstreak flies most actively in early morning and evening, often in shade.

29, 461, 646 **Atala**
(*Eumaeus atala*)

Description: 1¾" (44 mm). *Large. Above, male black
with shiny blue dusting around FW cell,*
similar *blue dots along HW margin; female
similar* with less blue, larger dots, and
FW iridescent green except for black veins
and wide margin. Both sexes *below black
with 3 rows of metallic blue dots on HW
and 1 coral-colored spot* near bright *coral-
red abdomen.* Fringes black.

Similar Species: Cycad Butterfly is larger and has white
fringe; range does not overlap in U.S.

Life Cycle: White egg, taller turban than most
hairstreaks; laid singly or in small
clusters. Host plants cycads (*Zamia,*
especially *Z. integrifolia*). Caterpillar,
¾–1" (19–25 mm), brick-red with 2
rows of yellow spots; feeds in groups.
Chrysalis at first shows dots of
caterpillar, then becomes mottled
reddish-brown.

Flight: Multiple broods; virtually year-round.

Habitat: Brushy wood edges and hammocks.

Range: Extreme S. Florida and Bahamas.

Once common in the subtropical wilds
of southernmost Florida, this species
was abundant in what is now
downtown Miami. Urbanization and
probably hurricanes, fires, and
competition from another species for
the host plant, all played a part in its
decline. The Floridian race was
considered extinct until a few small
colonies were found in the early 1960's.
Atala visits flowers of Spanish needles
and scrub palmetto with a slow and
lazy flight.

462 **Cycad Butterfly**
(*Eumaeus minijas*)

Description: 1¾–2" (44–51 mm). *Large. Black,
scaled with iridescent green above on male,
blue-green on female,* more on FW than

HW; row of green spots around HW margin, paler and larger on female. Black *below, with 3 rows of green spots around HW outer half,* pale on female; *orange spot near orange-tipped abdomen. Wing fringes pale whitish.*

Similar Species: Atala smaller, with black fringe.
Life Cycle: Caterpillar feeds on fronds of cycads (*Zamia,* including *Z. loddigesii*).
Habitat: Forests, trails, clearings with cycads.
Range: South America rarely to S. Texas.

Closely related and similar to the Atala, the Cycad visits the United States infrequently, perhaps carried by tropical storm winds.

413 Marsyas Hairstreak
(*Pseudolycaena marsyas*)

Description: 1¾–2″ (44–51 mm). *Large. Above, shiny sky-blue* with broad, black, inwardly suffused margin by *slightly drawn out FW tip. Below, pale grayish with black lines and small black dots* with white edges; sometimes also purple or bluish tints. 1 tail on HW.
Life Cycle: Unknown. Females have been observed laying eggs on young croton leaves (*Croton niveus*) in Mexico.
Flight: Autumn in Trinidad and Tobago.
Habitat: Scrub country to forest edges.
Range: Argentina north to northern Mexico, straying to Texas.

This butterfly is not uncommon in its range, but it is usually a solitary flier, with a fast dashing flight. It is very fond of taking nectar from white flowers. The Marsyas Hairstreak is the only member of its genus to enter our range, and may in fact be the sole member of the genus worldwide.

385 **Zebina Hairstreak**
(*Thereus zebina*)

Description: 1⅛–1¼" (28–32 mm). *Above, male bright, shiny blue with gray stigma,* black border, *small, rust-colored spot near tail. Female pale bluish-brown with row of dark spots along HW margin,* and 1 prominent orange spot near tail. Below, male dark gray, female light gray; both have black and white postmedian line fairly straight on FW, irregular on HW, small blue patch at base of tail with orange-capped black spots on either side.

Habitat: Low tropical dry forests and scrub country.

Range: South America to extreme S. Texas.

Common and widespread in the New World Tropics, the Zebina Hairstreak seldom crosses our border from Mexico. The females look like many other hairstreaks at first glance, but are distinguished by subtle differences. The males, however, are quite unlike other North American hairstreaks. Both sexes closely resemble another tropical species, the Spurina Hairstreak (*T. spurina*); only experts can distinguish them, and then with difficulty. In contrast, a third member of the genus, the Palegon Hairstreak (*T. palegon*), looks entirely different. It is milky sky-blue above and beautifully banded below with violet-gray and buff-orange.

460, 685 **Great Purple Hairstreak**
"Great Blue Hairstreak"
(*Atlides halesus*)

Description: 1¼–1½" (32–38 mm). Above, *male brilliant iridescent blue* with blackish margins and stigma, 1 or 2 HW tails, and a few blue marks by HW margin. *Female* similar but *blue duller, more restricted* to base of wings. *Below, uniform purplish-gray* except for *red at base of wings and metallic blue and green spots.* Abdomen below bright coral-red.

Similar Species: Atala greener, lacks tail. White M Hairstreak smaller, lacks bright orange spot on abdomen.

Life Cycle: Egg laid on mistletoes (*Phoradendron*), which parasitize various trees including oaks, walnuts, cottonwoods, sycamores, and junipers. Caterpillar green with darker back stripe, yellowish stripe low on sides; covered with inconspicuous orange hair. Chrysalis mottled dull black and brown; overwinters.

Flight: Typically 2 broods; February–April, July–October, with late summer brood more common. Additional broods may occur in southernmost part of range.

Habitat: Widespread in varied settings close to mistletoe-clad trees: oak savannahs, conifer flats, watersides.

Range: Coast to coast, mostly in South but residing or wandering north to Oregon, Illinois, and New York.

Strangely, this large, showy blue butterfly has long been referred to as a "purple" hairstreak. The northern limit of its breeding range is uncertain in many areas, partly because of its tendency to wander far from home.

389 White M Hairstreak
(*Parrhasius m-album*)

Description: 1⅛–1½" (28–32 mm). *Long-tailed. Above, bright iridescent blue with wide black margins;* male brighter blue with FW stigma. *Below,* both sexes have thin but prominent *white and black bordered postmedian band forming distinct "M" near 2 HW tails,* prominent reddish spot and blue area near tails.

Similar Species: Great Purple Hairstreak larger, similar above but black below; has bright orange spot on abdomen.

Life Cycle: Caterpillar, to ¾" (19 mm), downy,

light yellow-green, with dark, dull green back stripe and slanting side bars. Brown chrysalis overwinters, possibly in litter below host oak trees, (*Quercus*) which include live oak (*Q. virginiana*).

Flight: 3 broods in South; February–March, May–July, August–October. 2 broods in North; May–June, July–August.

Habitat: Around oaks: moist, open meadows and hills near woods; in northern areas, forest clearings and trails.

Range: Iowa and Connecticut south to Texas, Florida, and South America.

The White M Hairstreak gets its name from the white-bordered pattern below. A fast and erratic flier, it seems to be more abundant in the southeastern part of its range than farther north or west.

412 Aquamarine Hairstreak
(*Oenomaus ortygnus*)

Description: 1⅛" (28 mm). *Above, male brilliant intense greenish-blue* with black margins and stigma. *Female similar but duller, blue restricted* to wing bases. *Both sexes below pale rosy-gray, with few black median spots, and some greenish-blue patches on HW near 2 tails.*

Similar Species: White M Hairstreak has similar blue above, but has fine black lines forming pattern on HW below.

Life Cycle: Unknown. Host plant in Mexico is a member of the pawpaw family (*Annona globiflora*).

Flight: Year-round in Tropics.

Habitat: Low tropical dry forests and thornscrub.

Range: Widespread from central Mexico to South America, very rarely straying into S. Texas.

This exceedingly attractive hairstreak is only marginally part of the North American fauna. It is the only member of the genus *Oenomaus*.

423 Jade-blue Hairstreak
(*Arawacus jada*)

Description: 1″ (25 mm). With or without 1 HW
tail; *small droplike extension of HW* at
HW outer angle (tornus). Sexes similar
but male has small black stigma; *above,
shiny sky-blue with broad brown triangle on
FW tip,* small black marginal spots on
HW. *Cream-colored below with series of
long yellowish-tan bands* and orange spot.

Life Cycle: Unrecorded. Caterpillar feeds on a
relative of the potato plant (*Solanum
umbellatum*) in Mexico.

Flight: Possible stray at anytime; year-round in
Tropics.

Habitat: Tropical and subtropical forest edges
and scrub lands.

Range: Central America; uncommon in N.
Mexico, straying very rarely to S.
Arizona.

For many years this species was believed
to be a member of the genus
Dolymorpha. It is the only hairstreak in
our range with a cream-colored
underside.

449, 452 Brown Elfin
(*Incisalia augustinus*)

Description: ¾–1⅛″ (19–28 mm). Above, warm
brown to grayish, often orange to rust-
brown, especially in female. Chocolate-
brown *below; outer half purplish, light
brown, or mahogany.* Wing margins
lightly scalloped, usually checkered
black and white (older individuals may
lack checkering).

Similar Species: Other elfins have some contrasting gray
below. Pine elfins have zigzags below.

Life Cycle: Greenish egg. Caterpillar dull yellow-
green, turns intense green following
early molts; mature caterpillar green
with red and yellow bands. Feeds on
flowers, buds, and seed pods of host
plants, including blueberries

(*Vaccinium*), bearberry (*Arctostaphylos uva-ursi*), and azalea (*Rhododendron*) in East; dodder (*Cuscata*), California lilac (*Ceanothus*), salal (*Gaultheria*), apples (*Malus*), and madrone (*Arbutus*) in West.

Flight: 1 brood; April–May in East, February and March–early June westward, depending on elevation.

Habitat: Open woodland glades, pine barrens, bogs, chaparral and shrubby forest margins; also deserts, mesas, and parks.

Range: Widely distributed in Canada, ranging south to California, New Mexico, Michigan, and Virginia.

This species is the most common and widespread North American elfin, possibly because its caterpillars can feed on a great array of plants. Adults usually take nectar from plants in the heath family, especially bearberry.

451 Early Elfin
(*Incisalia fotis*)

Description: ⅞–1" (22–25 mm). *Above, male gray-brown*, with stigma; *female slightly reddish-brown*. Both sexes gray-brown *below with much hoary frosting, especially beyond white-edged postmedian line*, HW darker within line; indistinct submarginal row of black specks accompanied by pale yellow-orange parallels on HW margin.

Similar Species: Brown Elfin is browner below, does not have white-edged postmedian line. Moss Elfin similarly patterned but richer chestnut below.

Life Cycle: Early stages undescribed. Caterpillar feeds on cliff rose (*Cowania mexicana* var. *stansburiana*) and blends well with its flowers and leaves.

Flight: 1 brood; March–May.

Habitat: Desert mountains and canyons.

Range: E. Mojave Desert of California from Panamint to Providence Mountains, east to Utah, N. Arizona, W. Colorado, and New Mexico.

This early flying butterfly often appears before the snow has melted from the bleak foothill canyons. It occurs in scattered local colonies, and usually is not common. The differences between the Early and Moss elfins are not fully understood and some consider them the same species.

30, 450 Moss Elfin
(*Incisalia mossii*)

Description: ⅞–1″ (22–25 mm). *Above, male grayish-brown,* occasionally with rust area by HW outer angle (tornus), has stigma; *female more reddish-brown. Below 3 forms* for both sexes: red-brown (Colorado); purple-brown (Vancouver Island to San Francisco); ocherous (Big Sur and Sierra Nevada); *all darker within and lighter outside white postmedian line.* Usually white frosting outside line on HW, row of submarginal fine black dots, and row of marginal dark blotches; occasionally marks below on HW base.

Similar Species: Brown Elfin less contrasting below. Early Elfin grayer overall. Hoary Elfin grayer, lacks white line below.

Life Cycle: Pale blue-green egg laid on host plants, stonecrops and sedella (*Sedum* including *S. spathulifolium* and *S. lanceolatum*). Mature caterpillar, to ½″ (13 mm), variable: green with red and pink patterns, red, or yellow; all covered with short golden-brown hair. Purplish-brown chrysalis, ⅜″ (10 mm), overwinters in ground debris.

Flight: 1 brood; February–June, depending on locality.

Habitat: Canyons, steep rocky slopes, mossy bare summits and ridges in West; brushy foothill ravines, sagebrush hillsides, and flats eastward.

Range: Vancouver Island south along coast to
S. California and Arizona in Cascades
and Sierra Nevada east to SE. British
Columbia, Montana, and Idaho; also
Wyoming and Colorado around Front
Ranges.

The easily overlooked Moss Elfin stays
close to its host plant, flying erratically
and close to the ground, often in
inaccessible areas. Males come to damp
earth, perching on low shrubs or on the
ground. Females are more reclusive,
remaining higher up the slopes. A very
few populations in or near the San
Bruno Mountains of San Francisco may
represent a unique subspecies and are
officially protected as an Endangered
Species by the federal government.

453 Hoary Elfin
(*Incisalia polios*)

Description: ¾–1″ (19–25 mm). Dusky, gray-
brown above. *Below, outer FW and HW
clouded silver-gray,* bordered inwardly by
dark broken line, and dark brown
toward body.

Similar Species: Brown Elfin similar in pattern but
reddish-brown, not gray. Moss Elfin
also redder, with white line between
light and dark areas below.

Life Cycle: Immature caterpillar rosy to yellow,
turning green when older. Young
caterpillars well camouflaged on flowers
of host plant, bearberry (*Arctostaphylos
uva-ursi*). Sparsely hairy brown chrysalis
overwinters.

Flight: 1 brood; March–early June, mostly
April in East, May in West.

Habitat: Dry, open, often rocky areas. Local in
heathlands and barrens; also coastal
lowlands to windswept Appalachian
highlands. Generally above 8000′
(2440 m) in Rockies but near sea level
in East and Northwest.

Range: E. Alaska east to Nova Scotia, south

to Washington, New Mexico, and
New York.

Harbingers of spring, elfins usher in
balmy weather after long cold winters.
The Hoary and Brown elfins appear .
first, flitting low among heath scrub.

---

454 Frosted Elfin
(*Incisalia irus*)

Description: $7/8-1\frac{1}{4}''$ (22–32 mm). *Large. Male
gray-brown above; female reddish* overall or
in patches. Male has long thin FW
stigma. *Below HW colors do not greatly
contrast;* silver-gray dusting along
margin toward base overlays chestnut-
to gray-brown; single obscure brown-
black band bisects FW and HW below.
Short tail stump near HW tip; checkered
margins.

Similar Species: Other elfins have either more contrast
below with gray or reddish outer areas,
or lack short tail stump. Henry's Elfin
has tail stumps but is contrasting.

Life Cycle: Yellowish-green caterpillar feeds on
flowers and fruits of lupines (*Lupinus*)
and false indigo (*Baptisia tinctoria*).
Chrysalis weaves loose thread cocoon in
duff and leaf litter and overwinters.

Flight: 1 brood; late April–May, earlier
southward.

Habitat: Open second growth woods, roadside
areas near host plants; pine barrens, and
open brushy fields.

Range: SE. Canada and New England south to
N. Gulf States, west to Michigan and
Illinois; also isolated populations in E.
central Texas.

Although they are weak fliers, Frosted
Elfins appear to be efficient colonizers
that establish small dispersed
populations. However, they are never
abundant and are usually quite local.
An isolated Texas population is
sometimes thought a separate species.

455, 456 **Henry's Elfin**
(*Incisalia henrici*)

Description: ⅞–1⅛″ (22–28 mm). Above, dark
grayish-brown with reddish scaling,
especially on female; no FW stigma
on male. *HW below bicolored with deep
brown-black base contrasting with dull
yellow-brown and blue-gray outer area;*
prominent HW lobe at base; *short tail
stumps* usually present, longer in Florida
populations.

Similar Species: Frosted Elfin grayer above and below,
not as markedly 2-toned below.

Life Cycle: Caterpillar red to brownish-green, with
lighter oblique side dashes; eats flowers
and bores into fruit of blueberries
(*Vaccinium*), redbud (*Cercis canadensis*),
huckleberry (*Gaylussacia*), wild plum
(*Prunus*), and Texas persimmon
(*Diospyros texana*). Chrysalis
overwinters.

Flight: 1 brood; April–May, March southward.

Habitat: Coastal plain, brushy areas, acid scrub,
open forests, piedmont mountains, pine
barrens, barrier islands; occasionally
damp powerline cuts and other
woodland openings.

Range: Wisconsin, Michigan, Quebec, and
Nova Scotia south to Nebraska,
Illinois, Texas, and Florida.

Where several species of elfins are found
together, Henry's Elfin is likely to be
the least numerous. This species often
alights on a twig several feet high,
unlike the other elfins, which tend to
perch on or near the ground. Although
Henry's Elfin dwells in a variety of
habitats, it remains little known.

457 **Bog Elfin**
(*Incisalia lanoraieensis*)

Description: ⅝–1″ (16–25 mm). *Smallest elfin.* Dull
gray-brown above, darker below. *HW
below scalloped with dark, smudged, zigzag*

bands. Wing margins checkered black and white.

Similar Species: Eastern and Western Pine elfins much larger with crisper dark markings below.

Life Cycle: Caterpillar feeds on needles of black spruce (*Picea mariana*).

Flight: 1 brood; May–June.

Habitat: Spruce-tamarack bogs and muskegs.

Range: SE. Quebec, New Brunswick, and Maine; old records for New Hampshire.

Bog Elfins dwell with Jutta Arctics and Bog Fritillaries—northern butterflies that push south in the cold acid bogs of the East. Along with Bog Coppers, Bog Elfins occur only in these northeastern mires, but the copper is far more widespread than the elfin. The Bog Elfin was discovered relatively recently and takes its species name from the town of Lanoraie, Quebec.

38, 459   Eastern Pine Elfin
(*Incisalia niphon*)

Description: ¾–1¼″ (19–32 mm). Above, male dark brown; female tawnier. Below, mottled gray and brown; *FW cell below has 2 brown-black bars* (1 in middle, 1 at end); checks and chevrons *on HW below* and shallowly indented *black and white crisscrossing triangles,* faint hoary border. Both wings have checkered margins, appearing scalloped.

Similar Species: Western Pine Elfin has 1 black bar on FW cell, deeper HW checks and chevrons; lacks hoary border. Bog Elfin much smaller, markings blurrier.

Life Cycle: Caterpillar, to ⅝″ (16 mm), transparent green with longitudinal white stripes and white "collar" behind head; feeds on young leaves of scrub pine (*Pinus virginiana*), pitch (*P. rigida*), jack pines (*P. banksiana*), and other hard pines. Chrysalis, ⅜″ (10 mm), brown; overwinters.

Flight: 1 brood; March–early June, varying
with latitude.

Habitat: Pine and pine-oak woods, woodland
borders, roadsides, glades, and old
fields where young pines grow.

Range: Nova Scotia west to Alberta, southeast
to Texas and Gulf States.

The adult Eastern Pine Elfins emerge a
little later than other elfins; they take
nectar from wild plum, lupine,
everlasting, dogbane, and other
wildflowers. A stiff whack with a stick
on the branches of young pines will
send females spiraling into the spring
sunlight. Males often patrol from
shrubs a little distance from the pine
hosts. An adaptable species, the Eastern
Pine Elfin will colonize places where
native pines are absent, utilizing
ornamental species instead.

458 Western Pine Elfin
(*Incisalia eryphon*)

Description: ¾–1¼″ (19–32 mm). Above,
chocolate-brown; female warmer
orange-brown. Mottled gray and brown
*below; 1 bar across end of FW cell; deeply
indented white chevrons* overlay gray and
reddish-brown mottling on HW; no
hoary border. Margins of both wings
checkered, appearing scalloped.

Similar Species: Eastern Pine Elfin has 2 bars across FW
cell below and often hoary border. Bog
Elfin (in Maine) smaller, with dingier
colors and markings.

Life Cycle: Egg chalk-white. Mature caterpillar
rich velvet-green with cream-colored or
yellowish stripes; feeds on young shoots
of many pines, including lodgepole
(*Pinus contorta*) and ponderosa pine (*P.
ponderosa*). Chrysalis dark brown;
overwinters.

Flight: 1 brood; late May–July.

Habitat: At 6000–10,000′ (1830–3050 m);
along canyon bottoms and streamside

glades, spruce bogs, meadows in pine forests.

Range: British Columbia south to California, east to Alberta, Michigan, and Maine; south in Rockies to New Mexico.

Western Pine Elfins perch on small trees and shrubs. Adults seek nectar from pussy willows, pussy-paws, buckthorn, wild rose, lupine, and many other flowers. This species has recently been reported from the extreme Northeast in black spruce bogs, far away from its ordinarily western range. This is the only area in which the generally similar Western and Eastern Pine and Bog elfins all fly in proximity. The Western Pine Elfin seems more restricted to natural habitats than the eastern species.

Sandia Hairstreak
(*Sandia mcfarlandi*)

Description: 1–1⅛" (25–28 mm). Tailless. *Above, male gray-brown with dull golden tones* and somewhat darker margin; FW has stigma; *female rust-brown.* Both sexes *below bright light green with thin, prominent white and black postmedian line* running about a third in from margin, inwardly bordered by rust-brown broadly on FW.

Similar Species: Juniper Hairstreak and Xami Hairstreak have tails. Canyon Green Hairstreak lacks reddish below.

Life Cycle: Host plant is bear grass (*Nolina microcarpa*). Caterpillar variably colored pink to maroon or green, blending in with flowers and green seed capsules of host plant.

Flight: Multiple broods; mid-February–early July.

Habitat: Dry slopes and flats where bear grass is profuse.

Range: New Mexico (mostly east of Rio Grande) and W. Texas (Davis Mountains,

Big Bend National Park), south to Chihuahua, Mexico.

The Sandia Hairstreak takes its name from the mountains near Albuquerque, New Mexico, where it was first discovered in 1960. It can be very abundant, perhaps the most numerous spring butterfly in New Mexico. Adults usually fly in the morning and spend the night burrowed deep into the basal rosettes of bear grass.

440 Xami Hairstreak
(*Xamia xami*)

Description: ⅞–1″ (22–25 mm). Male *above light golden-brown with dark margins* and FW stigma; female similar but darker with wider margins. Both sexes *below light yellow-green; HW has thin white postmedian band with 2 fine toothlike projections* extending outward toward tail and some black dots on blue field near tail.

Similar Species: Juniper Hairstreak reddish above, darker green below. Olive Hairstreak redder above, deeper green below with 2 black and white lines across HW (not 1).

Life Cycle: Egg pale green, becoming dull white. Caterpillar yellow-green with reddish marks. Chrysalis pale reddish-brown to blackish-brown. Host plants in Mexico are succulents (*Echeveria, Sedum allantoides, S. texana*); host plants probably similar in North America.

Flight: Multiple broods; early April, early July, early October in Arizona; June–December in Texas, peaking in autumn.

Habitat: Dry, rocky areas.

Range: Mexico to S. Texas and north to S. Yavapai County, Arizona.

This species is quite rare and local in North America, although common around succulents in Mexico City gardens.

446 Thicket Hairstreak
(*Mitoura spinetorum*)

Description: 1–1¼" (25–32 mm). *Large.* Male *above steel blue with wide, dark, inwardly diffuse margins* and stigma; female similar but with wider margins and blue restricted toward body, especially on FW. Both sexes *below deep red-brown to slate-brown with prominent black-edged white postmedian line, forming "W" near HW tail;* row of submarginal dots along HW margin and black and orange spot on bluish field.

Similar Species: Johnson's Hairstreak dark brown all over above, not dull blue.

Life Cycle: Egg laid on dwarf mistletoes (*Arceuthobium*), parasites on certain conifers. Mature caterpillar green marked with red, white, and yellow; many bumps and ridges. Chrysalis brown; overwinters.

Flight: At least 2 broods in California and Oregon; March–September. 1 brood in Wyoming and Colorado; June–July.

Habitat: Coniferous forests or woodlands, usually in clearings, or small canyons, and desert canyons in Southwest.

Range: British Columbia to Baja California along Cascades and California ranges, east through Great Basin and Rockies to SE. Alberta, Colorado, and New Mexico.

Duller and deeper blue than the blues, the Thicket Hairstreak takes nectar at buckwheat and composite flowers. Males fly about summits and perch on shrubs, waiting for females to pass.

Johnson's Hairstreak
"Mistletoe Hairstreak"
(*Mitoura johnsoni*)

Description: 1–1⅛" (25–28 mm). *Above, male chocolate-brown* except orange-brown by tail; *female red-brown* except brown on margins and near tail. *Both sexes below brown,* thin white postmedian line; HW near tail has few black dots and bluish and orange scales.

Similar Species: Thicket Hairstreak deep blue above. Nelson's Hairstreak smaller, redder above and below.

Life Cycle: Host plant is western dwarf mistletoe (*Arceuthobium campylopodum*), a parasite on western hemlock. Caterpillar dark green with yellow, white, and red diagonal markings. Dark brown chrysalis overwinters, probably in forest litter beneath conifers.

Flight: Usually 1 brood; June–July. Possibly partial 2nd brood in August.

Habitat: In or near dense forest, especially western hemlock or red fir; occasionally open, drier woodlands, including digger pine.

Range: SW. British Columbia south in Coast Ranges almost to San Francisco Bay, and in Cascades and Sierra Nevada to Yosemite; an isolated population in Blue Mountains of E. Oregon.

Johnson's Hairstreak is named for the pioneering Washington zoologist, O.B. Johnson. It was considered one of North America's rarest butterflies until, a few decades ago, immense numbers appeared far south of its main range near Napa Valley, California. Since then this hairstreak has again been rare. Other western hairstreaks in the genus *Mitoura* undergo similar population fluctuations from time to time. Adults usually remain high in evergreens, where mistletoe grows, occasionally descending to take nectar from Pacific dogwood, Oregon grape, ceanothus, and pussy-paws.

447, 448 Nelson's Hairstreak
"Incense Cedar Hairstreak"
(*Mitoura nelsoni*)

Description: ⅞–1" (22–25 mm). *Above, male dark grayish-brown with rust-brown spots* near HW tail or over central parts of wings to varying degree; has stigma; *female rust-brown* over most of wings except for bases and margins. Both sexes *below brown, often with pinkish-red or purplish tint and variable white or brown postmedian band,* jagged to straight, prominent to muted. Near tail, a few dots, blue and orange field.

Similar Species: Thicket Hairstreak deep blue above. Johnson's Hairstreak larger, less reddish above and below; range does not overlap.

Life Cycle: Egg green; laid on incense cedar (*Libocedrus decurrens*) and other cypresses (Cupressaceae). Mature caterpillar bright green with yellow diagonal markings, covered by many bumps and ridges; blends with foliage. Chrysalis brownish-black; overwinters.

Flight: 1 brood; May–July.

Habitat: Forests and groves of host conifers from sea level to mid-elevation mountains.

Range: Pacific Coast from British Columbia to Baja California, east into N. Idaho, E. Oregon, W. Nevada.

The butterfly long known as Nelson's Hairstreak probably includes other separate but poorly distinguished species—all with host plants other than incense cedar. Muir's Hairstreak (*M. muiri*), named after the naturalist John Muir, occurs in central California; its host plant is Sargent cypress (*Cupressus sargenti*). Barry's Hairstreak (*M. barryi*) of Oregon has as host plant western juniper (*Juniperus occidentalis*). Three populations in the Pacific Northwest and northern Rockies use as host plant western redcedar (*Thuja plicata*) and may be the same species.

441 Juniper Hairstreak
(*Mitoura siva*)

Description: ⅞–1⅛" (22–28 mm). *Above. male dark grayish-brown with rust-brown* spots near HW tail or over central parts of wings to varying degree, has stigma; *female rust-brown* except for wing bases and margins. Both sexes *usually intense green below with prominent white postmedian line.* bordered inwardly with black and brown; near tail a few vague black dots lie in violet field. *Some populations purplish-brown below.*

Similar Species: Nelson's Hairstreak resembles brown forms but smaller, more red above and below; ranges do not overlap. Olive Hairstreak has 2 white lines on HW below; range barely overlaps. Loki Hairstreak has more prominent black spots on HW beneath.

Life Cycle: Green egg laid on young growth of host junipers (*Juniperus osteosperma. J. californica. J. scopulorum*). Mature caterpillar green with irregular yellow markings; has lumps and bumps. Dark brown chrysalis overwinters.

Flight: 1 brood; April–May in California, May–June in Great Basin and north. 2–3 broods in Southwest; June–July, sometimes August–September.

Habitat: Open, scrubby woodlands, arid lands, rocky outcrops, canyons, and scarps.

Range: SE. Washington south to S. California, east to S. Saskatchewan, Nebraska, and W. Texas.

The Juniper Hairstreak is a difficult hairstreak complex, possibly involving more than one species. Some populations in southeastern Washington, northeastern Oregon, Nevada, Utah, and parts of southern California are lilac-brown below, while the usual form is bright green.

444, 445   **Loki Hairstreak**
"Skinner's Hairstreak"
(*Mitoura loki*)

Description: ⅛–1″ (22–25 mm). *Above, male dark grayish-brown with rust-brown* spots on HW near tail, or over central parts of wings to varying degree, has stigma; *female usually rust-brown* in central parts of wings. Both sexes *below, green with prominent white and brown postmedian line; HW has dark marks near base and* full submarginal row of *bold black dots* parallel to margin.

Similar Species: Juniper Hairstreak has fewer and vaguer black dots near margin of HW below.

Life Cycle: Egg light green; laid on California juniper (*Juniperus californica*). Caterpillar rich green with white to yellow irregular markings; camouflaged on foliage. Mottled brown chrysalis overwinters, presumably in duff beneath shrubby host plant.

Flight: 2, sometimes 3 broods; March–May, June–July, and partial 3rd brood August–October.

Habitat: Dry woodland between true desert and mountain chaparral.

Range: Mountains of S. California south into Baja California.

The related Tecate Cypress Hairstreak (*M. thornei*) flies on Otay Mountain, California, where its caterpillars feed on tecate cypress (*Cupressus forbesii*). These hairstreaks, which are mostly red-brown and lack green, were considered a population of the Loki Hairstreak.

443   **Olive Hairstreak**
(*Mitoura gryneus*)

Description: ⅛–1″ (22–25 mm). Male *above dark brown with* variable amounts of *orange-brown or golden-brown, margins dark brown;* has small stigma. Female often has more orange or gold over brown.

Both sexes *below bright green with prominent white postmedian line* inwardly edged with red-brown and line made up of *2 white marks on HW near base, 1 atop other;* a few submarginal black spots and faint orange spot on bluish field by tail. Summer individuals darker.

Similar Species: Juniper Hairstreak has only 1 white-edged line below; range barely overlaps on W. Great Plains. Hessel's Hairstreak has white dot in FW cell beneath and brown patches outside HW white line beneath.

Life Cycle: Pale green egg. Host plants eastern redcedar (*Juniperus virginiana*), southern redcedar (*J. silicicola*) in Florida, and perhaps other junipers. Mature caterpillar deep green with yellow tint and pale green diagonal lines high on sides. Dark brown chrysalis overwinters.

Flight: 2 broods; April–May, July–August.

Habitat: Drier hillsides, rocky bluffs, and old fields with redcedar established or colonizing.

Range: W. Ontario south to Nebraska and W. Texas, east to Atlantic from S. New England to central Florida.

While most of its near relatives range across the western states, the Olive Hairstreak is common in the East. The decline of agriculture in New England has benefited this butterfly; it spreads among redcedars that grow in abandoned old fields, although further woodland growth decreases the available habitat. Urbanization also displaces the Olive Hairstreak, which is now absent from Staten Island, New York, and is rare on Long Island.

7, 41, 442 Hessel's Hairstreak
(*Mitoura hesseli*)

Description: ⅞–1″ (22–25 mm). Above, male dark brown; female somewhat reddish.

Below, deep green to bluish-green; FW has 1 white submarginal band; *2 white HW bands,* bordered by brown; *white dash nearest tail concave outwardly; brown patches* outside HW postmedian band. *FW cell below has 1 or 2 white dots.*

Similar Species: Olive Hairstreak has orange or golden tints above; below, white dash near tail concave inwardly, has no blue, and lacks brown patches outside HW band and dots within FW cell.

Life Cycle: Egg, to $\frac{3}{256}''$ h $\times$ $\frac{3}{128}''$ w (0.3 $\times$ 0.6 mm) green; becomes yellow-white. Caterpillar, to $\frac{5}{8}''$ (16 mm), bluish-green with white marks. Chrysalis, to $\frac{3}{8}''$ (10 mm), dark brown. Host plant is Atlantic white-cedar (*Chamaecyparis thyoides*).

Flight: 2 broods; May and July.

Habitat: Bogs and swamps close to white-cedar.

Range: Mostly along Atlantic from New Hampshire and Massachusetts south to North Carolina, along Gulf in Florida.

The rather rare Hessel's Hairstreak and the more common Olive Hairstreak represent a classic example of sibling species: 2 species which are nearly indistinguishable in appearance, but are biologically different. While the Olive Hairstreak is a widespread colonizer feeding only on redcedars in nature, Hessel's Hairstreak confines itself to inaccessible white-cedar swamps. Hessel's Hairstreak is named after Sidney Hessel, one of the first lepidopterists to recognize the difference between these biological species.

---

431 Bramble Green Hairstreak
(*Callophrys dumetorum*)

Description: $\frac{7}{8}$–$1\frac{1}{8}''$ (22–28 mm). Tailless. *Above, gray to dark gray-brown; female often has some reddish-brown,* especially over outer half of wings. *Below, green,* except for gray-brown area of FW disk; *white postmedian band* of irregular spots and dashes or band nearly absent.

Similar Species: Other green hairstreaks very similar; see individual entries.

Life Cycle: Egg deep green. Mature caterpillar bright green or red, covered with minute brown hair, yellowish lines along back, and white on sides. Chrysalis dark brown with black speckles; overwinters in litter beneath host plants, mainly deer weed (*Lotus scoparius*) and California buckwheat (*Eriogonum fasciculatum*) in California.

Flight: 1 brood; principally April–May, varying with elevation to February–June.

Habitat: Wastelands, dry washes, rocky hills, chaparral, and moist foothill canyons; maritime evergreen forest openings in Northwest.

Range: Puget Sound south to Baja California, east into Cascade and Sierra Nevada mountains.

The Bramble Green Hairstreak, a common spring butterfly, is often the only green hairstreak in the lush, low elevation forests of western Washington and Oregon, but it occurs far to the south as well. As a group, the green hairstreaks of the genus *Callophrys* are difficult to identify; even specialists differ on exactly which species should be recognized. Most of these species seem to intergrade with each other to some degree in one place or another.

436 Desert Green Hairstreak
(*Callophrys comstocki*)

Description: ¾–1″ (19–25 mm). Tailless. *Light gray above. Below, gray-green to green* (summer brood more yellowish); *postmedian white band of broken bars,* bent at middle, and *edged inwardly with black; FW disk largely green on underside.*

Similar Species: Other green hairstreaks similar but
Desert Green separated by habitat, very
gray upperside, and strong broken
white line below.

Life Cycle: Formally undescribed. Caterpillars said
to resemble those of Bramble Green
Hairstreak. Host plant probably
sulphur flower (*Eriogonum umbellatum
subaridum*) and other buckwheats
(*Eriogonum*). Chrysalis may overwinter
twice in very dry years.

Flight: 2 broods, and in some years, depending
on rains, 3 broods; March–April,
June–July, August–September.

Habitat: Desert mountain ranges, rocky slopes,
canyons and washes.

Range: Largely northern and eastern Mojave
Desert in S. California, W. and S.
Nevada, NW. Arizona, SE. Utah, and
SW. Colorado.

In years of ample rainfall, the Desert
Green Hairstreak can be quite
common. Males perch on rocks or low
shrubs and behave very territorially.
Females, found less often, seem to stay
higher up the canyon walls close to the
host plants.

433 Alpine Green Hairstreak
(*Callophrys lemberti*)

Description: ⅞–1¼" (22–32 mm). Tailless. *Male
grayish above;* female occasionally more
golden-brown. *Below moss-green.*
covering much of FW; *broken postmedian
white line* usually convex and
incomplete.

Similar Species: All green hairstreaks very similar; see
individual entries.

Life Cycle: Unknown. Alpine hoary buckwheat
(*Eriogonum incanum*) reported host plant.

Flight: 1 brood; mid-June to mid-July.

Habitat: High elevations, especially alpine fell-
fields and rocky slopes.

Range: Crater Lake, Oregon, south in Cascades
and Sierra Nevada to central California,

east to Warner Mountains and W. Nevada.

The Alpine Green Hairstreak is unique in its genus because it lives so high in the mountains, frequently above timberline. Some consider this species to be merely a high mountain race of the White-lined Green Hairstreak, but the absence of intermediate populations, the choice of a different host plant, and other evidence favor treating it as a separate species.

432 Canyon Green Hairstreak
(Callophrys apama)

Description: ⅞–1⅛″ (22–28 mm). Tailless. *Above, grayish-brown with coppery tints, especially in female. Below, apple-green with white postmedian band* varying from well developed, prominent, and angled about middle to absent; *bordered inwardly with blackish-brown and then reddish-brown; considerable brown in FW center.*

Similar Species: All green hairstreaks very similar; see individual entries.

Life Cycle: Unrecorded. Adult found on wild buckwheat (*Eriogonum*), suggesting this may be host plant.

Flight: Number of broods uncertain; March–June or July and in Colorado late August, most commonly May–June.

Habitat: 6500–10,000′ (1983–3050 m) in mountains, canyons, and foothills; open ponderosa pine forest and meadow edges in W. Colorado.

Range: SW. Wyoming and S. Utah through Arizona, New Mexico, and Colorado south to Mexico.

Success in identifying green hairstreaks depends in part upon locality. In the Colorado Front Range, a green hairstreak without white lines below will be this species. The number of

options increases west of the
Continental Divide. Most butterfly
watchers will do best to satisfy
themselves with the determination
that a butterfly is a green hairstreak,
leaving finer distinctions to experts.

430 Immaculate Green Hairstreak
(*Callophrys affinis*)

Description: ⅞–1⅛″ (22–28 mm). Tailless. *Above,
male dark gray-brown*, sometimes with
reddish-brown; *female rust-brown* over
most of wings. Both sexes *below slightly
yellowish-green; postmedian white band
generally absent* or composed of a few
faint marks; *considerable grayish about
FW center*.

Similar Species: All green hairstreaks very similar; see
individual entries.

Life Cycle: Eggs laid on sulphur flower (*Eriogonum
umbellatum*). Mature caterpillar varies
from grass-green to dark red, with light
lines and ridges down sides.

Flight: 1 brood; April–July, mostly June.

Habitat: Sagelands, often in lower mountains,
plateaus, and canyons.

Range: E. British Columbia south through E.
Cascades, Columbia, and Great Basin
and Rockies to S. Nevada, east to
central Montana and central Colorado.

Scarce and extremely wary, this
hairstreak is difficult to observe. It is a
strong flier, aided by the nearly
constant winds of the sagelands. Males
tend to return to the same posts day
after day.

435 Bluish Green Hairstreak
(*Callophrys viridis*)

Description: ⅞–1⅛″ (22–28 mm). Tailless. *Gray-
brown above. Green with bluish cast below*
in both sexes, with *most of FW below*

green-scaled. White postmedian band below may vary from quite complete to fragmentary or nearly absent. *Antennae noticeably whitish, fringe broad and white.*

Similar Species: All green hairstreaks very similar; see individual entries.

Life Cycle: Green egg. Mature caterpillar variable; light green to pink, with ridges. Host plants are wild buckwheats (*Eriogonum latifolium* and *E. nudum*) and deer weed (*Lotus scoparius*). Chrysalis dark brown, bluntly rounded.

Flight: 1 brood; March–May.

Habitat: Coastal bluffs, sand dunes, sandy or stony shores and slopes.

Range: N. California coast south to Monterey, inland in a few canyons.

Some lepidopterists consider the Bluish Green Hairstreak a subspecies of the Bramble Green Hairstreak. If it is a subspecies, it may be analogous to the sea level populations found in the Pacific Northwest; however, these sea level populations lack the bluish hue of the Bluish Green Hairstreak. As early as 1956, Bluish Green Hairstreaks were reported disappearing from the San Francisco Bay area. Today, virtually all populations on the Bay's islands, hills, and shorelines have been eliminated as the natural habitat has given way to development.

437 White-lined Green Hairstreak
(*Callophrys sheridanii*)

Description: ¾–⅞″ (19–22 mm). Tailless. *Above, gray* usually without any brown tones. *Below, dark blue-green with variable but rather straight postmedian white line;* sometimes bordered inwardly with black—solid or broken, thick or thin; *FW center has some green;* grayish limited to strip along trailing edge of FW.

Similar Species: All green hairstreaks very similar; see individual entries.

Life Cycle: Egg very pale green, flattened, hemispherical; laid on host plant, wild buckwheat (*Eriogonum umbellatum* and other *Eriogonum*). Caterpillar light green to pink, covered with bunches of short stiff spines, 2 rows of white spots down back. Chrysalis brown; overwinters.

Flight: 1 brood; April–May.

Habitat: 6000–10,000′ (1830–3050 m) open hillsides, canyon slopes, washes, and sagebrush habitats.

Range: S. British Columbia southeast of Cascade Crest to S. Nevada, central Arizona; east to Saskatchewan, North Dakota, and E. New Mexico.

The White-lined Green Hairstreak may be recognized in the eastern part of its range by the white line below—its namesake characteristic. Further west, the white line breaks down and so does the species' distinctiveness. The White-lined is the only green hairstreak of the genus *Callophrys* that reaches the Great Plains, although it does so only at the western rim.

---

434 Miserabilis Hairstreak
(*Cyanophrys miserabilis*)

Description: ¾–1″ (19–25 mm). *Above, male deep blue with marginal black shading.* especially on the FW costa and outer margin; has stigma. *Female duller blue.* Both sexes *below intense green* with some gray on FW; *HW has white postmedian band fragmented toward costa and reddish-brown marginal patches.* Tailed.

Similar Species: Silver-banded Hairstreak is purplish above, has fuller white line below. Goodson's Hairstreak tailless, yellower green below.

Life Cycle: Early stages undescribed. Host plant is Mexican palo verde (*Parkinsonia aculeata*).

Flight: Multiple broods; April–December in S Texas.

Habitat: Tropical and subtropical forest edges and scrub.

Range: S. Texas south to Costa Rica.

This emerald and sapphire butterfly is an uncommon breeding resident in the lower Rio Grande Valley. However, many similar looking hairstreaks are common in the American Tropics. For many years this species was identified as *Strymon pastor*.

429 Goodson's Hairstreak
(*Cyanophrys goodsoni*)

Description: ¾" (19 mm). Tailless. *Above, male pale lavender to silvery blue* with broad black marginal shading; female duller, blue restricted toward body. *Below, yellow-green with white postmedian band usually restricted to small dash* near HW outer angle (tornus) and sometimes a small red-brown marginal spot.

Similar Species: Miserabilis Hairstreak tailed and brighter blue above, greener below with more distinct white postmedian band.

Life Cycle: Early stages undescribed. Host plant is bloodberry (*Rivinia humilis*).

Flight: Multiple broods; June-October in southern Texas.

Habitat: Tropical and subtropical forest edges and scrub.

Range: S. Texas south to Costa Rica.

Although an irregular breeding resident in southern Texas, this tropical hairstreak is more common than its look-alike, the Miserabilis Hairstreak. One place to watch for this stray is the vicinity of Pharr, Texas. Previously called *Strymon facuna*.

400, 534 Behr's Hairstreak
(*Satyrium behrii*)

Description: ⅞–1⅛" (22–28 mm). Tailless. *Above,
bright golden-brown with dark margins*
broadest along costa; male has stigma.
*Below, gray to brown with fine, white-edged
black speckles, mostly on HW;* speckles
scattered over inner ⅔ HW and in even
row on outer third with additional red
and black spot near HW outer angle
(tornus).

Similar Species: Somewhat resembles coppers but bronze
coloring, black costa, and dark
underside distinctive.

Life Cycle: Overwinters as egg on host plant,
antelope brush (*Purshia tridentata* and
P. glandulosa). Mature caterpillar green
with white, yellow, or dark green
stripes and bars on back and sides;
covered with minute yellowish hair.
Chrysalis light tan, mottled brown.

Flight: 1 brood; mainly June–July, sometimes
August.

Habitat: Dry slopes of mountains and plateaus,
foothill creeks and canyons, grass-shrub
steppe; associated with sagebrush,
pinyon-juniper woodland, and Joshua
tree savannah.

Range: S. British Columbia south along east
slope of Cascades and Sierra Nevada,
locally along N. Transverse Ranges of
S. California; east through Oregon and
S. Idaho to SE. Wyoming, Arizona,
and New Mexico to E. Rockies.

Not venturing far from its host plant,
the Behr's Hairstreak tends to be
somewhat local. The tone of the
underside ground color and the
boldness of the black markings vary
considerably. The smallest and palest-
marked individuals come from
California, the largest and darkest-
marked from British Columbia. Unlike
most hairstreaks, Behr's Hairstreak is
passive and usually can be approached
at close range. Adults congregate
especially at wild buckwheat flowers.

411 Sooty Hairstreak
(*Satyrium fuliginosum*)

Description: 1–1¼" (25–32 mm). Tailless. Plain
dark *gray-brown or chocolate-brown above.*
Gray to *gray-brown below,* with or
without 1 or 2 rows of black spots
thinly ringed with white, larger on
FW. *Rounded wings,* gray-brown fringes.

Similar Species: Female Common Blue generally has
white fringes, more white below.
Female High Mountain Blue has dark
spots in cell, more white below.

Life Cycle: Unrecorded. Females have been seen
laying eggs in litter beneath lupines
(*Lupinus*). Strong association also of
adults with lupines suggests this as host
plant.

Flight: 1 brood; mainly July, records from
May–August.

Habitat: Moderate elevations, always with lupine
and often with sagebrush; dry
mountainsides, stony screes, meadows
among pine forests, plateaus, rolling
grasslands, and roadsides.

Range: S. British Columbia and SW. Alberta
south in mountains to SW. Oregon,
central California, central Nevada, east
to SE. Montana and NW. Colorado.

This butterfly looks more like a drab
female blue than a hairstreak. Even
its affinity for lupines suggests a blue,
but its wing structure is distinctive.
When the Sooty Hairstreak first
emerges, its deep rich gray wings have
green and purple highlights.

404 Acadian Hairstreak
(*Satyrium acadica*)

Description: 1⅛–1¼" (28–32 mm). Cool *gray*-
brown with *orange spot on HW near tail
above. Pale silver-gray below* with coal-
black bands on FW and HW,
consisting of small, widely separate
round dots; blue spot and orange patch

or zigzag border on HW near tail.

Similar Species: California Hairstreak is more reddish above, less orange near tail below. Sylvan Hairstreak usually much paler below, with less orange near shorter tail. Banded Hairstreak darker below with squarish spots that are connected into bands. Sylvan Hairstreak has bluish cast above.

Life Cycle: Egg overwinters. Caterpillar grass-green with yellowish dashes on side between 2 white lines; feeds on willows (*Salix*).

Flight: 1 brood; late June–July.

Habitat: Damp fields, meadows, streamsides; in canyon bottoms and seeps westward.

Range: British Columbia east to Nova Scotia, south to Washington, Colorado, Iowa, and New Jersey.

The Acadian Hairstreak alights on swamp milkweed, white clover, and dogbane, its favored nectar sources. In the Northwest this species and the California Hairstreak cannot be distinguished reliably on the basis of their looks, but their choice of habitat helps—the Acadian lives near willows, while the California occurs near oaks.

399 California Hairstreak
(*Satyrium californica*)

Description: 1–1¼" (25–32 mm). *Above, ruddy to dull brown with large dark yellow to orange splotches near margins;* FW broadly rimmed dark brown. *Below, cream-brown to brownish-black* with small black dots on FW and HW midbands; orange spots and blue patch near tail.

Similar Species: Both Sylvan and Acadian hairstreaks have bluish-gray cast above and are lighter and grayer below; Acadian has more orange on HW below.

Life Cycle: Egg overwinters. Caterpillar duff-brown with white side triangles and gray oval spots on back; underneath dingy ivory-green; consumes leaves of ceanothus (*Ceanothus*) and oaks (*Quercus*). Chrysalis reddish-brown mottled with black spots; has light greenish wing cases.

Flight: 1 brood; late May–early August, varying with elevation.

Habitat: Dry open foothills; lower elevation canyons in Rockies and Cascades, chaparral and ponderosa pine zones in drier mountains.

Range: S. British Columbia south to S. California, east to Wyoming and Colorado.

California Hairstreaks often cluster at such roadside flowers as chokeberry and dogbane in dry dusty foothills, canyons, and chaparral. Although the Sylvan and Acadian hairstreaks look similar, the California normally sticks to parched open habitat while the other 2 species favor moister willow drainages. However, where habitats and ranges overlap, they may intermingle.

405, 535 Sylvan Hairstreak
(*Satyrium sylvinus*)

Description: ⅞–1¼″ (22–32 mm). Cool gray *above with blue-gray cast;* reddish-brown near tail, colors diluted on FW base and HW. *Whitewashed below,* with jet-black FW and HW band spots distinct from surroundings; *all marginal markings nearly absent;* discrete orange and blue spots below near tail. *Usually short-tailed* but in some populations long-tailed or tailless.

Similar Species: All other hairstreaks usually much darker below, with larger orange spots and longer tails. Acadian and California hairstreaks very similar in Northwest.

Life Cycle: Eggs laid singly on host plant willows (*Salix*). Mature caterpillar apple-green, with cream-colored side lines from head to tail and paler diagonal slashes.

Chrysalis pale green flecked brown and deep olive-green.

Flight: 1 brood; late May–August, slightly later dates at higher elevations.

Habitat: Foothills near rivers, mountain canyons, sometimes wet meadows; always near willows in and along drainage bases.

Range: British Columbia south to Baja California, east to central Montana and New Mexico.

Fiercely loyal to specific colony sites, the Sylvan Hairstreak stays near willows, occasionally flying to such nectar sources as milkweed, white sweet clover, teasel, everlasting, and thistles. Sylvan populations without tails are considered by some lepidopterists to be another species, known as the Dryope Hairstreak (*S. dryope*).

398 Edwards' Hairstreak
(*Satyrium edwardsii*)

Description: 1–1¼" (25–32 mm). Warm light gray-brown above and below. *FW and HW bands below consist of rows of discontinuous small, dark brown oval spots* more or less ringed with white. Margin above often has orange tinge; fairly confined orange and blue spots near tail below.

Similar Species: Banded Hairstreak has continuous FW band of connected rectangles. Acadian Hairstreak has smaller FW band spots on grayer ground color, more extensive orange below.

Life Cycle: Egg laid under next year's buds or rough twigs; overwinters. Brownish caterpillar has hairy warts; feeds on oaks, especially scrub oak (*Quercus ilicifolia*). Mottled chrysalis secured to dead leaf or debris by silken girdle.

Flight: 1 brood; June–July, usually slightly earlier than Banded Hairstreak.

Habitat: Dense scrub oak thickets among open

woods and sandy barrens.

Range: Saskatchewan east to Maritimes, south to Texas and upland Georgia.

Edwards' Hairstreaks haunt oak thickets in sandy barrens, the males jostling incessantly for favorite perches. Often this species can be found alongside Banded Hairstreaks, although Edwards' tends to stick close to oaks while the Banded uses many host plants.

394 Banded Hairstreak
(Satyrium calanus)

Description: 1–1¼" (25–32 mm). Above, jet black with sooty cast. *Below, male warm brownish-black; female slate-colored; FW midband consists of thin rectangles, continuous although sometimes twisted,* broken slightly if at all; band and other dashes white-outlined on outer edges. Blue patch near tail; orange on either side may extend along margins.

Similar Species: Striped and King's hairstreaks have broader bands. Hickory Hairstreak browner, band on FW below clearly broken. Edwards' Hairstreak has discontinuous band below.

Life Cycle: Egg overwinters. Caterpillar yellowish-green or brownish, with duller broad transverse side stripes and deep brown wide band along back. Host plants vary regionally, including walnuts (*Juglans*), hickories (*Carya*), and oaks (*Quercus*); shagbark hickory (*Carya ovata*) preferred in Connecticut. Hairy, mottled brown chrysalis holds fast with tight silken girdle.

Flight: 1 brood; late June–early July.

Habitat: Deciduous forests eastward; woodland clearings, edges, roadsides, railway cuts, power lines, city parks, and hot oak-lined canyons in West.

Range: E. Saskatchewan, S. Wyoming, and SE. Utah south to Gulf, east to Atlantic.

This commonest of dark eastern
hairstreaks clings to low leaves and
shrubs bathed in sunbeams and engages
all newcomers in territorial tussles.
Sometimes half a dozen may chase
around together before settling. They
can be found on their favorite nectar
plants, such as milkweed, dogbane,
daisies, and sumac. Banded Hairstreaks
and their allies are some of the most
variable butterflies; certain
identifications often can be made only
by experts. The Banded Hairstreak may
actually prove to be a group of sibling
species; southern populations may be
distinct. Formerly called *S. falacer*.

396 Hickory Hairstreak
(*Satyrium caryaevorus*)

Description: 1–1¼" (25–32 mm). Above, brown-
black with slate-colored cast. *Below, cool
soot-gray to slate,* overlaid with distinctly
bluish cast when fresh. *FW midband
white-outlined* on both inner and outer
edges, *broken with branches strongly offset,*
inflated near top (especially in male).
*Small orange-red area, large blue HW spot
below near tail.*

Similar Species: Banded Hairstreak warmer gray-brown
below, FW band bent but not broken
and offset. Striped and King's
hairstreaks have broader band
segments. Edwards' Hairstreak warmer
brown with less conspicuous,
discontinuous FW band.

Life Cycle: Grass-green caterpillar has straight dark
green band on back, dark green side
marks. Host plants are hickories
(*Carya*), also reported on ashes
(*Fraxinus*); perhaps other broadleaf
trees. Sparsely hairy chrysalis mottled
brown.

Flight: 1 brood; June–July.

Habitat: Mixed forest zone south of boreal
forests; deciduous woods, roadsides,
and edges of fields.

Range: Minnesota and Quebec south through Iowa and Connecticut to Kentucky and W. Pennsylvania; rarely farther south.

Easily overlooked, Hickory Hairstreaks are difficult to spot among hordes of Banded Hairstreaks although they too are sporadically very abundant. The Northeast was inundated with this butterfly in the early 1960's, and again in 1980. It is not known whether Hickory Hairstreaks expand their range after long retractive periods or undergo explosive local booms. This species seeks nectar from dogwood, milkweed, Queen Anne's lace, and many other flowers.

397 King's Hairstreak
(*Satyrium kingi*)

Description: 1–1¼" (25–32 mm). *Above, uniform dark brown,* tinted bronze-green when fresh. *Below, FW midband has 3 thin offset segments, blending* into slate-brown; blue and orange spots near tail below not extensive; HW margin iridescent. Wings more rounded than most in genus.

Similar Species: Striped Hairstreak has broader midband segments, far more broken and offset than King's, and bluer spot. Banded Hairstreak has narrower, less broken and offset band segments. Hickory Hairstreak has narrower band segments, larger blue spot.

Life Cycle: Egg overwinters. Green caterpillar bores into young leaf buds of horse sugar (*Symplocos tinctoria*) and flame azalea (*Rhododendron calendulaceum*).

Flight: 1 brood; May–June near coast, July–August inland.

Habitat: Coastal plains near moist swamplands; oak hammocks; local in upland deciduous woods.

Range: Virginia south to Mississippi and Florida.

King's Hairstreak and its principal host plant, horse sugar, occur together in 2 separate areas—the coastal plain and the base of the southern Appalachians. Undescribed until 1952, this species remains rather vaguely known. With care, King's Hairstreaks may be watched when they settle on broad sunlit leaves or come to take nectar at chinkapin flowers.

395 Striped Hairstreak
(Satyrium liparops)

Description: 1–1⅜″ (25–35 mm). Warm brown-black above, buff-brown below. Some populations have orange-red flush to FW above. *Very wide, disjointed dark bands, bordered by white, boldly mark underside.* Blue and orange near tail.

Similar Species: Banded, Hickory, King's, and Edwards' hairstreaks all have much narrower bands below.

Life Cycle: Grass-green caterpillar striped lengthwise with yellow. Host plants are oaks (*Quercus*), willows (*Salix*), hollies (*Ilex*), blueberries (*Vaccinium*), various plums (*Prunus*), hawthorn (*Crataegus*), and others, especially members of rose family (Rosaceae). Chrysalis mottled.

Flight: 1 brood; July–early August.

Habitat: Deciduous woodlands, old fields, hedgerows and thickets, disturbed spots; lower elevation mountain canyons in West.

Range: S. Alberta and Maritimes south to Colorado, Arkansas, Mississippi, and Florida.

The Striped Hairstreak has very broad tastes and an extensive range yet it is not common. Apparently this butterfly shifts its colonies about from year to year. Striped Hairstreaks remain high in the treetops deep within prickly hawthorn thickets or take nectar from milkweed or white dogbane.

428 Gold Hunter's Hairstreak
(*Satyrium auretorum*)

Description: 1–1¼" (25–32 mm). *Both sexes brown above;* female and some populations have rust-colored wash or highlights; some males blackish, all males have stigma. *Below, both sexes dull brown with rows of vague darker bars,* faintly edged outwardly with white. Small orange and larger blue patch near *very short tail* on HW below.

Similar Species: Mountain Mahogany Hairstreak grayer without rust-colored highlights, longer tail on female, white line below lacks adjacent dark bars. Hedgerow Hairstreak has much more rust color above, longer tails.

Life Cycle: Egg pale purple; overwinters on host plants, California scrub oak (*Quercus dumosa*) in south and interior live oak (*Q. wislizenii*) to north, as well as other oaks. Caterpillar feeds on spring growth; varies from apple-green to pale orange, with short dark orange hair arising from whitish points. Chrysalis typically pinkish-buff with brown speckles and blotches.

Flight: 1 brood; late May–July.

Habitat: Mountains and foothills in oak woodland and chaparral.

Range: N. California south in Coast Ranges and W. Sierra Nevada to SW. California and probably Baja California.

The Gold Hunter was named by the French naturalist Boisduval, who described these butterflies at the time of the California Gold Rush. The species' numbers fluctuate from year to year, but generally this butterfly is local and scarce. Adults can be seen taking nectar at buckeye, buckwheat, dogbane, and horehound flowers.

415 Mountain Mahogany Hairstreak
(*Satyrium tetra*)

Description: 1⅛–1¼″ (28–32 mm). *Male dark gray above.* with lighter stigma; *female dark grayish-brown above. Below. dusky gray* but lighter than above, with *indistinct HW postmedian line edged outwardly with white* which continues over cell as faint whitish dusting. Row of faint black markings along HW margin ending in small faint blue patch near tail, which is very short on male, longer on female.

Similar Species: Gold Hunter's Hairstreak more or less rust-colored above, with bolder dark markings below; short tails on both sexes. Gray Hairstreak silver-gray below with strong white line; prominent orange spot on HW above and below.

Life Cycle: Egg overwinters. Mature caterpillar pale silver-green with diagonal bluish bars along sides, and covered with minute orange hair; blends in with leaves of host plant, mountain mahogany (*Cercocarpus betuloides*). Chrysalis pale mahogany with heavy black blotches.

Flight: 1 brood; mid-June through July.

Habitat: Arid scrub chaparral.

Range: S. Oregon from the Siskiyous to the Warner Mountains, south through California mountains and foothills to Mexico; also extreme W. Nevada east of Lake Tahoe.

The Mountain Mahogany Hairstreak is always found in chaparral, and is thus most common in southern California. It tends to be rarer and more sporadic farther to the north. Adults visit flowers avidly, especially buckwheat blossoms. For many years this species was known as *S. adenostomatis*, a name which referred to a mistaken association with the plant chamise (*Adenostoma*).

42, 424 Hedgerow Hairstreak
(*Satyrium saepium*)

Description: 1–1⅛" (25–28 mm). *Both sexes above russet or auburn-brown* with narrow dark border. Dull brown beneath, often paler outside irregular black, white-edged postmedian line; faint orange and blue spots near *short HW tail;* tail longer on female.

Similar Species: Mountain Mahogany Hairstreak gray with no russet above. Gold Hunter's Hairstreak has less reddish above, broader dark markings below, still shorter tails. Behr's Hairstreak golden, not russet above, broader black margin. Other russet hairstreaks are green or reddish below, or tailless.

Life Cycle: Pale gray-green egg overwinters on host plants, including species of wild lilac (*Ceanothus cuneatus, C. velutinus, C. sanguineus, C. macrocarpus*) and occasionally mountain mahogany (*Cercocarpus betuloides*). Caterpillar, to ½" (13 mm), light green with light stripes on back and sides, chevrons between, and covered with short golden hair. Chrysalis, to ⅜" (10 mm), brown, peppered with black.

Flight: 1 brood; late May–early July at lower elevations, July–September in mountains, occasionally into October.

Habitat: Mountains and foothills, mostly in arid forests or scrub, including chaparral, flowery roadsides, oak woodlands, sagelands, ponderosa pine forests and canyon floors.

Range: Pacific States from S. British Columbia south to Baja California and down Rockies from Montana to Colorado, N. Arizona, and New Mexico.

Early lepidopterists named this species the Hedgerow Hairstreak because it frequents shrubs and thickets. Adults seek nectar at goldenrod, mountain lilac, and other flowers. Usually less numerous than Behr's Hairstreaks, Hedgerows may appear in large groups.

388   Southern Hairstreak
(*Euristrymon favonius*)

Description: 1–1¼″ (25–32 mm). Above, vivid
brown-black with bright or dull
yellowish-orange patches on all wings,
larger on female. Below, dull brown
with thin black and white median
band; *HW has very wide red-orange
marginal band* tapering toward FW (not
separated into crescents) and
broadening toward blue patch; *2 long
HW tails.*

Similar Species: Northern Hairstreak has smaller red-
orange band below divided into
crescents, also less orange above,
shorter tails.

Life Cycle: Egg overwinters on oak. Caterpillar
light green with narrow, dark green
back stripe, oblique brownish and
yellow side stripes. Feeds on various
oaks (*Quercus*), eating buds and young
leaves; pupates in leaf litter.

Flight: 1 brood; May–June.

Habitat: Wooded coastal areas, barrier islands,
oak hammocks.

Range: West Virginia and New Jersey south to
Louisiana and Florida.

This butterfly seems to replace the
Northern Hairstreak in the coastal
Southeast. Further inland, some
intermediately patterned individuals
have been observed. It is possible that
the Northern and Southern hairstreaks
may be a single species, having 2
separate and dissimilar subspecies. For
nectar, adults visit chestnut and
chinkapin blossoms.

387   Northern Hairstreak
(*Euristrymon ontario*)

Description: 1–1¼″ (25–32 mm). Gray-brown
above, with *variable orange patches on all
wings* (mostly absent in North, well-
developed South and West). Tan-brown

below; thin black and white HW midline forms large "W" near 2 tails; FW line straight, indistinct; slight orange marginal row on HW capped with black, split into nearly distinct crescents, runs into blue patch near tails; lower tail longer.

Similar Species: Southern Hairstreak has more orange above, longer tails, continuous orange band on HW below. In other northern hairstreaks, line on underside expanded into band or broken into spots. Soapberry Hairstreak darker brown, orange more extensive below.

Life Cycle: In southern parts of range host plants include live oaks (*Quercus virginiana. Q. laurifolia* and probably *Q. ilicifolia*); perhaps hawthorn (*Crataegus*) used in Northeast. Chrysalis found in loose net in leaf trash below trees. Early stages undescribed.

Flight: 1 brood; June–July in North, late April–June southward.

Habitat: Open woodlands and coastal barrens in Northeast; oak groves in South and West.

Range: S. Ontario and Massachusetts south to Georgia, west (south of Great Plains) to Texas, SE. Colorado, and W. Arizona.

A rare butterfly in the Northeast, the Northern Hairstreak tends to become progressively more common toward the Southwest. It is most likely to be found on the "hairstreak flowers" such as New Jersey tea, dogbane, and milkweeds, or flying about in open sunny patches. The bright orange so characteristic of the southern populations is gradually lost through Missouri and Arkansas, becoming absent in the North. Extreme southwestern adults are blanched below, with all markings very reduced, and shorter tails. A population in the Davis Mountains of Texas has virtually no orange above and is often considered the separate species, Poling's Hairstreak (*E. polingi*).

Soapberry Hairstreak
(*Phaeostrymon alcestis*)

Description:

1–1¼″ (25–32 mm). *Above, warm sienna-brown.* Brown with all markings crisp *below; HW black and white line forms "VW" near 2 tails;* heavy coral-red HW border of connected crescents extends to or along FW; *white bar cuts across FW and HW centers.*

Similar Species: Northern Hairstreak lighter above and below, usually with some orange above. Southern Hairstreak has orange above. Banded Hairstreak lacks FW and HW bars.

Life Cycle: Egg overwinters. Host plant is soapberry (*Sapindus saponaria*). Early stages undescribed.

Flight: 1 brood; April–July, peaking in May.

Habitat: Roadsides, canyons, and other dry open areas with soapberry bushes and trees.

Range: S. Kansas, Oklahoma, Texas, and west through Arizona mountains.

These butterflies often perch high up on soapberry trees. Eggs can be found by inspecting leaf axils and bud scars in the fall and winter. The records of chinaberry as a host plant probably stem from the local southwestern name of western chinaberry for soapberry.

391 Endymion Hairstreak
(*Electrostrymon endymion*)

Description: ¾–⅞″ (19–22 mm). Small. *Golden-orange above with brown* at base, costa, and margin. *Gray below with postmedian band black, white, and red;* thin, dark submarginal band and prominent red spot and blue patch by HW tail.

Similar Species: Olive Hairstreak green below. Red-banded and Dusky Blue hairstreaks both black and blue above.

Life Cycle: Unknown.

Flight: Multiple broods; April–May, August, and November–December.

Habitat: Tropical and subtropical forest edges and scrub.

Range: S. Brazil straying rarely into S. Texas.

The Endymion Hairstreak is one of many neotropical butterflies that may occasionally be found in the southernmost extremities of North America. Endymion was a shepherd in Greek mythology lulled into perpetual sleep by the moon—hardly a suitable namesake for this flighty and alert hairstreak.

421 Angelic Hairstreak
(*Electrostrymon angelia*)

Description: ¾–⅞" (19–22 mm). Small. *Above, tawny-brown, darker at base. Below, smoke-gray white postmedian line broken, lacking red edge.* Prominent orange and black spots on either side of blue patch near tail.

Life Cycle: Undescribed. Brazilian pepper (*Schinus terebinthifolius*) reported host plant.

Flight: Year-round in successive broods.

Habitat: Subtropical hardwood hammocks.

Range: S. Florida south through Caribbean.

A recent immigrant to Key West, this Caribbean hairstreak has now established a foothold as far north as Miami. After alighting on a shrub it often rubs its hind wings together repeatedly, a habit common to many gossamer wings but especially well-developed in this species. Angelic Hairstreaks take nectar from seagrape as well as Brazilian peppers and other flowers.

Amethyst Hairstreak
(*Chlorostrymon maesites*)

Description: ¾–⅞" (19–22 mm). *Small. Above, male brilliant iridescent blue-purple; female bright blue with dark FW tip* and small HW marginal spots. Both sexes *below bright green; black postmedian line* lacks white on FW, partially white-edged and fairly straight on HW with an angle near inner margin; *brown and gray marginal patch near 2 tails.*

Similar Species: Silver-banded Hairstreak larger, paler purple above; broad white line across underside. Telea Hairstreak redder purple; HW postmedian line forms "W".

Life Cycle: Unknown.

Flight: At least 2 broods; December– September, peaking in early June.

Habitat: Hardwood hammocks.

Range: West Indies to Miami, Florida.

Generally quite rare in North America, this exquisite butterfly appears with sparse regularity in the keys of southernmost Florida. It can be found perching high up in direct sunlight on guamachil, Brazilian pepper, and Spanish needles. The combination of intense deep purple and bright green on the Amethyst Hairstreak is unusual.

438 Telea Hairstreak
(*Chlorostrymon telea*)

Description: ¾" (19 mm). *Small. Male brilliant iridescent reddish-purple above; female duller, bluer. Below, chartreuse green* with black spots above violet and blue metallic patch near tail.

Similar Species: Amethyst Hairstreak not as bright.

Habitat: Lowland tropical dry forest, especially disturbed areas near rivers; scrub country and wood edges, especially near flowers.

Range: Brazil straying north to S. Texas.

In spite of its extreme beauty, the Telea rarely appears where it might be encountered and studied in North America. Adults are especially fond of visiting the flowers of tropical trees, particularly mango, cordias, and caesarias. For a long time, the Telea Hairstreak was considered a subspecies of the Amethyst Hairstreak. This brilliant hairstreak is rare in North America, but more common south of the border.

439 Silver-banded Hairstreak
(*Chlorostrymon simaethis*)

Description: ¾–⅞" (19–22 mm). *Small. Above, male dull iridescent purple,* margins are dark, widest at FW tip; *female's blue largely restricted to FW base,* much less iridescent. Both sexes *below bright yellow-green with silver-white postmedian band* on both wings, *forming "V" near HW tail;* terminal *orange-brown patch along entire HW outer margin.*

Similar Species: Both Amethyst and Telea hairstreaks smaller, brighter purple, lack broad white line beneath. Blue-and-green Hairstreak blue above, not purple, lacks white line below.

Life Cycle: Undescribed. Host plant is balloon vine (*Cardiospermum halicacabum* and *C. corindum*).

Flight: 2 broods; April–May, August–December, peaking in October.

Habitat: Arid thorn-scrub forest.

Range: S. California, S. Arizona and S. Texas into Baja California and South America; also in Antilles.

The Silver-banded Hairstreak flies swiftly and is difficult to follow. When it settles with wings closed on a green leaf (often a nettle), its chartreuse underside keeps it well hidden. Uncommon and local throughout most of its range, this species is well established in extreme S. Texas, as well as Baja California and S. Florida, but occurs only as a rare visitor to southernmost Arizona and California.

___

409, 481   Early Hairstreak
(*Erora laeta*)

Description: ¾–1″ (19–25 mm). Tailless. *Above, steel-blue with black* tips and margins; male less blue than female. *Below, bluish-green with broken row of reddish, white-rimmed spots;* white fringe sparsely orange-dusted.

Life Cycle: Unrecorded. Beech (*Fagus*) and hazelnut (*Corylus*) are suspected host plants.

Flight: 1 brood; March in South through late May–late June northward, partial 2nd brood in August.

Habitat: Edges of mature beech-maple woods, dirt roads and shaded patches among woodlands, mature deciduous forests in Canada.

Range: Michigan through E. Canada and New England, south in Appalachians to S. Carolina.

Few North American butterflies have attracted as much attention as this rare species, whose habits are shrouded in mystery. Although classic lore places Early Hairstreaks in mature beech forests, not all habitats conform, and individuals appear from time to time in varied wooded sites. The Early Hairstreak may turn out to be far less rare than its reputation suggests once its behavior is fully known. Adults will alight on bare ground, at times remaining so still that they may be trod upon by the unwary. It is possible that Early Hairstreaks live high in the woodland canopy, descending within viewing range only occasionally.

410, 484 Arizona Hairstreak
(*Erora quaderna*)

Description: ¾–1″ (19–25 mm). Tailless. *Above,
male bronze-brown* with blue confined to
HW margin, *female deep violet-blue with
black tips and margins. Below, smoky
olive-green crossed by rows of rust-colored,
white-rimmed spots;* rust-colored fringe.
Blue and green scales lost as butterfly
ages, leaving brownish ground
color.

Similar Species: Blue-and-green and Silver-banded
hairstreaks both brighter, tailed.

Life Cycle: Unknown.

Flight: 2 broods; April and late June–July.

Habitat: Open oak-pine forests, generally at
5500–7500′ (1678–2288 m) in
Arizona.

Range: Arizona, New Mexico, and Texas south
to Mexico.

Denizens of hot canyons and backroads,
Arizona Hairstreaks gather in April and
June around damp spots and flowers,
favoring ceanothus, cherries, and bear
grass. Although this butterfly resembles
the Early Hairstreak, its habitat, range,
and habits are utterly different. The
Arizona Hairstreak is encountered in far
greater numbers. Both species share the
unusual trait of the females being
apparently more numerous than the
males. However, this statistic probably
reflects behavior and sampling bias
rather than actual sex ratio.

393 Red-banded Hairstreak
(*Calycopis cecrops*)

Description: ¾–1″ (19–25 mm). Brown-black *above,
pale iridescent blue patches* near tails on
male, covering much of HW above on
female, sometimes also on FW base.
Gray-brown *below with bright jagged
tricolor banner of white, black, and red,
with flame-red innermost and broad* on

both FW and HW; between 2 tails *biggest HW spot more black than red.*

Similar Species: Dusky Blue Hairstreak has narrow red stripe below, especially on FW; HW spot between red is more red than black.

Life Cycle: Pearl-white, dimpled egg laid in leaf litter. Caterpillar pale yellow upon hatching, darkening with age; has thick brown hairy coat and blue-green back stripe. Chrysalis chestnut-brown, mottled black. Host plants include dwarf sumac (*Rhus copallina*), croton (*Croton*), and possibly waxmyrtle (*Myrica*).

Flight: Multiple broods. 3 in central Atlantic States; April–May, July, September–October.

Habitat: Open countryside, woodland margins, fields, and barrens.

Range: Ohio and New Jersey south to Texas and Florida; rarely farther north.

Especially common in the coastal southeast, the Red-banded Hairstreak becomes sparser in the mid-Atlantic states and the Ohio Valley, and is merely a visitor further north. The flowers of the host plant and of other sumacs attract the adults as nectar sources. Especially active at dusk, these butterflies lurk among leaves until disturbed.

392 Dusky Blue Hairstreak
(*Calycopis isobeon*)

Description: ¾–⅞" (19–22 mm). *Small.* Brown-black *above; blue at bases of outer HW* more prevalent in female. Dark gray-brown *below with tricolored banner: thin flame-red innermost.* followed by black and white lines; *red bleeding* past black and white into tail area on HW; between tails *biggest HW spot, much more red than black.*

Similar Species: Red-banded Hairstreak has broad red which does not bleed into black and white part of HW band below, also HW spot has more black than red. 2 overlap in S. central states only.

Life Cycle: Unrecorded.

Flight: Multiple broods, approximately March and November.

Habitat: Tropical and subtropical forest edges and scrub.

Range: S. and E. Texas south to Panama; rarely extending to Kansas.

From the large, primarily tropical genus *Calycopis*, only the Dusky Blue and Red-banded hairstreaks reach North America. In Central America large numbers of blue-winged hairstreaks of this and other genera often fly together and it is easy to confuse them.

Clytie Hairstreak
(*Ministrymon clytie*)

Description: ¾–⅞" (19–22 mm). *Above, male pale blue* over most of HW and lower portion of FW; black toward FW tip; *female darker blue at base,* outer portion of wings black. *Below, postmedian line rather broadly colored with red,* heavily scaled with white outside line. *Reddish dashes prominent near HW base,* orange and black spot and blue patch near tail.

Similar Species: Leda Hairstreak has redder markings below.

Flight: Multiple broods; year-round, straying to U.S. at any time.

Habitat: Lowland mesquite thorn scrub; often near irrigation ditches with lantana.

Range: S. Arizona and S. Texas south into Mexico.

Little known in North America, the Clytie Hairstreak can be fairly common in the limited number of places it lives. The spring brood is more deeply colored than summer individuals.

407 Leda Hairstreak
(*Ministrymon leda*)

Description: ¾–⅞″ (19–22 mm). *Small. Above, blue at base, gray-black outwardly;* male blue paler, may cover most of HW. *Below,* gray crossed by *jagged black postmedian line* with or without red outline; black-centered red spot and blue patch may or may not be present near 2 tails; faint spots at wing base.

Similar Species: Clytie Hairstreak has more prominent, less red postmedian line and basal spots below.

Life Cycle: Egg pale green; laid on host plant, mesquite (*Prosopis juliflora*). Mature caterpillar apple-green with pale yellow chevrons above and short, thick brown hair all over. Chrysalis pale dull green-brown with brown and black marks.

Flight: At least 2 broods; May–July or August, September–October or even November.

Habitat: Thorn scrub with host plant, often in sandy washes and flats, foothill canyons, and drier mountain forests.

Range: Eastern Mojave Desert in New York Mountains of California, south and east to deserts of S. California, Baja California, and S. Arizona.

The Leda Hairstreak has 2 distinct forms, with and without red markings. This kind of dimorphism may be due to different seasonal conditions, such as moisture variation, or it may be a balanced, genetically fixed set of frequencies within the population. Although the exact cause of the dimorphism remains uncertain, lepidopterists know that the 2 forms definitely belong to the same species because eggs of one have produced the other.

408 Larger Lantana Butterfly
(*Tmolus echion*)

Description: ⅞–1" (22–25 mm). *Above, male deep iridescent blue-purple,* restricted on FW to base by *dark diffuse border. Female above dullish sky-blue, foggy at borders.* FW blue more extensive than in male; diffuse darker spot at end of FW cell. Both sexes *below light grayish with rust-red spots forming short broken row near base and full postmedian row,* continuous on FW and *much broken and offset on HW.*

Life Cycle: Unknown. Host plants are species of lantana (*Lantana*) and many other plants.

Flight: Year-round in Tropics.

Habitat: Weedy habitats, such as roadsides and fields around lantana.

Range: Hawaii; rarely in extreme S. Texas.

This species is common in Hawaii, where it was introduced in 1902 in an effort to control the spread of weedy lantanas, which the caterpillars eat. However, the Larger Lantana's presence created additional problems because the caterpillars also consume peppers, eggplants, potatoes, and other garden crops. Ironically, lantana is one of the best nectar flowers for the adult butterfly. A related species, the Azia Hairstreak (*T. azia*), tends to wander slightly more frequently to Texas and the Florida Keys. This very small hairstreak is dark grayish above and the female has bluish-white on the outer half of its hind wings; below, both sexes have a bright red postmedian line. This species feeds on woody leguminous plants throughout much of Latin America and enters our southernmost areas occasionally in the spring.

Sonoran Hairstreak
(*Hypostrymon critola*)

Description: 1" (25 mm). *Male above bright blue-lilac, purplish especially surrounding large black FW stigma,* with narrow black margins; female has paler, more restricted blue above. Both sexes *below light gray with fine dark striations* evenly shading in most of wings except for *unmarked submarginal strip,* vague orange and black spot near 2 tails.

Similar Species: Tails and male stigma distinguish it from blues, striations below from other hairstreaks.

Life Cycle: Unknown. Close association of adults with salt-loving shrub (*Maytenus phyllanthoides*) in Baja California suggests it may be host plant.

Flight: April–May in Arizona; possibly year-round in Mexico.

Habitat: Desert scrub, usually alkaline; coastal scrub in Mexico.

Range: Huachuca and Patagonia Mountains of S. Arizona; also in Sonoran Desert of Mexico.

While common in parts of Baja California, this lovely hairstreak is generally found in our area only infrequently. Recent records suggest it may be resident in Arizona.

---

386 Gray Hairstreak
"Common Hairstreak"
(*Strymon melinus*)

Description: 1–1¼" (25–32 mm). *Above, deep slate-gray with orange spot* on HW; female browner. *Below, dove-gray;* straight, thin FW and HW red and black midband lined with white; bold orange and blue patches above tail, black spot at HW outer angle (tornus). Abdomen has orange sides.

Similar Species: Avalon Hairstreak sandy-buff below with reduced bands and orange. Other similar hairstreaks are browner or lack bright orange spot above.

Life Cycle: Egg pale green. Caterpillar variable, usually grass-green to translucent green, with white to mauve diagonal side stripes; various host plants are nearly 50 different plants in over 20 families; host plants include corn (*Zea mays*), oak (*Quercus*), cotton (*Gossypium*), strawberry (*Fragaria*), and mint (*Lamiacea*), legumes and mallows preferred. Chrysalis brown with copious black mottling.

Flight: Variable, number of broods increasing southward: 2 in North, 3 or more in South; April–October.

Habitat: Open deciduous woods, coastlines, roadsides, chaparral, old fields, parks, vacant lots, and other open spaces.

Range: British Columbia to Maritimes, and south to Baja California, Florida; also to Venezuela and Colombia.

Absent only from the far North, the Gray Hairstreak is one of the most generally distributed butterflies. The caterpillar is known regionally as the "cotton square borer," and has upon occasion damaged commercial bean, hops, and cotton crops. Gray Hairstreaks now occur in all the California islands, having colonized Catalina Island, stronghold of the Avalon Hairstreak, in 1978. By 1979 both species from Catalina showed atypical variation, implying possible hybridization between them. Until now, the Gray Hairstreak has appeared to exclude the Avalon Hairstreak from the other California islands; therefore, scientists and conservationists are watching the Gray Hairstreak's influx to Catalina with apprehension.

414 Avalon Hairstreak
(*Strymon avalona*)

Description: ¾–1″ (19–25 mm). *Above, brownish-gray* often with whitish-blue suffusion; orange capped black spot on HW. Pale whitish to sandy-buff *below with frost-white outer third* beyond midband on both FW and HW; thin, vague FW and HW *median band lightly red and black, scalloped on HW* into 3 or 4 distinct lobes; slight trace of orange above tail, modest black spot at HW outer angle (tornus).

Similar Species: Gray Hairstreak more slate-colored above, gray below, orange spots more prominent.

Life Cycle: Egg pale green. Caterpillar dusky gray-green, with fainter back and side stripes; burrows deep into buds of scrubby lotuses (*Lotus argophyllus* and *L. scoparius*); populations around NW. isthmus of Catalina Island feed on buckwheat (*Eriogonum*). Translucent light brown chrysalis flecked with darker brown.

Flight: Multiple, indistinct broods, possibly continuous; mainly February–October, peaking in March.

Habitat: Steep brushy slopes along dry coastline; rarer in interior.

Range: Catalina Island, off S. California coast.

This and the Uncompahgre Fritillary have the most restricted ranges of any Nearctic butterfly species. Avalon Hairstreaks swarm in large colonies over much of Catalina Island, drinking nectar from many flowers on this desert island, especially in the morning, and perching and mating during the hot afternoon. The Avalon Hairstreak's close relative, the Gray Hairstreak, seems to exclude this species from other Channel Islands, and may endanger it through competition or genetic swamping. The name *avalona* comes from the town of Avalon on Santa Catalina.

390 Mexican Gray Hairstreak
(*Strymon bebrycia*)

Description: ⅞–1⅛″ (22–28 mm). *Male gray-brown above with* a few dark marginal spots and *small red spot by tail;* has stigma. *Female brown above with dark patches* like male's stigma but larger, *HW has some light blue* overscaling at center and around HW marginal spot row; *1 spot by tail has prominent red.* Both sexes light gray *below with HW postmedian line red and white.* usually lacking any black; black-centered *red spot and blue patch near tail;* faint white markings along HW margin.

Similar Species: Gray Hairstreak male has bigger orange-red spot on HW above, female lacks blue above; both sexes have black as well as red edging to white line below.

Life Cycle: Unknown. Host plant possibly croton (*Croton*).

Flight: Multiple broods; February and September–December in U.S.

Habitat: Unrecorded.

Range: Costa Rica north to extreme S. Texas.

This tropical butterfly wanders fairly regularly into Texas. The species has also been known as *S. buchholzi.*

Alea Hairstreak
(*Strymon alea*)

Description: ⅞–1″ (22–25 mm). *2 seasonal forms. both dark gray above with black spots near tail:* summer form paler toward margin of HW above; winter form pale only around spots. *Below. summer form* even gray, *black and white postmedian line edged inwardly with red.* bright orange cap to black spot on HW; *winter form darker gray inside postmedian line. lacks red on line. much lighter beyond.* faint orange cap to black spot near tail.

Similar Species: Columella Hairstreak has postmedian

band below composed of broken spots.
Yojoa Hairstreak has white stripe on
HW below.

Life Cycle: Unknown. Host plant is myrtle croton
(*Bernardia myricaefolia*); caterpillar eats
buds and blossoms.

Flight: Multiple broods; February, April–May,
June–July, and September–December.

Habitat: Desert scrub and semi-arid areas,
especially along rivers and near flowers.

Range: Mexican lowlands north to Austin,
Texas.

The Alea Hairstreak displays marked
seasonal dimorphism. The summer
form was named *S. laceyi* after the 19th-
century Texas naturalist H.G. Lacey,
but the name *S. alea* is an older one
that has taken precedence.

419 Yojoa Hairstreak
(*Strymon yojoa*)

Description: ⅞–1⅛" (22–28 mm). *Above. dark gray-
brown with a few black marginal spots by
HW tail;* male has dark stigma; female
more rounded wings. *Below. gray with
fine black and white postmedian line*
offset but rather straight and *set well out
toward margins; white marbling. especially
on HW* as postbasal strip and beyond
postmedian band; orange and black
spots at either side of tail with blue
patch between. Abdomen yellow-
tipped.

Similar Species: Gray Hairstreak has red along line
beneath, orange HW spot above.
Others lack white marbling below.

Life Cycle: Unknown. Host plant in Mexico is
hibiscus (*Hibiscus tubiflorus* and
cultivated varieties).

Flight: Multiple broods throughout the year.

Habitat: Tropical mountain forests, plantations,
fields among deciduous forest, and arid
scrub.

Range: Central America and Mexico straying to
extreme S. Texas.

Although common in the Tropics, this species occurs very rarely in North America.

---

406 Columella Hairstreak
(*Strymon columella*)

Description: ⅛–1" (22–25 mm). *Above, male brown or gray-brown* with prominent dark stigma, a few *black spots near HW tail; female* similar with *some light blue* often on rear half of HW. Both sexes pale *gray to brown below,* with vague HW postmedian band of broken spots, a few dark spots at costa and cell, fine white markings over HW outside of postmedian band, and *prominent black spot near tail* with or without prominent orange.

Similar Species: Reddish Hairstreak has orange spot on HW above; female Reddish lacks blue on HW above. Mexican Gray Hairstreak grayer, with red along HW submarginal line below, and orange on HW above.

Life Cycle: Egg pea-green. Caterpillar becomes dark green with darker back stripe, covered with translucent brown hair; feeds on alkali mallow (*Sida hederacea*), in California probably other mallows (*Sida*). Chrysalis pink-brownish yellow with brown flecks and green lines.

Flight: Year-round in Texas and Florida. In California 2 broods; March–April, August–October.

Habitat: Alkaline flats, often around agriculture in California and Texas.

Range: S. California, S. Arizona, S. Texas, and S. Florida, south to South America and Caribbean.

While the Columella Hairstreak is common in parts of Florida and Texas, it is quite rare in California. Butterflies from Florida are browner below with the white markings of the hind wing especially prominent beyond the

postmedian line; they also have a
prominent red spot near the tail below.
Texan hairstreaks are similar but lack
the prominent orange, as do California
broods, which are light gray below. A
very similar species, the Limenia
Hairstreak (*S. limenia*), was recognized
in Florida in 1972. So far it has been
found on Big Pine Key and Key West,
where it flies March–May and in
December, visiting flowers of Spanish
needles. The Limenia Hairstreak differs
from the common Columella in having
a more prominent hind wing
postmedian band, more prominent
black spots in the hind wing cell, and a
very prominent black spot below in
from the tail.

420 Reddish Hairstreak
(*Strymon rufofusca*)

Description: ⅞" (22 mm). Tailed. *Above.* male dull,
tawny-*brown* with small dark stigma;
female rounder, darker brown with
dark spot-row along HW margin; *both
sexes have red-capped spot by tail. Below.*
both sexes tan with broken thin *red
postmedian line and 2 fine rows of white
dashes* outside lines; small mostly black
spot near tail.

Similar Species: Gray and Mexican Gray hairstreaks
grayer above. Most others have white
and/or black postmedian line, not just
reddish.

Life Cycle: Unknown. Host plants in Mexico are
mallows (*Malvastrum coromandelianum*
and probably also *Sida*).

Flight: Multiple broods; March, September,
and November–December.

Habitat: Deciduous forest clearings, tracks, and
edges.

Range: Brazil north to S. Texas; also St.
Vincent in the Lesser Antilles.

The Reddish Hairstreak is widespread
in the American Tropics. It enters our

area only occasionally, along the southernmost borders.

416, 510 Cestri Hairstreak
(*Strymon cestri*)

Description: ¾–1″ (19–25 mm). Tailless. *Chocolate-brown above with bluish overtones.* subtle in male, brighter in female; large, dark stigma on male; black spots around HW margin on both sexes. *FW distinctively convex below tip.* FW cocoa-brown *below; HW complexly patterned with alternating. blurred bands of brown and cream-colored spots.* 1 distinct, round black spot near HW outer angle (tornus).

Similar Species: Smaller Lantana Butterfly bluer above, more vaguely patterned below and has prominent dark spot along HW costa below near base.

Life Cycle: Unknown.

Flight: Throughout the year.

Habitat: River bottomlands, patches of brush, canals, and ditches; waterside woodlands.

Range: Central America, straying to S. Texas.

The Cestri Hairstreak's status in North America remains vague. First found and subsequently rediscovered around Pharr, Texas, it may be rare and regular or a random visitor to the state.

417 Smaller Lantana Butterfly
(*Strymon bazochii*)

Description: ⅞–1″ (22–25 mm). Tailless. Male *above has FW dark gray-brown with dark. prominent stigma; HW pale blue* with dark margin along costa and dark dot row along outer margin; HW outer angle (tornus) somewhat pointed. Female similar above except blue slightly more lilac and HW tornus more rounded.

Both sexes *below soft gray mottled with soft brown; HW has dark spot at base near costa and whitish streak* running from base to middle of outer margin.

Similar Species: Cestri Hairstreak less blue above, more distinct pattern below, has dark spot near lower angle of HW margin, and lacks spot at HW costa.

Life Cycle: Unrecorded. Host plants in Texas include lippia (*Lippia alba* and *L. graviolens*) and lantana (especially *Lantana camara*); in Hawaii 2 members of the mint family (*Hyptis pectinata* and *Ocimum basilicum*), as well as lantana.

Flight: Probably 2 broods in Texas; March–May, October–December.

Habitat: Scrubby second growth in seasonally dry tropical zones; gardens and weedy areas in Hawaii.

Range: Extreme S. Texas south to Brazil and Antilles; introduced in Hawaii.

In southern Texas the Smaller Lantana Butterfly is resident and sometimes common. Along with the Larger Lantana Butterfly, it was intentionally introduced as a natural control agent for weedy lantanas in Hawaii.

384 Acis Hairstreak
(*Strymon acis*)

Description: 1″ (25 mm). *Steel-gray above*, with red spot near shorter tail, rust-colored spot near longer tail; *no blue. Below, clear gray crossed by broad white, black-edged lines* (1 on FW, 2 on HW), *1 large red-orange patch near long tail, 1 along lower margin;* small white dots near HW base.

Similar Species: Martial Hairstreak has blue above, more rust-colored HW spot below, and lacks white dots near HW base below. Gray Hairstreak has less white on lines below, which are also less solid and straight.

Life Cycle: Wild croton (*Croton linearis*) is host

plant. Otherwise unknown.

Flight: Several broods; year-round.
Habitat: Deciduous woodland edges and gardens around lakes, beaches, and hammocks.
Range: S. Florida south throughout Antilles.

In Jamaica, the Acis Hairstreak is said to associate with the elegant Silver-banded Hairstreak. Several subspecies occur on the Antillean islands. The Florida subspecies, Bartram's Hairstreak (*S. a. bartrami*), was named after the pioneer naturalist of the Southeast, John Bartram.

383 Martial Hairstreak
"Long-tailed Hairstreak"
(*Strymon martialis*)

Description: 1" (25 mm). *Above, pale blue* with blackish-brown FW tips and costa and gray-brown HW costa (female darker); few marginal black dots near *2 tails and 1 rust-colored spot at HW outer angle (tornus). Below,* light gray with black and white postmedian band connected to submarginal markings through *prominent orange field* which extends up toward tornus.

Similar Species: Leda and Clytie hairstreaks have spotty bands below, lack broad orange field. Acis Hairstreak has blue extremely understated or absent above; broader lines and redder HW spot below.

Life Cycle: Dull green caterpillar covered with short white, stiff hair; feeds on trema (*Trema floridana*).

Flight: Multiple broods; virtually year-round.
Habitat: Secondary growth near shore.
Range: Florida as far north on coast as Sarasota, south to Antilles.

The Martial Hairstreak flies commonly but locally in southern Florida. Adults visit Spanish needles, bay cedar, and buttonwood mangrove flowers.

418 White Hairstreak
(Strymon albata)

Description: ⅞–1⅛" (22–28 mm). *Largely white above with brown tip and base;* large black stigma on male. *Below, tan within and whitish outside dark postmedian line,* with vague dark markings in white areas and small black and orange spot near tail.

Similar Species: No other S. Texan hairstreak has heavy white scaling.

Life Cycle: Host plant in Mexico is Indian mallow *(Abutilon incanum)*.

Flight: Summer and fall in U.S.

Habitat: Forest edges and second growth in tropical dry forest; scrub country around watercourses.

Range: Costa Rica straying north to S. Texas.

Only on very rare occasions has the White Hairstreak been seen north of Mexico. Adults take nectar from cordias and caesarias.

403 Coral Hairstreak
(Harkenclenus titus)

Description: 1–1¼" (25–32 mm). Tailless. *Male has dark, pointed, triangular wings; brown above,* unspotted or with light reddish spots along HW margin. *Female has rounded wings, light brown above,* usually with reddish spots along HW margin, blurs along FW margin. Some populations heavily suffused with orange above in both sexes. Below, both sexes warm gray-brown with or without conspicuous rows of black, white-rimmed spots across wings or vaguer black dots; always *prominent row of large, bright coral-red spots along entire HW margin below.*

Similar Species: No other hairstreaks combine the row of coral spots with the absence of tails.

Life Cycle: Egg overwinters. Downy caterpillar yellowish-green, pinkish in middle of back; feeds on developing fruits of plums and wild cherries (*Prunus*), also western serviceberry (*Amelanchier alnifolia*).

Flight: 1 brood; June–August, mostly in July.

Habitat: Meadows, brushy clearings, roadsides, watercourses; usually mountain canyons in West; very often near wild cherry clumps.

Range: British Columbia east to Maritimes, south to E. central California, NE. Arizona, Texas, and Georgia; absent from far Southwest, Gulf Coast, Northwest Coast and far North.

The Coral Hairstreak displays a high degree of variation in its appearance and habitat. Common in meadows in the Northeast, rather rare and local in western canyons, it flies swiftly but dallies at flowers. Its favored eastern nectar source is butterfly weed, while in the Rockies adults often settle on bee plant. The unusual genus name *Harkenclenus* is an amalgam of the names of the late Harry Kendon Clench, a noted hairstreak specialist.

505 Western Pygmy Blue
(*Brephidium exilis*)

Description: ⅜–¾″ (10–19 mm). *Tiny. Bicolored above:* white-fringed chocolate-brown with *ultramarine blue inward.* female larger than male and less blue. *Below:* gray-brown blending to bluish-gray at base; tiny iridescent blue-green centered *black spots on HW margin;* white striations across wings.

Similar Species: Eastern Pygmy Blue fringes darker, bases of wings brownish below; ranges usually do not overlap.

Life Cycle: Egg aquamarine. Caterpillar pale green to cream-white, with yellow stripes on back and sides, and tiny brown bumps overall; feeds on plants of goosefoot family, including pickleweed (*Salicornia ambigua*), saltbush (*Atriplex*), and pigweed (*Chenopodium*).

Flight: Continuous broods in S. Texas and
California, peaking in late summer and
autumn.

Habitat: Lowland, often disturbed places and
coastlines: washes, marshes, alkali flats,
railroad tracks, and vacant lots.

Range: E. Oregon, California, and Great Basin
south to Texas and South America, east
to Nebraska and other plains states.

This smallest western butterfly is often
abundant, but nevertheless usually
passes unnoticed because of its
diminutive size and slow flight. Despite
its minuteness and seeming fragility,
the Western Pygmy Blue emigrates
northward each year from its year-round
southern homeland. Northern records
probably represent these emigrants and
their summer offspring rather than
resident colonies.

506 Eastern Pygmy Blue
(*Brephidium isophthalma*)

Description: ½–¾" (13–19 mm). *Tiny. Male
uniform brown above. female reddish;* blue
absent or merely a dusting. *Below.*
gray-brown with dark fringes and little
or *no white at base;* blue and black spots
on HW margin; white striations across
wings.

Similar Species: Western Pygmy Blue has whitish
fringes, bases of wings whitish on
underside; ranges usually do not
overlap.

Life Cycle: Pale greenish caterpillar feeds on the
salt marsh plants, glasswort (*Salicornia*)
and saltwort (*Batis*).

Flight: Multiple broods; peaking in February,
April, June, and September.

Habitat: Salt marshes and tidal flats.

Range: Coastal Alabama, Georgia, and Florida.

The smallest eastern butterfly is not as
widespread as its western counterpart,

lacking the strong emigratory and colonizing tendency of the Western Pygmy Blue. At times it is rare or nearly absent, then suddenly for a short while this species becomes very common. This blue hugs the coastline, flying low and slowly over its salty habitat.

508 Cassius Blue
(*Leptotes cassius*)

Description: ½–¾″ (13–19 mm). *Tiny. Male light blue above* with thin black border; some males have white on HW. *Female white above.* often with blue near border; broad brown-black margins and inner banding and black HW spots. *Both sexes translucent. with underside markings showing through.* Both sexes pale white *below. with brown bands* profusely scalloping FW and inner HW, *fading away into pale patches* in basal wing halves; spotted brown borders, 2 marginal HW spots.

Similar Species: Antillean Blue has dark markings beneath, more in short bars than extended bars; usually only 1 black HW spot. Marine Blue purplish-blue above, complete bold brown scalloping below.

Life Cycle: Egg pastel green changing to ivory. Young caterpillar dark yellow, becoming speckled green suffused with russet; feeds on flowers of leadwort (*Plumbago*), milk pea (*Galactia*), lima bean (*Phaseolus limensis*), rattlebox (*Crotalaria*), and other legumes. Chrysalis attached to lower leaf surfaces by silken girdle.

Flight: Multiple broods; most months where resident, summer and autumn where emigrant.

Habitat: Fields, roadsides, and parks.

Range: Resident in S. Florida and S. Texas, emigrating through Kansas and Missouri; south to South America.

Very common in the southern part of
its range, the Cassius Blue flies in
almost any sunny spot. Adults seem to
prefer flowers of trees and shrubs.
Although this species maintains
northern populations during moderate
weather, it still succumbs to frigid
midwestern winters.

487 Marine Blue
(*Leptotes marina*)

Description: ⅝–1" (16–25 mm). *Above, male uniform
light purple to lavender-blue,* without
white inner HW margin; *female dull
violet,* with broad brown border, and
brownish suffusion to bases of FW and
HW. Blanched bluish or brownish-
white *below, entirely scalloped with pale
brown crescents;* HW spots iridescent
pale blue to silver-colored.

Similar Species: Cassius Blue paler blue not purplish
above, scalloping stops halfway down
wings leaving blank area. Marine and
Cassius overlap in central states.

Life Cycle: Egg green, ridged white. Caterpillar
variable, from pure green to rich
brown. Short-haired, dark yellow
chrysalis has paler wing cases. Pea
family host plants include alfalfa
(*Medicago sativa*), sweet pea (*Lathyrus
odoratus*), false indigo (*Amorpha*), and
others; also leadwort (*Plumbago*).

Flight: Multiple, continuous broods in Tropics
and frost-free temperate regions;
peaking February–November in
Southwest. Summer emigrant in
central and mountain states.

Habitat: Weedy, open sites; commonly along
watercourses.

Range: Resident in S. Texas to central
California; summer emigrant to
Illinois, and Nebraska; also American
Tropics.

Like Cassius Blues, Marine Blues fly
northward in summer, dying back in

fall and winter to frost-free strongholds. Extensive ornamental plantings of leadwort and wisteria in California bring legions of these shimmering purple-blue butterflies to city and suburban streets and parks. East of the Rockies, they follow canals and streams far into the plains and foothills. Like many emigrant butterflies, Marine Blues exploit ephemeral habitats—margins, flats, streambeds, and wastes, where weedy annuals spring up between frosts and floods.

---

485 Blackburn's Bluet
"Hawaiian Blue"
(*Vaga blackburnii*)

Description: ⅞–1″ (22–25 mm). *Above, male deep lilac-blue with thin dark margin; female blue restricted to wing bases, has wider, more diffuse dark margins. Both sexes below unmarked pale green* (some gray on FW).

Similar Species: Greater Lantana Butterfly has no green below.

Life Cycle: Originally fed chiefly on the Hawaiian koa tree (*Acacia koa*); has since adopted several introduced shrubs, including acacia (*Acacia*). Also uses mamake (*Pipturus albidus*) and other tropical plants.

Flight: Year-round in successive broods.

Habitat: Koa forests on hills and mountains; kipukas (islands of older vegetation among lava fields); lowlands and occasionally gardens.

Range: Restricted to Hawaiian Islands.

Blackburn's Bluet and the Kamehameha Butterfly make up the entire native butterfly fauna of Hawaii. Specialists believe that the Bluet's nearest relatives are found thousands of miles to the west, on the Bonin Islands. The Bluet has proved to be very adaptable, and remains abundant

despite development and the
introduction of non-native species to
the islands.

Cyna Blue
(*Zizula cyna*)

Description: ¾–1″ (19–25 mm). Above, *fragile*
slender FW and rounded HW both
translucent lilac-blue, with broad,
indistinct darker margins. Below, ash-
gray, peppered with small black spots.

Life Cycle: Unrecorded.

Flight: Several broods; March–September in
U.S.

Habitat: Variety of open sites, including
scrubby edges and disturbed urban
habitats.

Range: New Mexico, S. Texas, and Arizona
south to South America.

For a long time, this extremely frail-
looking butterfly was erroneously
considered an African import. An
apocryphal story relates that Cyna Blue
caterpillars were brought to the United
States in camel fodder in the 1800's by
Jefferson Davis, then Secretary of War.
The camels were to be used as pack
animals. Since then, this butterfly has
been reevaluated as a native species after
all. The Cyna Blue is a very weak, slow
flier.

480 Miami Blue
(*Hemiargus thomasi*)

Description: ¾–1⅛″ (19–28 mm). *Above*, bright
blue with dark, narrow, scalloped
margin; *female has black-centered orange
spot at base of HW*. *Below*, black
chevrons along FW margin, dots cover
HW; *broad inner white diffuse band*
crisply divides wings into separate dark
areas, with *light and dark strongly*

contrasting; sparkling blue-black spots at HW base.

Similar Species: Similar small blues lack HW orange spot above.

Life Cycle: Host plants include the legumes cat's-claw (*Pithecellobium*) and gray nicker (*Guilandina*).

Flight: Multiple, overlapping broods; year-round.

Habitat: Open sunny areas.

Range: S. Florida and Antilles.

This locally numerous butterfly is fond of visiting flowers, particularly Spanish needles. Its appearance changes with the seasons, becoming more blue in winter and spring than in summer and fall, when black encroaches from the wing margins. Such seasonal pattern variation may be caused by changes in humidity, temperature, or the length of the day.

488, 504 Antillean Blue
(*Hemiargus ceraunus*)

Description: ¾–1″ (19–25 mm). *Above. male lavender-blue* with narrow black border, *female dusky brown-black.* washed steel-blue at wing bases. Drab gray-brown *below with 2 rows of whitish arrowheads* inside margins; thin, broken, white-outlined FW and HW black bars cross wing centers; *1 or 2 prominent. coal-black HW spots on margin usually lack any orange.*

Similar Species: Miami Blue has orange HW spot above. Reakirt's Blue has inflated black FW spots below. Cassius Blue brown-banded, not black-barred, below.

Life Cycle: Egg pale gray-blue. Caterpillar deep green, yellow, mauve, or dull red with yellow or green tiger stripes and covered with silver-white hair; consumes buds and young leaves of numerous legumes, including locoweed (*Astragalus*), mesquite (*Prosopis*), beans

(*Phaseolus*), and partridge pea
(*Chaemaecrista*). Bright green chrysalis
turns brown with age.

Flight: Multiple broods; continuous in
American Tropics, March–November
in Southwest, year-round in Texas and
Florida.

Habitat: Fields, roadsides, and sunny spots;
principally lower deserts and foothills
westward, sporadically into mountains.

Range: S. California east through Nevada, S.
Utah, and New Mexico; Texas;
Alabama, Georgia, and Florida south
into Mexico and Antilles. Absent from
central Gulf States.

Three separate races of this species reach
into the United States from the
Antilles, eastern Mexico, and Baja
California. The Southwestern
population usually has only 1 eyespot
beneath, while the Central American
subspecies has 2 eyespots. The
population found in southeastern states
is nearly purplish above, and has more
contrasting buff-colored markings
below and may show an orange spot
below. Because the Antillean Blue's
caterpillars feed on a wide array of
legumes, the species is capable of
inhabiting many kinds of warm places.

474, 503 **Reakirt's Blue**
"Solitary Blue"
(*Hemiargus isola*)

Description: ¾–1⅛″ (19–28 mm). *Above,* male
lilac-blue; female dusky with bluish
interior; on both *HW spots lack orange.*
Cream-brown, flushed with white *below;*
prominent series of inflated round, coal-
black FW spots rimmed by white,
continuing around HW margin with
several small and 2 or 3 larger spots.

Similar Species: Antillean Blue lacks row of spots on
FW below. Miami Blue contrasting
light and dark below. Silvery Blue

much larger, lacks prominent spots at HW outer angle (tornus).

Life Cycle: Undescribed. Caterpillar feeds on flowers and fruits of legumes, including mesquite (*Prosopis*), indigos (*Indigofera*), ornamental acacias (*Albizzia*), and others.

Flight: Many overlapping broods in South; July–October in North.

Habitat: Fields, gardens, and open areas where legumes and weedy nectar sources abound.

Range: Southern states from S. California to Mississippi; north to Nevada, Wyoming, Saskatchewan, and Michigan; south to Costa Rica.

Largely inhabiting the American Tropics, Reakirt's Blue becomes much rarer and more sparsely distributed to the North and East. Resident southern populations invade the North annually in highly variable numbers, depending on climate and other conditions. Although they persist well into autumn, these blues lack the ability to overwinter in the North, and they die back each year. Because they tend to be loners, they are sometimes called "Solitary Blues."

401, 402, 490 **Eastern Tailed Blue**
(*Everes comyntas*)

Description: ¾–1″ (19–25 mm). Above, male bright silver-blue with thin dark margin and orange and black HW spots near *threadlike tail;* female slate-gray and black shot with blue. *Grayish-white below with distinct curved rows of gray-black spots* becoming hazier toward borders; *conspicuous orange black-edged spots above HW tail.* Both sexes have white fringe.

Similar Species: Western Tailed Blue larger, paler, with fewer and less distinct spots beneath.

Life Cycle: Eggs laid in flower buds and stems.

Caterpillar variable, often dark green and downy with obscured brown and lighter side stripes. Host plants are many legumes, especially clovers (*Trifolium*), slender bush clover (*Lespedeza*), beans (*Phaseolus*), tick trefoil (*Desmodium*), wild pea (*Lathyrus*), and others. Caterpillar overwinters inside pea and bean pods. Chrysalis buff-colored.

Flight: 3 broods in North, probably more in South, often overlapping; 1st flight begins in early spring.

Habitat: Disturbed sites: fields, gardens, powerline cuts, railroad lines, and crop fields.

Range: S. Canada to Central America, covering entire area east of Rockies; more spottily west to Pacific at low elevations.

One of the East's most abundant butterflies, the low-flying Eastern Tailed Blue readily adapts to human activities; roadsides and rights-of-way create new suitable habitats for its leguminous host plants. Color patterns vary seasonally, spring females bearing much more blue than those of later summer. Populations west of the Rockies may have been introduced after people altered the natural landscape.

471 Western Tailed Blue
(*Everes amyntula*)

Description: ⅞–1⅛" (22–28 mm). Above, bright lavender-blue with very narrow dark margin; female more brown and black than blue. *Below, chalk-white, with grayish markings hazy and reduced,* sometimes absent; *orange spot above threadlike HW tail inconspicuous,* sometimes absent.

Similar Species: Eastern Tailed Blue grayer beneath with more prominent dark and orange spots.

Life Cycle: Egg pale green. Caterpillar varies from straw-color to jade-green; has transverse mauve and maroon side slashes, and short fine white hair; feeds on locoweed (*Astragalus*), peas (*Lathyrus*), and vetch (*Vicia*). Mature caterpillar overwinters in seed pods of host plants. Chrysalis, white haired, dingy white or brownish-yellow; egg-shaped.

Flight: 2 staggered broods at lower elevations and latitudes; January–September, peaking in May and June in California. 1 brood at higher elevations and farther north; June–August.

Habitat: Moist meadows, canyons, and along roadsides, sandy clearings, and forest margins.

Range: Alaska south to Baja California and Mexico, east across Canada to N. central states, and south in Rockies to Arizona and New Mexico.

This butterfly largely replaces the Eastern Tailed Blue in the West and in mountains. Offshore, it often abounds on the California Islands in far greater numbers than are normally seen on the mainland. As they drink, tailed blues twitch their hind wings in a way that calls attention to their threadlike tails, possibly distracting predators from attacking the body.

477, 482 Spring Azure
(*Celastrina ladon*)

Description: ¾–1¼″ (19–32 mm). *Spring brood deep silvery violet-blue above.* female has coal-black FW border; *below. slate-gray. black checkered border. and variable black HW interior spotting* with marginal spotting small and fading toward FW. *Summer brood blanched violet-blue above.* with pale white basal half to FW and most of HW; female black bordered and much whiter; *below. washed out* pale white with faint markings.

Similar Species: Sooty Azure similar below but blackish above with little blue. Arrowhead Blue has white arrowhead in HW cell below.

Life Cycle: Egg pea-green; laid in flowers and buds. Caterpillar highly variable; usually cream-colored, daubed rosy, checkered on dusky back and green side slashes. Many flowers are host plants, especially dogwoods (*Cornus*), viburnum (*Viburnum*), ceanothus (*Ceanothus*), blueberries (*Vaccinium*), black snakeroot (*Cimicifuga racemosa*), and meadowsweets (*Spiraea*). Plump, golden-brown chrysalis overwinters.

Flight: Multiple broods on East Coast; March to mid-April, May–June, mid-July to August. Fewer northward, perhaps more to South; 1–2 broods in West.

Habitat: Open decidous woods, roadsides, and brushy areas from sea level through mountains; clearings, glades, and many other places in and near woodlands.

Range: Alaska east across Canada, and south through entire U.S. to Mexico and Panama in mountains.

Widespread and common in early spring, Spring Azures signal the return of warm weather. This species presents a complex set of identities to entomologists; only recently was the Sooty Azure recognized as a separate species. Genetic and biological evidence suggests several distinct butterflies still remain within the broad definition of the Spring Azure. A darker, bluer, highly polymorphic first brood produces chrysalises which may overwinter or emerge the same year. Chrysalises from the first of the paler, single form, late spring and summer broods will hatch several weeks later, with the last ones of late summer overwintering. The caterpillars of different broods feed on different plants depending upon what is flowering at the time. For many years the Spring Azure was known by the scientific name *C. argiolus,* which is now considered a related Old World species. Current nomenclature reestablished the 18th-century name.

464, 509 Sooty Azure
"Dusky Azure"
(*Celastrina ebenina*)

Description: ¾–1¼" (19–32 mm). *Above, male blackish-brown, with occasional blue* scaling at wing bases; *female lustrous, pale gray-blue,* black-bordered, with extensive whitish suffusion. *Below,* both sexes pale ash-colored to bluish-white, HW irregularly black dotted, with *dusky marginal spots large and distinct along entire border.*

Similar Species: Spring Azure similar below but much bluer above. Female Eastern Tailed Blue has tails and orange spots.

Life Cycle: Egg greenish, ribbed. Mature caterpillar uniform whitish blue-green, faintly yellow tiger-striped (not checkered like Spring Azure); feeds on goat's beard (*Aruncus dioicus*). Chrysalis brown, flecked black; overwinters.

Flight: 1 brood; April–early May.

Habitat: Rich, moist, deciduous forests, especially shaded northern slopes where goat's beard grows.

Range: Illinois and W. Pennsylvania south to Missouri and North Carolina.

Only recently distinguished from the widespread Spring Azure, with which it flies, the Sooty Azure is much rarer and more local. These blues imbibe moisture from damp sand and dew and may take nectar at violet and geranium flowers. In most blues the male is blue and the female grayer; the Sooty Azure reverses this pattern.

489 Sonoran Blue
(*Philotes sonorensis*)

Description: ¾–⅞" (19–22 mm). *Above,* male silvery pale sky-blue, FW has *2 prominent orange spots by margin,* some crisp black dots beyond cell and black margin at tip. Female slightly darker silver-blue, with black spots and even more prominent orange spots on both FW and HW. *Below, both sexes have orange spot on FW; HW lacks any red or orange,* only brownish-gray with light gray over-scaling and a few black spots.

Life Cycle: Egg light green with white patterning. Caterpillar pale green to patchy red; feeds on (and sometimes burrows into) succulent foliage of host plants, stonecrops (*Dudleya*); live-forever (*Sedum purpureum*) suspected. Chrysalis light brown; overwinters amid litter or under stones near host plant.

Flight: 1 brood; typically February–April but into May or June at higher elevations up to about 5000′ (1525 m).

Habitat: Lower mountain canyons and dry washes, rocky slopes, and chaparral; always near host plant.

Range: Santa Clara and Stanislaus counties in California, south to Baja California in mountains, and east to Mojave and Colorado deserts.

Flying in the earliest days of spring, the Sonoran Blue occurs sporadically and rarely in northern California. Unlike its close relatives, the Sonoran Blue rests with its wings partially open. Most of the time, it is seen patrolling near host plants, often along narrow canyon bottoms, rocky slopes, or along the bases of cliffs or ledges. Its iridescence is so reflective that in flight the Sonoran Blue looks almost like a flashing neon light.

486, 532 Square-spotted Blue
(*Euphilotes battoides*)

Description: ¾–1″ (19–25 mm). Above, male
bright blue, with wide or thin black
margins on HW sometimes giving way
to row of small spots; sometimes orange
submarginal shading on HW. *Above,
female* dark brown or, rarely, with blue
and *usually with* prominent or faint *HW
orange submarginal band.* Both sexes
below light grayish, with black spots
sometimes quite large, *may be square*
but not always; *orange submarginal HW
band bounded on both sides by row of black
dots,* outer row never encircled by any
metallic scaling. Fringes often
checkered.

Similar Species: Often difficult to distinguish from
other species in genus, especially
Dotted Blue, as well as Pale and Rita
blues; usually Square-spotted Blue
purpler, more heavily spotted below.

Life Cycle: Egg pale white; laid deep in opening
flower buds of host plant, wild
buckwheat (*Eriogonum*) or shrubby
perennials. Caterpillar resembles color
of blossoms it eats, ranging from
turquoise to yellow to pink; pupates in
litter or soil beneath host plant.

Flight: 1 brood, exactly when host plant
flowers; March–September, any 1
population flying for about a month.

Habitat: Extremely variable; sea level to over
10,000′ (3050 m), often drier areas.

Range: S. British Columbia to Mexican border,
and east to W. Montana and
Wyoming, SE. Colorado, and central
New Mexico.

All *Euphilotes* are intimately associated
with their buckwheat host plant; the
adults habitually take nectar from the
flowers and also roost on them at night.
Most of the many geographic subspecies
in this genus of blues are more precisely
host plant subspecies. The 4 species of
Euphilotes are notoriously difficult to
distinguish from one another without

resorting to dissection of genitalia. Not only do the species vary widely from one general area to another, but the different species in the same area also tend to closely resemble one another. The Square-spotted and Dotted blues are nearly indistinguishable.

470 Dotted Blue
(*Euphilotes enoptes*)

Description: ¾–1″ (19–25 mm). Above, male bright blue, with wide or thin black margins on HW sometimes giving way to row of small spots; sometimes orange submarginal shading on HW. *Above, female* dark brown or, rarely, with blue and *always with* prominent or faint *HW orange submarginal band,* occasionally with blue toward base of wings. Both sexes *below light grayish, with black spots; orange submarginal HW band bounded on both sides by row of black dots,* outer row never encircled by any metallic scaling. Fringes often checkered. (Some females have extensive blue scaling.)

Similar Species: Often cannot be distinguished from Square-spotted, Pale, and Rita blues, but Rita and Pale blues paler, less heavily marked.

Life Cycle: Egg blue-green. Caterpillar cream-white with pinkish and brown markings. Chrysalis pale chestnut-brown. Many wild buckwheat species (*Eriogonum*) are host plants.

Flight: 1 brood, exactly when wild buckwheat flowers; March–October, for a month or less.

Habitat: Extremely variable; sea level to high mountains; often drier areas.

Range: Entire W. U.S.; also in extreme S. Alberta and Saskatchewan, over most of Wyoming, Colorado west of plains, extreme NW. New Mexico.

An early spring form from the Mojave Desert of California, westernmost

Colorado Desert, and westernmost
Nevada has at times been considered a
separate species, the Mojave Blue (*E.
mojave*); it is associated with the
buckwheat *Eriogonum pusillum*. Unlike
other *Euphilotes*, the Mojave female has
much blue above, and is one of the few
in its genus to feed on an annual species
of wild buckwheat instead of
perennials.

469 Pale Blue
(*Euphilotes pallescens*)

Description: ¾–⅞″ (19–22 mm). Male above
violet-blue; narrow dark margin,
sometimes broken into rows of dots on
HW. *Female above* dark brown or gray-
brown, sometimes with flush of
grayish-blue on FW base; *orange
submarginal band* sometimes has small
marginal dark spots. Both sexes *below
light gray to white, with small black dots;
thin or wide orange submarginal band on
HW*. Outermost row of black dots on
HW never has any surrounding
metallic scales.

Similar Species: Square-spotted, Dotted, and Rita blues
difficult to distinguish, but Pale Blue
generally lighter, less heavily marked
below.

Life Cycle: Egg pale green; laid in opening flower
buds of wild buckwheat (*Eriogonum
kearneyi, E. microthecum,* and *E.
plumatella*). Caterpillar ivory-white
turning orange-yellow; camouflaged
among flowers. Chrysalis orange-
yellow; overwinters on pebbles or twigs
at base of host plant.

Flight: 1 brood; 1 month or less, late July–
October, most records for August
and September.

Habitat: Juniper woodlands, arid slopes, and
sandy desert washes.

Range: California along western border of
Mojave Desert, across Nevada to W.
Utah.

Female butterflies from the Stansbury
Mountains of Utah have gray on the
base of the fore wings. This feature is
less or not at all apparent in individuals
from other areas. The Pale Blue often
flies with Acmon or Lupine blues
around the same buckwheat flowers.
The usually rarer Pale Blues fly much
faster and more erratically than the
more common blues in the genus
Euphilotes.

Rita Blue
(*Euphilotes rita*)

Description: ⅞–1" (19–25 mm). *Above, male* light
silver-colored to bright blue with dark
margins and small marginal spots on
HW; *often submarginal orange band.
Female* brown above with *HW
prominently marked with submarginal
orange,* extending well onto FW in
some individuals. Both sexes *below
whitish to gray-brown with many small
black dots; HW orange submarginal band*
sometimes extending over FW margin
as well; outer row *HW black dots lack
metallic scaling.*

Similar Species: Square-spotted and Dotted blues
darker, more heavily marked beneath.
Pale Blue lighter, with lighter
markings below. Orange-bordered Blue
has metallic scaling around HW black
dots below.

Life Cycle: Eggs laid on wild buckwheats
(*Eriogonum effusum, E. flavum, E.
leptocladon, E. racemosum, E. wrightii*).

Flight: 1 brood, coinciding with blooming
period of host plant; late June–
September.

Habitat: Deserts, juniper-pinyon woodlands, and
rolling prairie grasslands.

Range: Nevada, Utah, Arizona, Colorado, S.
Wyoming, and W. New Mexico.

The Rita Blue is the only *Euphilotes*
blue whose range extends east of the

Rockies into the Great Plains. Others in the genus occur in the mountains and deserts west of the plains. A good place to see the Rita Blue is at the mouth of Madera Canyon in the Santa Rita mountains of Arizona. Some lepidopterists also consider the southern California population of the Pale Blue to be Rita Blues.

465 Small Blue
(Philotiella speciosa)

Description: ½–⅝" (13–16 mm). *Very small.* Above, male uniform light blue with dark margins; female uniform brown. Both sexes white to gray *below; bold black dots on FW; HW has bold black dots mostly in an irregular row* or only a few tiny black dots. *Wings rather long and narrow.*

Similar Species: Other small blues have either orange bands or metallic submarginal spots on HW below; Small Blue has neither.

Life Cycle: Egg minute, light-colored; laid singly. Caterpillar apple-green with short white hair and rosy or purplish marks on back; rests within bract around stem of host plant. Principal host plant is punctured bract *(Oxytheca perfoliata)*; also trilobia *(O. trilobata)* and kidney-leaved buckwheat *(Eriogonum reniforme)*. Chrysalis rust-brown.

Flight: 1 brood; on deserts mostly April–early May, in Sierra Nevada foothills early May.

Habitat: Deserts; also sandy washes and some areas bordering chaparral or arid woodlands.

Range: Chiefly W. Mojave and Colorado deserts of S. California; also California from San Joaquin Valley south along western edge of Sierra Nevada and east to White Mountains into W. Nevada.

At the height of the brief desert wildflower season, the Small Blue may be found flying almost invisibly close to the desert floor. Small Blues from the Sierra Nevada foothills, found near Mariposa and at Briceburg on the Merced River, are among the rarest North American butterflies. They are nearly immaculate white on the hind wings below and males have wide borders on the upperside.

476 Arrowhead Blue
(*Glaucopsyche piasus*)

Description: 1–1¼" (25–32 mm). *Large*. Above, male dullish deep-blue to violet-blue; wide brown margins are diffuse inwardly. Female blue much restricted by broader margins. Both sexes *below, pale gray to dark brownish-gray* with black dots; *across HW row of white arrowheads,* inward-pointing (sometimes fused together) between crisp black median spots and dark ground color marginal spots. *1 large white arrowhead crosses HW disk below. Fringes checkered.*

Similar Species: Spring Azure dark form lacks large white arrowhead in HW cell below.

Life Cycle: Pale egg laid on various species of shrubby perennial lupines (*Lupinus,* including *L. albifrons* and *L. excubitus*). Caterpillar variable: blue-green, greenish-white, yellow-brown, with reddish or whitish back stripe and side bars. Chrysalis overwinters.

Flight: 1 brood; April–May in lowlands, June–July in mountains at 8000–10,000′ (2440–3050 m).

Habitat: Mountain and foothill canyons and grassy slopes near lupines.

Range: Extreme S. British Columbia and Saskatchewan south to S. California, N. Arizona, and New Mexico, and east to westernmost South Dakota, Nebraska, and Front Range of Colorado.

For unknown reasons, many areas with lupines lack this uncommon butterfly.

The wing color of the underside is palest in California and Oregon mountain populations, where the white arrows are almost lost in the light ground. Dark individuals occur in the Great Basin and the Rockies, with the darkest of all coming from the lowland foothills around Los Angeles. This race may now be extinct.

468 Silvery Blue
(Glaucopsyche lygdamus)

Description: 1–1¼" (25–32 mm). *Above, male silver-blue with narrow black margins; female dark with diffuse, wide dark margins.* sometimes *blue restricted* to base, or wings totally brown. Both sexes pale gray to dark brownish-gray *below, usually with 1 bold, crooked row of black rounded dots, each ringed with white* (missing in northern populations); very few other dots, and sometimes a bluish cast to wing bases.

Similar Species: Common Blue lighter beneath, spots less contrasting with rims, usually with a marginal row of spots. Reakirt's Blue much smaller, with prominent spots at HW outer angle (tornus).

Life Cycle: Egg laid on various hosts in legume family (Fabaceae): deer weed (*Lotus scoparius*), lupine (*Lupinus*), wild pea (*Lathyrus*), vetch (*Vicia*), locoweed (*Astragalus*), and others. Mature caterpillars variable: green to tan with darker (often reddish) back stripe and lighter, oblique dashes on sides. Pale brown chrysalis marked with small black dots; overwinters.

Flight: 1 brood; about 1 month, precise time varying with altitude and latitude—March–July depending on location.

Habitat: Widespread from sea level to above timberline: mountain meadows, open woodlands, brush, disturbed or burned areas, canyons, seeps, and streamsides.

Range: E. Alaska and Nova Scotia south to
Baja California, central Arizona, New
Mexico, Oklahoma, Alabama, and
Georgia.

The slow-flying Silvery Blue is among
the first species to appear in spring. It
is easily recognized by the small
number of dots on the underside. Over
10 geographic races are identified; the
shade of blue, depth of iridescence, and
size of spots vary dramatically from area
to area. These blues can stand chilly,
windy weather in the early spring but
never linger into late summer.

475 Xerces Blue
(*Glaucopsyche xerces*)

Description: 1⅛–1¼" (28–32 mm). *Extinct.* Above,
male lilac-blue, female brown. *Below,
extremely variable:* grayish-brown with
*either white splotches or white-ringed black
dots,* dots usually rather reduced
and surrounded broadly by white,
occasionally large and bold.

Similar Species: Overall pattern resembles Silvery Blue,
but black spots smaller with less white
around them.

Life Cycle: Egg pale green; laid on lupine (*Lupinus
arboreus*) and deer weed (*Lotus scoparius*).
Caterpillar pale green with yellow
markings and covered by white pile.
Chrysalis pale gray-green.

Flight: 1 brood; March–April.

Former Habitat: Coastal sand dunes of San Francisco,
California.

Former Range: Upper San Francisco Peninsula from
about North Beach to Presidio and
south along coast to Lake Merced
district.

The Xerces Blue disappeared forever in
1943 when an expanding military
facility claimed the land where the last
colony lived. Its memory is recalled in
the name of the Xerces Society, a

worldwide group devoted to the
conservation of rare insects through
recognition and protection of their
unique habitats.

473, 496, 538 Northern Blue
(*Lycaeides argyrognomon*)

Description: ⅞–1¼" (22–32 mm). Above, male
bright silvery purple-blue with narrow
dark border; female gray-brown with
greater or lesser rows of orange spots
around margins above. Dirty white to
light fawn *below; black line along extreme
outer margins thin, becoming inflated at
veins forming distinct triangular spots;*
submarginal row of silvery blue-green,
orange, and black spots below
somewhat pale and reduced, especially
in West.

Similar Species: Orange-bordered Blue usually has
brighter orange above and heavier black
marginal line below; some populations
very difficult to distinguish. Shasta
Blue smaller, much duskier. Acmon
Blue male has orange on HW above;
female lacks orange on FW above.
Euphilotes blues' habitats rarely
overlap.

Life Cycle: Pale gray-green egg laid singly on host
plants: lupines (*Lupinus*), crowberry
(*Empetrum*), laurel (*Kalmia*), and
Hudson Bay tea (*Ledum palustre*) among
others. Caterpillar variable, from pea-
green to pale brownish, with obscure
light side stripes.

Flight: 1 brood; June–August, later at higher
elevations.

Habitat: Cool zones in western mountains; open
areas, heaths, and bogs in northern
coniferous and mixed forests.

Range: Holarctic: in North America, Alaska,
and Yukon east to Minnesota and
Maritimes; British Columbia south to
central California in mountains and
coastal bogs; Pacific Northwest to S.
Colorado in Rockies.

The Northern Blue is the butterfly hikers often stir from muddy spots along the Cascade Crest Trail. The relationship between the Northern and Orange-bordered blues is extremely complex. Variability is the rule in these 2 butterflies. In some areas they are easily separated while in others, such as the Colorado Rockies, the 2 species are similar to all appearances. Genitalic dissection is then the only way to make definite identification. Novelist Vladimir Nabokov, an accomplished and respected lepidopterist, devoted many years of meticulous study to Northern Blues and their allies.

472, 500, 501   Orange-bordered Blue
"Melissa Blue"
(*Lycaeides melissa*)

Description: ⅞–1¼" (22–32 mm). Above, male vivid silver-blue or dark blue with very narrow black margins. *Female* slate-colored gray-brown above, sometimes shot with blue; *has orange margins. Below* dusky, whitish to tan; solid, bold black line along extreme margins does not form distinct bulges at veins; black markings crisp, solid; *extensive orange flanks iridescent blue-green HW spots.*

Similar Species: Northern Blue usually has less and paler orange, fewer sharp black spots, and thinner marginal line below; thicker at veins. *Euphilotes* blues are smaller, lack iridescent spot on HW below.

Life Cycle: Egg pale green with frosty white ridges. Caterpillar green covered by delicate brown pile, darker green above, with faint oblique side stripes. Host plants vary geographically and altitudinally, and include lupine (*Lupinus*), alfalfa (*Medicago sativa*), crazyweed (*Oxytropis*), and wild licorice (*Glycyrrhiza*); lupine preferred eastward. Overwinters in East as egg and very

young caterpillar. Chrysalis is green
with many translucent yellowish spots.

Flight: 2 broods; May to mid-June and July–
August. Apparently 3 broods in
lowland West, especially in alfalfa
fields.

Habitat: Open sunny areas, such as lupine stands
and dry mountain meadows in West,
and sand barrens and dry open woods
in East.

Range: Mid-Canada south through the Sierra
Nevada and Rockies to Mexico, east to
W. Great Plains States and Manitoba;
Minnesota and Michigan east to New
Hampshire in spotty colonies.

Along with sulphurs and whites,
lowland western Orange-bordered Blues
swarm over alfalfa fields, where their
caterpillars eat flowers and young
leaves. The northeastern subspecies,
known as the Karner Blue, was named
by novelist Vladimir Nabokov. It is a
protected insect in New York State.
Limited to such places as the Albany
Pine Bush and other sand barrens, the
Karner Blue survives in isolated
colonies across the northern Midwest
and Northeast.

463, 478, 497 **Greenish Blue**
(*Plebejus saepiolus*)

Description: ⅞–1¼" (22–32 mm). *Above. male
silver-blue* (sometimes faintly greenish)
with dark margins. sometimes with
marginal dot row on HW. *Female 2
forms: brown* form with or without
submarginal orange on HW, or HW
and FW; also *blue* form (not as blue as
male). Both sexes usually have *dark spots
at ends of cells on FW and HW above.
Below male silver-white to gray.* often
with blue tinge at base; *female browner;
both sexes have black dots on FW and HW
of nearly equal size.* those on HW base
have more black than surrounding

white; *any orange inconspicuous,* limited to outer dot rows, usually only 1 faint rust-colored spot near HW outer angle (tornus).

Similar Species: Common Blue often lacks FW cell bar, HW cell bar below often whitish, FW spots below larger than those on HW. Silvery Blue has fewer spots below.

Life Cycle: Formally undescribed. Eggs laid in clover flowers (*Trifolium monanthum, T. longipes, T. wormskioldii,* and probably others). Caterpillar greenish or reddish; overwinters half grown.

Flight: 1 brood in some areas, 2 in others; late spring and early summer, mostly June and July.

Habitat: Sea level to high mountains and high latitudes in wet areas: bogs, meadows, grassy slopes, roadside ditches.

Range: Alaska south to mountains of S. California; east through S. Canada to Great Lakes, Gaspé, and Maine; also to W. South Dakota and Nebraska, Colorado west of plains, and N. New Mexico.

In almost any moist, clovery meadow within its range, the Greenish Blue abounds in summertime. Its taste for clovers no doubt has enabled it to extend its range, for the Greenish Blue may now be found along roadsides and in other disturbed sites, where European white clover has been used as a ground cover. Adults take nectar at bistort, asters, and other flowers. There is great variability in coloring both within and between populations.

493 San Emigdio Blue
(*Plebulina emigdionis*)

Description: ⅞–1⅛" (22–28 mm). *Above, male deep lilac-blue with wide orange-brown margins blurring* into blue; HW margin has diffuse black dots on orange submarginal band also blurring inwardly into blue. *Female bluish-brown*, bluer toward body and *prominent orange* submarginal band on HW and suggested on FW, extending inward along veins. *Below*, both sexes pale gray-brown with many black spots, bold on FW, smaller on *HW with outermost row of dots heavily encrusted with metallic blue-green* scales on yellow-orange field.

Similar Species: Male Orange-bordered Blue lacks dots on margin above; female has sharper orange bands above. Male Acmon Blue has sharp orange HW margins above; female lacks orange on FW above.

Life Cycle: Eggs laid on hoary saltbush (*Atriplex canescens*) and a legume (*Hosackia purshiana*). Caterpillar variable, from blue-green to gray or brown with black speckles and blotches; overwinters partially grown. Chrysalis green or straw-yellow.

Flight: Usually multiple broods (perhaps dependent on summer rains); late April–May, late June–early July, and August–early September.

Habitat: Usually dry riverbeds and stream courses.

Range: S. California from S. San Joaquin Valley and Mojave Desert south to Victorville area.

Colonies of this rather rare blue are few in number and widely scattered. But hoary saltbush, the host plant, is common throughout the West, including many areas of the Mojave Desert where the San Emigdio Blue is absent. This species may be a Pleistocene relict, which was once more widespread when the desert held large lakes of melted glacial water. Under the current hotter and drier conditions, the blue has withdrawn to a few suitable localities.

44, 467 Common Blue
(*Icaricia icarioides*)

Description: 1–1⅜" (25–35 mm). *Above, male silver-blue to violet-blue,* with dark margins, usually without FW cell mark. *Female completely brown or with blue restricted* to wing bases, with or without FW cell mark. *Below, both sexes pale* or silver-gray to cream-tan or brownish with *black spots on FW, black or white spots on HW;* sometimes marginal crescent row (often very faint) and orange absent or limited to a few rust-colored *HW spots usually smaller than FW spots, prominently encircled with or replaced by white.* Bar near HW cell below more white than black.

Similar Species: Silvery Blue has fewer, usually more prominent black spots below. Greenish Blue has HW spots below as large as those on FW and a touch of orange at HW outer angle (tornus).

Life Cycle: Egg delicate green; laid on lupine (over 40 species or forms of *Lupinus* recorded as host plant, but usually only one used in a given area; always hairier lupine preferred). Caterpillar, to ⅜" (10 mm), green and covered with short white hair, many diagonal marks along sides; overwinters half grown. Chrysalis, to ⁵⁄₁₆" (8 mm), green with red-brown on abdomen; adult emerges in a few weeks.

Flight: 1 brood; 1 month or more from April in S. California to August in mountains of California and in Rockies.

Habitat: Sea level to over 10,000' (3050 m), including coastal sand dunes, mountains, valleys, meadows, streams, sagelands, and roadsides. Always close to lupines.

Range: British Columbia to S. California, and east to W. Saskatchewan, the Dakotas, Nebraska, W. edge of Colorado plains, and south to E. New Mexico.

This highly variable species has over a dozen named subspecies, yet many populations have individuals resembling several varieties. One of these, the Mission Blue, is restricted to the San Bruno Mountains of San Francisco; it has been listed as an Endangered Species.

499 Shasta Blue
(*Icaricia shasta*)

Description: ⅛–1" (22–25 mm). *Above, male blue-violet or silvery-lilac blue;* dark margins become marginal dot row on HW. *Female brown or blue,* has wider margins, especially on FW; *HW margin has dot row often capped inwardly with some orange crescents. Both sexes often have black bar at end of FW cell* and/or HW cell also (more common on female). Below, both sexes are gray with much white overscaling, dots on HW browner than on FW; *HW below has prominent row of metallic bluish-green encircled marginal dots, inwardly bordered by* distinct row of *orange crescents, in turn bordered inwardly by row of white triangles* pointing toward brownish dots.

Similar Species: Northern Blue larger, less dusky. High Mountain Blue lacks orange row. Acmon Blue has bright orange, more distinctly marked below.

Life Cycle: Undescribed. Host plants include lupine (*Lupinus lyalli*), locoweed (*Astragalus spatulatus* and *A. calycosus*), and clover (*Trifolium dasphyllum*).

Flight: 1 brood; mainly July, varying with altitude.

Habitat: Chiefly high mountains, often near or above treeline; as low as 7000' (2135 m) in the Sierra Nevada, and 3000' (915 m) in Oregon. Alpine fell-fields, arctic-alpine ridges, meadows near streams, and forest clearings; also sagelands, prairie and pinyon-juniper woodlands.

Range: Eastern slopes of central Oregon Cascades and California Sierra Nevada; east to extreme S. Saskatchewan and south through Dakotas, W. Nebraska, central Colorado, Great Basin, and Rockies.

The Shasta Blue is at home as high as 13,000′ (3965 m), fluttering weakly over rock and cushion plants. The lower prairie forms are more rapid and erratic fliers. Both tend to live in colonies. There is some geographic variation and individual populations can be variable as well. Males from the Sierra Nevada have broader borders, while those from above treeline in Colorado vary in color from clear to gray-blue. Females from the prairies can be almost as blue as males, and those of Mt. Charleston near Las Vegas, Nevada, are very blue with prominent bright orange crescents above. The common name for this species comes from Mt. Shasta, California.

494, 531	**Acmon Blue** "Emerald-studded Blue" (*Icaricia acmon*)
Description:	¾–1″ (19–25 mm). *Above. male bright lilac-blue with narrow black margins; HW has pinkish-orange submarginal band* surrounding small marginal black dots. *Female dark brownish; blue. if present. limited* to wing bases (mostly FW); *red-orange submarginal band more prominent. Below.* both sexes white to gray with many small black dots; orange submarginal band prominent on HW (never on FW); *outermost row of HW black dots have metallic green caps.* Fringes are white; *FW below has black dot in middle of cell.*
Similar Species:	*Euphilotes* blues lack metallic scaling on HW below and are confined to buckwheat host plants. Lupine Blue almost indistinguishable, generally larger with more and brighter orange markings, female often blue. Orange-

bordered Blue male lacks orange on HW above; female has orange on both HW and FW above. Shasta Blue duskier, with less orange.

Life Cycle: Egg pale green; laid on wild buckwheat (*Eriogonum*), locoweed (*Astragalus*), bird's-foot trefoil and deer weed (*Lotus*), lupine (*Lupinus*), other legumes, and knotweed (*Polygonum aviculare*). Young caterpillar overwinters; mature caterpillar dirty yellow, covered with fine white hair, green back stripe, and various side markings. Chrysalis brown with green abdomen.

Flight: Multiple broods; February–October, according to locale.

Habitat: Virtually any habitat in West except driest deserts, dense forests, and urban areas, sea level to 10,000′ (3050 m); rarer at higher elevations.

Range: Canada to Mexico; Pacific Coast east to Saskatchewan and Dakotas, western edge of Great Plains, and W. Texas. Isolated populations in Minnesota, Nebraska, and Kansas.

Nearly the most ubiquitous western blue, this species flies close to the ground between flower visits. Spring brood females may be quite blue. Almost any crowd of blues at a mountain mud puddle will include some Acmon Blues. As they drink they twitch their hind wings, and in so doing, flash their emerald scales in the sunshine. This immediately distinguishes them from the similar, smaller buckwheat blues in the genus *Euphilotes*. The Acmon Blue was one of the few species of butterflies that lived on the slopes of Mt. Saint Helens before the volcanic eruption of 1980 destroyed this habitat. Versatile and quick to colonize, this species will probably return to the mountain soon.

495 Lupine Blue
(*Icaricia lupini*)

Description: ¾–1⅛" (19–28 mm). *Male purplish-blue above with broad, diffuse dark borders* and broad, bright orange-red band around HW margin. *Female slate-brown or blue above,* also with orange band. *Below,* both sexes silver-gray with curved rows and clusters of *black dots, those nearest margin often connecting with margin; orange submarginal band often capped with metallic blue-green flecks on HW below.*

Similar Species: Acmon Blue very similar, usually has narrower dark margin above, more separate near-marginal dots below. *Euphilotes* blues lack green flecks. Orange-bordered Blue male has no orange band above; female orange-banded both on HW and FW above (not just on HW).

Life Cycle: Caterpillar feeds on wild buckwheats (*Eriogonum,* including *E. fasciculatum, E. umbellatum,* and *E. ovalifolium*).

Flight: 1 brood; May–August, mostly June–July.

Habitat: Moderate elevations in chaparral, mountain canyons, and forest openings.

Range: Central Cascades of Oregon south through Sierra Nevada and Coast Ranges into ranges of S. California, east to Nevada.

The Lupine Blue's common and scientific names are misleading, for while several other species of blues employ lupines as the host plants for their caterpillars, this one feeds only on wild buckwheats. The Lupine Blue strongly resembles the Acmon Blue, and only recently have they been recognized as separate species. A population of blues from the Olympic Mountains of Washington was formerly named to this species but is now known to belong to the Acmon Blue instead.

533 Orange-veined Blue
(*Icaricia neurona*)

Description: ¾–⅞" (19–22 mm). *Above, brown with wing veins outlined in orange* (more on FW than HW and more pronounced on male than female); HW has prominent orange submarginal band enclosing marginal row of brown points. Below, pale brownish-gray with many small neat black spots, prominent orange submarginal band between marginal and submarginal rows of dots, marginal dots finely encircled with metallic scales.

Life Cycle: Egg light gray-green, matches color of host plant wild buckwheat (*Eriogonum wrightii*). Mature caterpillar green with broad white diagonal bands on sides, darker stripe down back, and abundant short white hair. Chrysalis green with yellowish abdomen and blue-gray wing cases.

Flight: 1 or 2 broods; early June or May and August, depending upon area.

Habitat: Arid slopes or on exposed summits at 4000–8000' (1220–2440 m) in proximity to host plant.

Range: Mountains of S. California, from the southernmost Sierra Nevada south through Tehachapi Mountains, Mt. Pinos district, and San Gabriel, San Bernardino, and San Jacinto ranges.

This distinctively marked blue is found in small numbers only where the ground-hugging host plant grows, and flies only a few inches off the ground. Along with the San Emigdio Blue, the Hermes Copper, and the Avalon Hairstreak, the Orange-veined Blue is part of southern California's endemic fauna. No other state has as many full species of butterflies entirely confined within its borders.

479 Yukon Blue
(*Vacciniina optilete*)

Description: ¾–⅞" (19–22 mm). *Above, male uniform bright purplish-blue* with fine black margins; *female has diffuse, wide brown margins, purplish-blue confined to wing bases.* Both sexes *below, gray with bold black spots* except for some subtle marginal markings; *1 prominent orange spot* between marginal and submarginal rows near HW outer angle (tornus).

Similar Species: Western Tailed Blue tailed. Northern Blue has row of orange spots, not just 1 prominent spot. Greenish Blue paler both above and below.

Life Cycle: Not completely known. Caterpillar overwinters; host plant is grouseberry (*Vaccinium myrtillus*).

Flight: 1 brood; July.

Habitat: Subarctic taiga and tundra, often in or around sphagnum bogs; restricted to areas where host plant occurs.

Range: Yukon River valley of E. Alaska, S. Yukon, and extending east to Hudson Bay around Churchill, Manitoba. Also European Alps, arctic Europe, Asia, and Japan.

In the short span of an arctic summer, Yukon Blues emerge in small numbers with the more numerous Northern, Greenish, Silvery, and Western Tailed blues. While a number of sulphurs, satyrs, and brush-foots have circumpolar distribution, the Yukon and Northern blues and the American Copper are the only gossamer wings which occur both in the Old World and the New World.

483, 502, 507 High Mountain Blue
"Arctic Blue"
(*Agriades franklinii*)

Description: ¾–1" (19–25 mm). *Above, male lustrous grayish-blue, sometimes slightly greenish.*

with dark FW margin and often HW marginal dot row; *female, dark to reddish-brown,* sometimes with pale blue and white markings. *Both sexes have small dark bar,* often surrounded by a few white scales, *at end of FW cell, and often on HW also. Below,* both sexes have many whitish spots, sometimes with pupils, on dark grayish-brown field with much white scaling; *white spots may dominate HW, or appear 2-toned* with inner half white-spotted on dark field and outer half with fine dark marks (sometimes lacking) on white field. Lacks any prominent orange. HW fringes usually checkered.

Similar Species: Greenish Blue deeper blue above, pale below with black spots (instead of reverse). Common Blue larger, bluer above, paler below with spots less dominant.

Life Cycle: Early stages poorly known. Egg laying has been observed on shooting star (*Dodecatheon*) and rock-jasmine (*Androsace*) in alpine area and diapensia (*Diapensia lapponica*) in arctic area. Blueberry (*Vaccinium*) also a suspected host plant.

Flight: 1 brood; June—July or August.

Habitat: Arctic tundra, subarctic and subalpine forests, high elevation slopes, ridges, and mountain meadows; sometimes wet prairies.

Range: E. Alaska to Ellesmere Island and Labrador, south to central Sierra Nevada of California, high mountains of central Arizona and New Mexico, and Black Hills of South Dakota.

There is great variation in the amount of blue on males, red-brown on females, and of overall spotting. Some lepidopterists consider that the High Mountain Blue belongs to one or another Old World species—either the Glandon Blue (*A. glandon*) or the Arctic Blue (*A. aquilo*).

METALMARKS
(Riodinidae)

Approximately 1000 species
worldwide, 90% in the New World;
about 24 in North America. North
American metalmarks are small
butterflies, with wings spanning ⅝–2"
(16–51 mm). Most are subtly colored
brown, gray, or rust, or are otherwise
dark, although brilliant tones and
patterns characterize tropical members
of this family. These butterflies get
their name from the shiny metallic
markings of many species; however,
some lack these markings completely.
Most wings also have black spots and
checkered patterns. Although they
superficially resemble checkerspots and
crescentspots, metalmarks have a more
angular shape and softer hues. In
common with the gossamer wings and
snouts, male metalmarks have reduced
fore legs. At rest, most species perch on
the undersides of leaves, and hold the
wings out flat, although some keep
them open at about a 45° angle.
The eggs are more or less flattened and
turban-shaped. Most eggs bear a
network of fine lines on the surface, but
there is a certain amount of variation
among different species. Caterpillars are
also flattened and plump. The downy
chrysalises are attached with silk to leaf
litter or to the stem of the host plant.
Adult metalmarks do not wander and
are quite specific in their ecological
requirements, haunting a variety of
sunny places. A number of swamp and
dune species are endangered by
drainage and development. North
American metalmarks are rather poorly
known; much more research remains to
be done.

549, 552 Little Metalmark
"Virginia Metalmark"
(*Calephelis virginiensis*)

Description: ⅝–¾" (16–19 mm). *Small. Above,
uniform bright rust* to orange, *with*
marginal and inner *bands of conspicuous,
glistening silver-green spots.* Below, dull
orange, flecked with black or lead-gray,
not much paler than above. *Fringes
uncheckered.*

Similar Species: Northern Metalmark larger, with
wavier outline. Swamp Metalmark
larger, with more pointed wings and
less red.

Life Cycle: Yellow thistle (*Cirsium horridulum*) is
host plant in Texas.

Flight: 3 broods in mid-Atlantic states; May,
July, September. Individual broods less
defined farther south; every month in
Florida.

Habitat: Open grassy fields, pine savannah, salt
marsh meadows, and wood margins.

Range: Virginia to Florida, west to Texas and
Arkansas.

Distributed sporadically in all of the
Gulf States, the Little Metalmark
becomes progressively more local to the
north. It is especially fond of pine
flats, woodland edges, and damp
habitats.

542 Northern Metalmark
(*Calephelis borealis*)

Description: 1–1¼" (25–32 mm). Dingy orange-
brown *above with wide central dark band*
bisecting both wings; *has relatively*
indistinct but *uninterrupted line of
metallic marks.* Below, light yellow-
orange, often flushed with red, has
lead-gray spotting. Fringes slightly
checkered.

Similar Species: Swamp Metalmark brighter reddish
above, has less checkered fringes. Little
Metalmark smaller, with much more

rust above, wings more rounded.

Life Cycle: Caterpillar stout, whitish-green with black spots on back, and long plumelike hair on back and sides; eats ragwort (*Senecio obovatus*); overwinters in leaf litter when ⅔ grown.

Flight: 1 brood; late June–early August.

Habitat: Dry, open woods and hilly meadows, especially shale and limestone outcroppings.

Range: S. New York and N. New Jersey to Virginia and Kentucky in Appalachians, west through Ohio to Indiana and Missouri.

The Northern Metalmark is largely geographically and ecologically isolated from the Swamp Metalmark, although they may overlap in the upper Midwest. Fond of drier habitats, Northern Metalmarks occur in widely dispersed, local, sparse colonies. Like many metalmarks, they land on the undersurfaces of leaves, with wings held out flat.

545 Fatal Metalmark
"Dusky Metalmark"
(*Calephelis nemesis*)

Description: ¾–1″ (19–25 mm). Usually *dull dark gray-brown above;* hazy darker band crosses middle of FW and HW; *FW margin straight, often concave; fringes uncheckered.* Below, yellow-orange, with diffuse lead-black spots. Metallic markings typically inconspicuous both above and below.

Similar Species: Lost Metalmark usually markedly 2-toned, with convex outline and checkered fringes.

Life Cycle: Dark gray caterpillar studded with silver lumps, has long grayish-yellow hair; some hairs form archway along back, others lie along surface. Host plants include bush sunflower (*Encelia californica*) and mule fat (*Baccharis*

glutinosa). Chrysalis dirty yellow,
sparsely matted with yellow hair; hangs
by silken girdle.

Flight: 2–3 broods in California; May–June,
August–October in coastal areas.
Multiple broods in Texas most of year.

Habitat: Chaparral canyons beside rivers within
arid regions.

Range: S. California and Baja California
through N. Mexico, east through
Arizona to E. Texas.

Fatal Metalmarks display considerable
geographical variety; even within a
population, ground colors may range
from deep orange to chocolate-brown.
The pattern variation shown by many
metalmarks contributes to the difficulty
of identifying species.

544, 546 Lost Metalmark
(*Calephelis perditalis*)

Description: ⅝–⅞" (16–22 mm). *FW broad with
rounded borders.* Dull chocolate-brown
above with broad, diffuse, blacker
central band; *narrow rows of metallic
markings; fringes checkered.* Dirty gray-
brown to tan, and *unmarked below.*

Similar Species: Fatal Metalmark has uncheckered
fringes and FW straight or gently
rounded, not markedly convex.
Rawson's Metalmark larger, with much
brighter orange beneath; not as strongly
2-toned above.

Life Cycle: Host plants are snakeroots (*Eupatorium
odoratum,* possibly also *E. serotinum* and
E. betonicifolium).

Flight: 4–5 broods; year-round, most abundant
in autumn.

Habitat: Moister pockets among arid lands.

Range: S. Arizona, S. Texas, and Mexico.

One of our more subtly colored species,
the Lost Metalmark stands in contrast
to the many brilliant South American
metalmarks.

39, 547   Wright's Metalmark
(*Calephelis wrighti*)

Description: ¾–1" (19–25 mm). *Above, uniform dull red-brown* with smoky-white sheen, especially when fresh; has blurred dark markings and shimmering silver *metallic markings; lacks black central bands. Contrasting FW fringes* strongly checkered and conspicuous. Below, dull gray-orange with small black dots and bands of silver markings.

Similar Species: Other southwestern metalmarks roughly similar but less uniformly colored, and lack prominent fringe checkering.

Life Cycle: Caterpillar, to ⅞" (22 mm), greenish-gray with white bumps, black squares along sides, and long, threadlike hair arching over back and sides; eats outer layers of sweetbush stems (*Bebbia juncea*); pupates in ground debris.

Flight: 2–3 broods; early spring, June–July, October; exact dates depend upon summer rains.

Habitat: Near watercourses in arid deserts, dry canyons, and flats.

Range: Colorado Desert and S. California, N. Baja California, Mexico, and SW. Arizona.

This species was named for W. G. Wright, who wrote a monumental work published in 1905 entitled *The Butterflies of the West Coast of the United States.*

551   Swamp Metalmark
(*Calephelis muticum*)

Description: ⅞–1⅛" (22–28 mm). Wings angular. *Above, uniform deep mahogany* with fine, blacker scalloping; *lacks wide, central darkened bands;* silver-green *metallic markings form continuous band.* Below, yellow to orange-brown with black to lead-gray flecks.

Similar Species: Northern Metalmark similar but darker and duller above, more rounded; occurs in drier habitats. Little Metalmark smaller, with more rounded wings and more red.

Life Cycle: Sculptured egg tucked under leaves of young swamp thistle (*Cirsium muticum*), upon which caterpillar feeds. Caterpillar pale green with plumelike hair, few or no black spots along back; overwinters partially grown. Chrysalis has silken girdle, surrounds itself with fuzzy hair of caterpillar woven into loose mat.

Flight: 1 brood; July–August.

Habitat: Wet meadows, swamps, and bogs.

Range: W. Pennsylvania across N. central Great Lakes States to SE. Minnesota; recorded once in New England.

Swamp and Northern metalmarks are best distinguished by habitat and geography. The Swamp Metalmark is found in more northern and wetter areas. Although basically they do not overlap, together the ranges of the Swamp, Northern, and Little metalmarks cover most of the United States east of the Mississippi. This pattern suggests that these 3 metalmarks may have developed relatively recently from a common ancestor.

543 Rawson's Metalmark
(*Calephelis rawsoni*)

Description: ¾–1" (19–25 mm). *Above, dull reddish-brown to dark brown* with central *black band indistinct;* fringes moderately checkered; *undulating FW margin squared off* near center; heavy, *wavy inner row of large metallic markings.* Beneath, uniform light tan with well defined black dots and lead-gray metallic markings.

Similar Species: Fatal Metalmark smaller, rounder, and

with more conspicuous bands above.
Other *Calephelis* metalmarks of the same
range nearly indistinguishable from
Rawson's.

Life Cycle: Host plants in composite family
(*Eupatorium havanense* and *E. greggii*).
Flight: Probably 2 broods; June–November.
Habitat: Desert washes and moist areas in arid
lands.
Range: S. central Texas west to Big Bend
National Park; also Edwards Plateau.

In the past, Rawson's Metalmark has
been mistakenly identified as *Calephelis
guadeloupe*, another species that is part
of the nearly indistinguishable
southwestern *Calephelis* metalmarks.
The related Freeman's Metalmark (*C.
freemani*) closely resembles Rawson's.
Found in the Davis Mountains of
Texas, it was named for the noted
skipper biologist H. A. Freeman.

---

550 Arizona Metalmark
(*Calephelis arizonensis*)

Description: 1⅛″ (28 mm). *Above,* light brown to
cinnamon brown, striated with fine
blackish lines and crossed by a *fuzzy,
irregular dark brown median band; whitish
patch near tip of each wing* and 2
submarginal metallic lead-colored lines.
Below, uniform ocher with black
speckles and *2 strong silver or lead-colored
submarginal metallic lines.* Light brown
fringe is checkered with white.
Similar Species: Lost Metalmark more 2-toned, lighter
checkered fringes. Fatal Metalmark
smaller, has somewhat less pronounced
dark band above. All of these closely
related metalmarks are difficult to
distinguish.
Life Cycle: Unreported.
Flight: Perhaps 2 broods; February–March and
sometimes autumn.
Habitat: Mountain canyons.
Range: Mountains of S. Arizona.

Autumn broods seem to be paler, with less defined markings than those of spring. The Arizona Metalmark has been found in the Baboquivari, Santa Catalina, Santa Rita and Patagonia mountains. It was previously confused with the Fatal Metalmark. A similar Mexican species, the Nogales Metalmark (*C. driesbachi*), has been seen at Nogales, Arizona. It is darker, more reddish, and more dully marked than the Arizona Metalmark.

541, 548 Red-bordered Metalmark
"Schaus' Metalmark"
(*Caria ino*)

Description: ¾–1″ (19–25 mm). *Male above dark blackish-brown,* sometimes with greenish bar along costa; both wings have wide, deep *rust to maroon borders; below brick-red,* darker along margin, peppered with iridescent steel-blue spots. *Female above grayish-brown,* marked with duskier squares and *rust borders;* below, orange-brown with conspicuous silver or steel-blue spots.

Similar Species: Female resembles *Calephelis* metalmarks but has steel-colored spots all over wings below (*Calephelis* have only 2 marginal rows of spots).

Life Cycle: Host plant is hackberry (*Celtis pallida*). Caterpillar spins silken shelter and later pupates in it.

Flight: Continuous broods; year-round except January and February northward.

Habitat: Thickets and thorn forests.

Range: Yucatan and Oaxaca, Mexico, north to S. Texas.

The Red-bordered Metalmark inhabits thorn forests in a somewhat patchy arc from southern Texas south to the Yucatan, Mexico, as well as a separate region in western Mexico.

491 Blue Metalmark
(*Lasaia sula*)

Description: ¾–1¼" (19–32 mm). FW long and triangular, pointed. *Male above iridescent, dull blue-green* with black and white margins; on both wings several thin, black, broken bands give appearance of checkering. *Female above dingy* gray-brown, heavily checkered and scalloped by broken black lines and whitish squares; medium-sized black ovals dot outer wings just inside black and white margins. *Both sexes below banded* dusky light brown and white.

Similar Species: None in North America. Blues have more rounded wings, lack spotting above.

Life Cycle: Unreported.

Flight: Perhaps 3 broods; April, June, and August–November in S. Texas.

Habitat: Subtropical forest edges, scrubby open areas.

Range: Honduras north to extreme S. Texas around Pharr and Brownsville.

Another bluish metalmark, the Narses Metalmark (*L. narses*), has a large, dark patch near the center of the fore wing. In the past, this butterfly has been seen in extreme southern Texas, although its regular occurrence there is doubtful; it resides farther south.

678 Pixie
(*Melanis pixe*)

Description: 1⅝–2" (41–51 mm). *Elongated FW coal-black above,* interrupted by brilliant *yellow-orange tip;* red spots at wing bases. HW also black above with red spot near abdomen and *flashing row of red marginal* spots. Similar below with more red spotting at base.

Life Cycle: Unreported. Caterpillar feeds on guamuchil tree (*Pithecellobium dulce*) of legume family.

Flight: January and July in Texas.
Habitat: Gardens, ornamental plantings, and thorn scrub with host plant in Texas.
Range: Panama north to Chiapas along West Coast; also extreme S. Texas south to Yucatan.

This unmistakable butterfly is a member of the primarily South American genus *Melanis,* which contains some of the most brilliantly colored butterflies in the New World Tropics. Its bright spots against the dark background may mimic distasteful tropical moths, protecting this species from attackers.

555 Zela Metalmark
(*Emesis zela*)

Description: 1⅛–1½″ (28–38 mm). FW pointed, triangular in male, slightly hooked in female. *FW above grayish-brown with pinkish cast* varying to russet; female marked with broken black bands more strongly than male. *HW above has broad yellowish-orange patch* toward FW, *bleeding into deeper gray beyond;* black marks in interrupted bands more conspicuous than on FW. Below, similar but paler and blurred; subtle metallic markings.
Similar Species: Ares Metalmark generally similar but browns and oranges darker and more intense; metallic spots, not bands.
Life Cycle: Unreported.
Flight: At least 2 broods; March–April and June–August.
Habitat: Dry canyons, creek bottoms, and nearby back country roads.
Range: Central America north to central and S. Arizona in mountains.

Male Zela Metalmarks defend territories in open areas along creeks and canyons. The Falcate Metalmark (*E. emesis*) is a smaller species which rarely strays into the lower Rio Grande Valley of Texas from Mexico. Warmer brown above, its wings have broad gray bands and 2 or 3 white spots toward the slightly hooked fore wing tip. Falcate Metalmark caterpillars feed on the legume *Caesalpina mexicana,* primarily an ornamental plant in Texas.

553, 554 Ares Metalmark
(*Emesis ares*)

Description: 1¼–1½" (32–38 mm). FW pointed, triangular. *Above, deep grayish-brown,* with distinct blackish spotting; *fringes strongly checkered. HW bicolored:* half toward FW rosy-orange, lower half dark brown with more distinct marginal spotting and inner banding than on FW. Metallic markings vague. Below, similar but pale.

Similar Species: Zela Metalmark usually smaller and paler, with markings in bars rather than spots, HW mostly orange.

Life Cycle: Unknown.

Flight: 1 brood; August–September.

Habitat: Hot, dry canyons and washes.

Range: Mountains of S. Arizona to Sierra Madre Occidental of Mexico.

Ares and Zela metalmarks have often been confused but their differences, although subtle, are fairly constant. In flight and at rest, these metalmarks resemble certain day-flying moths, but can be quickly distinguished from moths by their clubbed antennae.

569, 570 Mormon Metalmark
(*Apodemia mormo*)

Description: ¾–1¼" (19–32 mm). *Above, brightly patterned,* varying geographically; FW outer half and base coal-black to ash-gray; inner half nut-brown to russet-orange; *4 black-bordered, pearl-white squares boldly mark inner FW area,* with others forming 2 indistinct bands of tooth-shaped spots outwardly; fringe strongly checkered. HW ground color and spotting similar, russet area usually confined to broad swath above outer wing margin. Below, FW pale brown; large white squares dot margin and interior; HW variable (ash-gray to white, or russet) with very large, puffed, white squares and polygons. Metallic markings obscured by white.

Similar Species: Gray Metalmark smaller, much grayer above, and more subtly marked below.

Life Cycle: Pale pinkish egg. Mature caterpillar dark gray-violet on back and sides, lighter below; each segment studded with several dark-based clusters of short hair. Host plants are a wide variety of buckwheats (*Eriogonum*). Mottled brown chrysalis hidden in leaf litter or sometimes within hollow plant stalk.

Flight: 2 broods in lower areas, 1 brood northward and at higher elevations; principally March–June, August–October. In arid situations, timing of 2nd brood depends upon summer rainfall.

Habitat: From beach dunes to mountains, typically in dry, often rocky, washes and slopes.

Range: SE. Washington, E. Oregon through Idaho and central Rockies to W. Texas; across Great Basin to Pacific Coast south to Baja California and NW. Mexico.

The unmistakable, common, and swift-flying Mormon Metalmarks perch vertically, either head up or head down, in blazing sunshine. Their flashing colors match mottled desert rocks and sand, making flying adults difficult to track. A number of named subspecies reflect complex patterns of geographic variation. One of these, the very red Lange's Metalmark (*A. m. langei*), has been designated an Endangered Species, and survives precariously in its isolated coastal dune habitat east of Antioch, California. Federal measures are being taken to rescue this butterfly and several endangered wildflower species of the Antioch Dunes.

---

577 Gray Metalmark
(*Apodemia palmerii*)

Description: ¾–⅞" (19–22 mm). *Above, dusky brown-gray,* flushed with orange and slightly dappled with white; some species have orange margins and white fringes. *Below, pale peach* with extensive white overlay and scalloping; wavy, light brown line cuts across middle of FW and HW.

Similar Species: Mormon Metalmark larger, with less gray above, and more distinctive markings beneath.

Life Cycle: Egg flattened, light green, dotted with ridged hexagons. Mature caterpillar faded blue-green, with lemon-yellow stripes, and line of short tufts of white hair. Host plant is honey mesquite (*Prosopis juliflora*). Gray-green chrysalis has straw-colored wing cases.

Flight: 2 or 3 broods; April, June–August, October–November, time and length probably determined by summer rainfall.

Habitat: Semiarid deserts near honey mesquite.

Range: Desert mountains of SE. California, east through Arizona and New Mexico to W. Texas, and south to Baja California and central Mexico.

Gray Metalmark caterpillars spend much of their time in silken hideaways between sewn mesquite leaflets. They may occasionally emerge during winter for brief feedings, but do not mature until spring. Three small relatives of the Gray Metalmark dwell in south Texas. Hepburn's Metalmark (*A. hepburni*) has fewer reddish markings,

and lacks the former's marginal row of
white dots. The 2 species occupy much
of the same range in southern Arizona,
west Texas, and Mexico. Walker's
Metalmark (*A. walkeri*) is chiefly
Mexican, but has been found in
southernmost Texas. The pattern above
and below consists of shades of gray,
with no orange at all. The Narrow-
winged Metalmark (*A. multiplaga*) is
slate-gray above with prominent white
spots; it is lighter below with some
orange at the fore wing base. The fore
wings are narrow and curve inward
below the blunt tip.

Crescent Metalmark
(*Apodemia phyciodoides*)

Description: ⅞–1⅛" (22–28 mm). *Above, yellow-
orange, checkered* heavily with brown-
black; margins and costa dark; rough
band of deeper orange along FW and HW
centers. Below, FW flushed gray
outwardly, reddish inwardly; irregular
gray-black spotting. *HW below* ash-gray
with prominent, white midband; diffuse
white checks near body and again along
margin.

Similar Species: Small checkerspots and crescentspots
lack pointed FW. Crescent Metalmark
has no crescent-shaped spot on HW
margin.

Life Cycle: Unknown.

Habitat: Roadsides and openings among drier
mountainous scrub woodland.

Range: Chiricahua Mountains of SE. Arizona
and dry mountains in Chihuahua and
Sonora.

The Crescent Metalmark resembles a
small checkerspot or pearly crescent.
One of the rarest North American
butterflies, the Crescent Metalmark was
long known only from 2 specimens
collected in 1924 in the Chiricahuas.
Then, in 1979, it was rediscovered in

Mexico. Although this species survives there, it is unclear whether the Crescent Metalmark is now extinct in Arizona, simply difficult to locate there, or if the location of the early butterflies was incorrectly labeled.

591 Nais Metalmark
(*Apodemia nais*)

Description: 1⅛–1½″ (28–38 mm). Male has long, triangular wings; female's rounded. *Above, deep russet-orange* with faint grayish-brown cast, both wings profusely *blotched with black squares and bars,* forming 3 bands parallel to margin, becoming haphazard near body. Fringes heavily checkered. *Short white dash* along FW costa. Below, FW dull orange-yellow with black spots repeated; HW ivory-white, with black capped *orange spot row* along outer margin; several orange patches, outlined in black, toward base.

Similar Species: Chisos Metalmark generally similar but much paler above and below, with FW band, not spot, on costa.

Life Cycle: Caterpillar pale green, with bristles bunched into small clusters studding upper surfaces. Host plant is buckbrush (*Ceanothus fendleri*) in Colorado, probably also in Arizona.

Flight: 1 brood; June–early August, occasionally May and September.

Habitat: Brushy chaparral, foothills, and moister mountain canyons.

Range: Colorado south through Arizona and N. New Mexico to Chihuahua, Mexico.

A key characteristic that distinguishes the metalmarks from many other butterflies is their reduced fore legs. Rather different from most other metalmarks, Nais Metalmarks sometimes have well-developed fore legs, which are nonetheless too small to be used for walking. Adults visit dogbane and other flowers, and settle on damp ground. In moist mountain canyons they tend to occur singly or in pairs, although sometimes several share a common area.

Chisos Metalmark
(*Apodemia chisosensis*)

Description: 1⅛–1¼" (28–32 mm). *Above, light orange-brown* with broad *white band or patch* along outer central FW (especially in female), margin checkered; both wings have extensive black spots in rows toward margins, variable inward. *Below,* FW yellow-orange and flushed with white pattern of blotches repeated; *HW ivory-white, lacking orange* except as hint near marginal black spotting; has fairly distinct *row of median black egg-shaped spots.*

Similar Species: Nais Metalmark is deeper orange, lacks white FW band.

Life Cycle: Caterpillar constructs leaf shelter; feeds on wild cherry (*Prunus harvardii*).

Flight: 2 broods; May and early August; 2nd brood less reliable, may depend on rainfall.

Habitat: Agave scrub in mountains around 5000' (1525 m).

Range: Big Bend National Park in S. Texas; in Chisos Mountains above 1 mile (1.6 km) elevation; probably also neighboring Mexican mountains.

Desert butterflies must frequently undergo a dormant stage during the hottest, driest part of the summer. This summer hibernation or diapause is known as aestivation. The caterpillars of Chisos Metalmarks both overwinter and aestivate—those originating from the spring brood wait out the driest summer months in a rolled cherry leaf; the fall brood caterpillars spend the winter in the same manner.

SNOUT BUTTERFLIES
(Libytheidae)

Fewer than a dozen species worldwide; 2 in North America. Although Libytheidae is the smallest family of butterflies, it is represented on every continent populated by butterflies. Such a pattern suggests an ancient lineage; indeed, snouts are found in fossil shales some 30 million years old. Since many of these deposits contain impressions of hackberry leaves, it seems likely that North American snout butterflies have fed on hackberries at least since then.

Snout butterflies have extraordinarily long, beaklike palpi that make up the "snout," which has no obvious specialized function. Their fore wing tips extend well beyond the outer margins and are squared off. The fore legs of males are reduced, but females possess well-developed, functional fore legs. The 2 North American species are mottled soft rust, orange, brown, and black, but their Asian relatives are brightly colored. The cylindrical caterpillars are green with yellow streaks. They emerge from ridged, oval eggs and transform to hanging, angular, green chrysalises that overwinter.

366 Snout Butterfly
(*Libytheana bachmanii*)

Description: 1⅝–1⅞" (41–48 mm). *FW has bright tawny-orange patches on basal half, dark brown tips have clear white patches;* HW brown above and banded with orange. Below, FW has gray tip; HW of male mottled, purplish gray-brown, female all gray. *FW extended, squared at tip; HW margins scalloped. Prominent snout,* ¼" (6 mm).

Similar Species: Southern Snout paler with orange

patches more smeared; FW extends less; HW lacks scallops and is sandy-brown, not purplish or gray, below.

Life Cycle: Egg pale green. Caterpillar dark green with yellow stripes; hump behind head has pair of yellow-based, black tubercles. Feeds on hackberries (*Celtis*), and pupates before overwintering.

Flight: 3 or more broods; February or March in South, later farther north; year-round in extreme southern range.

Habitat: Wood edges, stream courses, and deciduous forests with hackberries; anywhere on emigration.

Range: States and provinces of Great Lakes, east to central New England; south through Rockies and east into Mexico, Arizona and S. California. Probably resident only in South.

With their squared-off fore wings and long, beaklike palpi, snout butterflies can be easily recognized. Like the Southern Snout, this species visits mud puddles and its favorite nectar sources, which include rabbit brush, peach, and dogwood. Both species emigrate in remarkable numbers.

365 Southern Snout
(*Libytheana carinenta*)

Description: 1⅝–1⅞" (41–48 mm). *Pale orange over much of HW and FW above,* with brown patches near HW base; FW tip has brown patches with white spots. Beneath, FW orange at base, HW mottled with pale sandy gray-brown areas and striations. *FW extended, squared at tip. Prominent snout,* ¼" (6 mm).

Similar Species: Snout Butterfly has brighter orange, more confined light patches; HW margins are scalloped and mottled purplish gray-brown beneath.

Life Cycle: Unrecorded. Host plants probably hackberries (*Celtis*).

Flight: Most of year.
Habitat: Canyons, partially wooded draws, hilly
           scrub, and desert washes.
Range: Arizona and Texas south to Paraguay,
           occasionally emigrating as far north as
           Kansas.

Both snout species take part in
emigratory flights of millions of
butterflies that carry individuals farther
north than they normally occur. These
mass flights seem to be related to
population pressures. No return
migration follows.

BRUSH-FOOTED BUTTERFLIES
(Nymphalidae)

Approximately 3000 species worldwide;
150–160 residing in or visiting North
America. A large, diverse family of true
butterflies, brush-footed butterflies
occur worldwide except at the polar ice
caps. They are generally some shade of
orange and medium-sized, with wing
spans of 1½–3″ (38–76 mm), although
there are both small and large brush-
foots in a wide spectrum of colors and
shapes. Their unifying characteristic is
the reduced fore legs of both males and
females. These vestigial fore legs are
useless for walking, and have given rise
to the common name of the family.
Brush-footed butterflies also have large,
prominent knobs on their antennae and
robust, furry palpi. This richly diverse
family incorporates the admirals,
fritillaries, checkerspots, crescentspots,
anglewings, leafwings, painted ladies,
tortoiseshells, and longwings.
Brush-foot eggs, caterpillars, and
chrysalises vary greatly. The host plants
include many plant families, ranging
from trees, shrubs, and vines to wild
and cultivated herbaceous plants.
Caterpillars are usually spiny, and the
thorny, angled chrysalises hang upside
down from silken pads. Most members
of the family overwinter as caterpillars
or chrysalises; tortoiseshells and
anglewings overwinter as adults.
Most of the species are powerful fliers
and possess strong territorial instincts.
The painted ladies include a number of
long-distance emigrants—several
reinvade the North each year. Many of
the individual species have extremely
wide ranges, and some are found
throughout the northern hemisphere or
may even be worldwide. As a rule,
adults seek nectar, sap, scat, and
carrion equally avidly. They often perch
with their wings half open.
Some authors treat the hackberries and

goatweeds (apaturines) and the longwings (heliconiines) as separate families, but they are included as Nymphalidae here. Although the satyrs (Satyridae) and monarchs (Danaidae) are also sometimes grouped with Nymphalidae, they appear as separate families in this guide. All fritillaries were formerly classified in the genus *Argynnis;* they are now divided into the greater fritillaries (*Speyeria*) and the lesser fritillaries (*Clossiana, Proclossiana,* and *Boloria*). Other genera of the Nymphalidae have been redivided and classified under new generic names.

19, 593 Gulf Fritillary
(*Agraulis vanillae*)

Description: 2½–2⅞" (64–73 mm). FW long, narrow. *Brilliant red-orange above* with a few black spots, black network along HW border, and cluster of tiny white spots near FW costa. Below, FW similar but with bright coral-pink base and metallic silver-white teardrops near tip; *HW below dull to rich brown or olive with many silver-white orbs and streaks.*

Similar Species: Mexican Silverspot is darker, broader with darker pink on FW beneath. Greater fritillaries (*Speyeria*) are often silver-colored below, but are duller and more black-spotted, with rounder wings; most do not occur with Gulf Fritillary.

Life Cycle: Egg oblong, ribbed, yellow. Caterpillar, to 1½" (38 mm), dark brown with rust-colored stripes, and 6 rows of branching black spines (2 on head are long and curve backward). Host plants are passion flowers (*Passiflora incarnata* and other species). Chrysalis, to 1⅛" (28 mm), long, curved; mottled brown and warty, resembling a dried-up leaf.

Flight: Several broods; early spring–winter in far South, summer in North.

Habitat: Subtropical forest edges, city gardens, canyons; open, sunny areas with abundant flowers.

Range: San Francisco Bay to Baja California; resident throughout southern U.S. into Mexico, emigrating northward into Great Basin, Rockies, and Midwest, Great Lakes and mid-Atlantic states.

Some lepidopterists regard the Gulf Fritillary and other longwing butterflies as a separate family, the Heliconiidae. However, they have much in common with the other brush-footed butterflies. As its name implies, this beautiful insect haunts the Gulf of Mexico, and may be seen flying far out over the water. Although it has silver spots like the true fritillaries, the Gulf Fritillary is not closely related to them. Significant emigratory flights of Gulf Fritillaries often take place from the Southeast. Colonization of the North is temporary, as neither the butterfly nor its host plants can withstand northern winters.

601 Mexican Silverspot
(*Dione moneta*)

Description: 2⅝–2⅞″ (67–73 mm). Wings broad; FW elongated into blunt, rounded point. *Tawny-orange above,* deep orange-brown at base, *heavily veined and marked with black.* FW beneath similar but paler, usually with dark pink at base; silver spots on FW tips. *HW below brown with very bold, elongated silver spots* and streaks.

Similar Species: Gulf Fritillary brighter red-orange, narrower wings, with brighter pink at base of FW beneath. Mexican Silverspot chunkier and darker overall. True fritillaries have rounder wings and spots, but are not in same U.S. range.

Life Cycle: Egg maroon. Caterpillar also maroon with 6 rows of spines, orange spots, and silver speckles. Chrysalis long, curved, brown, and warty, with wing cases projecting beneath. Host plants are passion flowers (*Passiflora*).

Flight: Several broods; most of year.
Habitat: Mountain chaparral and drier subtropical environments.
Range: Texas south to Peru.

The Mexican Silverspot has only rarely been found in North America, as a visitor to southern Texas. Recently it has become established there and may remain as a breeding resident.

592 Julia
(*Dryas iulia*)

Description: 3⅛–3⅝" (79–92 mm). *Wings long and narrow. Above, male clear bright orange,* has black stigma; *female duller orange-brown. Black bars* and black margins fairly pronounced or may be absent. Florida individuals have more black than those from Texas. Both sexes tan beneath with silver streak on HW costa.

Life Cycle: Egg yellow, oblong, and ribbed. Caterpillar light brown with spots and 6 rows of branching spines. Chrysalis warty, brown and gold. Host plants are passion flowers (*Passiflora*).

Flight: 2 or more broods; year-round where weather permits.
Habitat: Hammocks, islands, and gardens.
Range: S. Texas and extreme Florida; also well into American Tropics.

The only member of its genus, the Julia differs in several respects from the other longwings. It is sometimes extremely abundant in the Florida Keys. The female's duller coloring may protect its valuable egg load from attack by predators. The Julia is also thought to be distasteful because its caterpillars feed on poisonous passion flowers.

Adults are vigorous fliers but often stop to take nectar at flowers.

4, 34, 644 Zebra Longwing
(*Heliconius charitonius*)

Description: 3–3⅜″ (76–86 mm). *Wings long and narrow. Jet-black above, banded with lemon-yellow* (sometimes pale yellow). Beneath similar; bases of wings have crimson spots.

Life Cycle: Egg yellow, ³⁄₆₄″ h × ⁷⁄₂₅₆″ w (1.2 × 0.7 mm). Caterpillar, to 1⅝″ (41 mm), white; 6 dark-patched rows of black, branched spines. Feeds on passion flowers (*Passiflora*). Mottled brown chrysalis, to 1⅛″ (28 mm), has metallic spots on sides; spiny.

Flight: Multiple broods; year-round in Florida, except when colder weather occurs.

Habitat: Hammocks, thick woods, and forest edges.

Range: Resident from Texas to South Carolina, south through West Indies and Latin America, wandering to S. California, Great Basin, Colorado, and Great Plains.

Completely distinctive, the Zebra Longwing does not at all resemble the Zebra Swallowtail. The Zebra's usual flight is slow, feeble, and wafting, although it is able to dart quickly to shelter. Zebra Longwings roost communally at night, assembling at dusk. Hammocks and thickets throughout Everglades National Park are good places to see gatherings of these butterflies.

---

643 Crimson-patched Longwing
(*Heliconius erato*)

Description: 3–3⅜″ (76–86 mm). *Wings long, narrow, and rounded. Black above, crossed* on FW by broad crimson patch, and on HW by narrow yellow line. Below, similar but red is pinkish and HW has less yellow.

Life Cycle: Egg yellow, oval, and ribbed. Caterpillar studded with 6 rows of branched spines. Host plants are passion flowers (*Passiflora*).

Flight: Several broods; year-round, fluctuating in numbers.

Habitat: Forest openings and edges.

Range: S. Texas to South America.

The Crimson-patched Longwing is typical of a large number of similar butterflies limited to the American Tropics. All feed on passion flower foliage, which is suspected of rendering them distasteful. They form the basis of many large mimicry complexes, whereby butterflies and moths of several families fool their predators and gain collective protection. Longwings in the genus *Heliconius* live as adults for several months, feeding on pollen necessary for long-term egg production. Other butterflies are not known to utilize this form of concentrated protein. Flapping rather feebly, this species flies low over the ground along the edges of woods, showing off its crimson patches.

20, 626   Variegated Fritillary
(*Euptoieta claudia*)

Description: 1¾–2¼" (44–57 mm). FW somewhat elongated. Tawny-brown *above with zigzag black band* in middle of both wings, followed by blackish shadowy line, black dots, and finally 2 bands of black along margin; FW has several black circles and crescents. *Below, whitish-brown with FW orange on basal half* and pattern of upperside faintly repeated; *HW somewhat mottled or variegated,* with whitish veins.

Similar Species: Mexican Fritillary lacks dark markings on HW disk above.

Life Cycle: Egg cream-colored, ribbed; laid on various host plants, including violets and pansies (*Viola*), flax (*Linum*), passion flower (*Passiflora*), stonecrop (*Sedum lanceolatum*), moonseed (*Menispermum*), and plantain (*Plantago*). Caterpillar, to 1¼" (32 mm), white with red bands, black spines; red head has 2 very long black spines. Chrysalis, to ¾" (19 mm), pale shiny blue-green with black, yellow, and orange marks and gold bumps. Adult overwinters in extreme South, but not in North.

Flight: Continuous broods; spring–fall.

Habitat: Open areas such as grasslands, subtropical fields, and mountain summits and meadows; everywhere but deep forests.

Range: Resident from Arizona to Florida and southern plains, emigrating periodically to S. California and northward to SE. British Columbia, Northwest Territories, and Quebec.

The caterpillars of this species eat more different types of plants than those of almost any other butterfly except the Painted Lady, Spring Azure, and Gray Hairstreak. Both caterpillar and chrysalis are among the most beautiful of all in our range. The Variegated Fritillary has characteristics of both true fritillaries (*Speyeria*), whose caterpillars feed only on violets, and longwing fritillaries (*Agraulis*), whose caterpillars eat passion flowers. The caterpillars of the Variegated Fritillary thrive on both plants.

---

599 Mexican Fritillary
(*Euptoieta hegesia*)

Description: 1¾–2¼" (44–57 mm). Bright orange *above with black zigzag in middle of FW* (*but none on HW*), black dots inside from 2 rows of black lines on margins, black circles near FW base. Orange-brown *below, FW orange on basal half;* pattern of upperside repeated faintly below.

Similar Species: Variegated Fritillary duller orange with dark markings on HW disk above and longer wings.

Life Cycle: Egg brownish-black, taller than broad, and ribbed. Caterpillar maroon with silver bands and black spines; red head has 2 black horns; feeds on passion flower (*Passiflora foetida*) and turnera (*Turnera ulmifolia*). Chrysalis brown or black with gold and silver spots and short spines.

Flight: July–December in Texas. Year-round in Mexico.

Habitat: Subtropical open woodlands and fields, especially seasonally dry lowlands.

Range: Mexico north to S. Texas, S. Arizona, and rarely S. California; also Antilles.

The Mexican Fritillary is not a year-round resident of the United States except in southern Texas; it appears in California only as an immigrant.

323, 656, 657 **Diana**
(*Speyeria diana*)

Description: 3–3⅞" (76–98 mm). *Very large,* rounded. *Male black above* on inner ⅔, *outer third brilliant orange* with a few small black spots; tan-orange below, basal ⅔ of FW heavily black-marked, *1 row of silver spots on HW margin below. Female black above with outer third iridescent blue on HW and bluish-white spots on FW;* below, FW black with whitish and black spots, HW brownish-black with white postmedian and marginal lines.

Similar Species: Other fritillaries have numerous silver or yellowish spots below. Great Spangled Fritillary much less contrasting above than male Diana. Female Red-spotted Purple has longer, narrower wings, reddish spots below.

Life Cycle: Black caterpillar has many branching spines, orange at base; feeds on violets (*Viola*). Chrysalis mottled light brown and red.

Flight: 1 brood; mid-June to September.

Habitat: Deciduous and pine woodlands near streams.

Range: Maryland and W. Pennsylvania west to S. Illinois and E. Oklahoma, south to N. Louisiana, N. Georgia, and Carolinas.

Considered the most beautiful fritillary, this species has decreased in range due to cutting of forests, although it is still common in parts of the Great Smoky Mountains. The male and female look completely different. The blue female mimics the Pipevine Swallowtail, as do the Eastern Black Swallowtail, female dark form Tiger Swallowtail, and the Spicebush Swallowtail.

609, 640, 641 Great Spangled Fritillary
(*Speyeria cybele*)

Description: 2⅛–3″ (54–76 mm). Above, orange with 5 black dashes near FW base; several black dashes near HW base, irregular black band in middle of wing followed by row of black dots, plus 2 rows of black crescents, the outer in a line along margin. *Below,* FW yellowish-orange with black marks similar to upperside and a few silver spots near tip; *HW reddish-brown with silver spots on base* and middle of wing, and broad *yellow band and silver triangles* next to brown margin. Female darker above, especially at base. Western male brighter orange with more pointed FW; female straw-colored outwardly, black at base.

Similar Species: Other fritillaries have 1 or more black

spots on FW base below wiggly black lines.

Life Cycle: Tiny caterpillar overwinters after hatching from pale brown egg. Caterpillar black with branching spines that are orange at base; feeds on violets (*Viola rotundifolia*). Chrysalis mottled dark brown.

Flight: 1 brood; June to mid-September.

Habitat: Moist meadows and deciduous woods in East; also moist pine and oak woods, conifer forest openings, and wet meadows in West.

Range: S. British Columbia, S. Quebec and Maritimes south to central California, New Mexico, and N. Georgia.

The Great Spangled Fritillary flies swiftly but pauses to take nectar from black-eyed Susans, thistles, and other flowers. Females of this and most other fritillaries mate in June or July, but many of them disappear, perhaps hiding under leaves or bark, to reappear in late August and September, when they lay their eggs near violets. By this time, the shorter-lived males, which have emerged from chrysalises a few days or weeks earlier than females, are scarce. Eastern populations are large, rounded, tawny, and common. In the West, this species occurs more rarely; some lepidopterists make it a separate species, the Leto Fritillary (*S. leto*).

608, 617, 623 Aphrodite
(*Speyeria aphrodite*)

Description: 2–2⅞″ (51–73 mm). Above, orange with many black dashes and spots, *black dot near FW base above hind edge;* FW veins lightly scaled with brown. *Below,* FW resembles upperside but has silver spots near tip; *HW disk cinnamon to chocolate-brown with silver spots,* those near margin triangular, usually with narrow yellow band inside triangles.

Similar Species: Atlantis and most other fritillary males
have wing veins bordered with brown
on FW above. Great Spangled Fritillary
lacks black dot near lower FW base.
Aphrodite larger than other western
fritillaries and HW below usually has
narrower yellow outer band.

Life Cycle: Caterpillar overwinters after hatching;
feeds on violets (*Viola lanceolata, V.
fimbriatula, V. nuttallii, V. primulifolia*).
Caterpillar blackish-brown with black
bands and many spines (orange on
sides). Chrysalis brownish-black with
yellow wing cases and gray abdomen.

Flight: 1 brood; late June–September.

Habitat: Wooded areas; open deciduous woods
or coniferous woodlands; sometimes
moist prairie meadows.

Range: SE. British Columbia east to Nova
Scotia, and south to E. Arizona,
Nebraska, and N. Georgia.

Adults take nectar from thistles, rabbit
brush, dogbane, and other flowers,
sometimes congregating by the dozens
as they drink. In the Colorado foothills
during August and September,
females lay eggs under mountain
mahogany bushes and other places
where violets have long since dried up
and will not reappear until the next
year. The females may smell the violets'
dormant roots.

667 Regal Fritillary
(*Speyeria idalia*)

Description: 2⅝–3⅝″ (66–92 mm). *Very large.
Above,* both sexes have red-orange FW
with blue-black spots and white-dotted
black margin; *HW black with 2 rows of
light spots* (cream-white on female, outer
row rust-orange on male) and orange
at base. *Below,* FW similar to
upperside; *HW deep olive-brown with
many silver spots.* Female larger, darker
than male.

Similar Species: No other fritillary has very dark HW with white spots above.

Life Cycle: Eggs tan; laid in late summer. Caterpillar yellowish-brown with black blotches and lines, yellowish bands, and many spines (some silver on back and orange on side); overwinters. Host plants are violets (*Viola*). Chrysalis light brown with black spots.

Flight: 1 brood; June–early September.

Habitat: Wet meadows in woodland areas and moist tallgrass prairies, especially virgin grasslands.

Range: Manitoba and E. Montana east to S. Ontario and Maine, south to E. Colorado, N. Arkansas, and W. North Carolina.

The Regal Fritillary may one day be very rare and restricted because its natural grassland habitat is rapidly disappearing as land is plowed or developed. Adults take nectar from milkweeds and thistles.

604, 642 **Nokomis Fritillary**
(*Speyeria nokomis*)

Description: 2¾–3″ (70–76 mm). *Large*. Above, male brilliant orange with pattern of sparse black spots; female more or less yellowish-white on outer half of wing with many black spots and dark basal half. *Below*, FW pinkish-orange with black pattern and yellowish tip with white spots; *HW has many silver spots, pale yellowish band near margin;* HW disk color varies from yellow in male to olive-green in female in California and Nevada, to golden-tan or brown in both sexes in Colorado, Arizona, and New Mexico.

Similar Species: Great Spangled Fritillary male duller orange above with less heavy black spotting; female yellower, darker at base; both sexes have brighter submarginal yellow band below.

Life Cycle: Eggs laid near blue violet (*Viola nephrophylla*). Caterpillars overwinter in grass stems after hatching, mature caterpillar orange-yellow with transverse black stripes and many orange spines. Chrysalis black with center of wing cases orange, and orange stripes on abdomen.

Flight: 1 brood; late July to mid-September.

Habitat: Wet meadows, edges of beaver ponds, and grassy springs, mostly in mountainous wooded areas or canyons with pinyon pines and junipers.

Range: Nevada and E. California east to W. Colorado, south to Arizona, New Mexico, and Mexico.

Tucked away among vast desert landscapes, tiny pockets of lush vegetation at desert seeps nurture colonies of Nokomis. These big fritillaries rarely stray from their local, far-apart colonies. Distribution of Nokomis Fritillaries was probably more continuous during an earlier, moister climate; drying lands and water diversion could mean their extinction. A Mexican race with blue females is thought already to have disappeared from Arizona. Adults of both sexes seek nectar at thistles.

622 Edwards' Fritillary
(*Speyeria edwardsii*)

Description: 2⅜–2¾" (60–70 mm). *Large.* Above both sexes orange with black margin (heavier on female), black zigzag in middle, row of black dots beyond it, and black wavy bars near base. *Below, FW bright pinkish-orange at base,* yellowish outwardly with same black pattern as upperside, and *silver triangles in green border; HW bright olive-green with many large silver spots.*

Similar Species: Callippe Fritillary smaller with lighter borders above. Egleis Fritillary smaller, less orange above. Other fritillaries with green below are smaller, have smaller silver spots, lighter borders and/or lack pinkish FW bases below.

Life Cycle: Caterpillar feeds on violets (*Viola nuttallii*). Otherwise undescribed.

Flight: 1 brood; late June–early September.

Habitat: Grasslands, open pine forests; sometimes foothill chaparral and mountain canyons.

Range: S. Alberta to SW. Manitoba, south to N. New Mexico and W. Nebraska.

Of the western fritillaries, this is the only one whose appearance remains quite constant throughout its range. When Edwards' Fritillary stops for nectar at thistle or coneflowers, its green and silver wings stand out, but it is camouflaged among grasses.

605 Coronis Fritillary
(*Speyeria coronis*)

Description: 2–2¾″ (51–70 mm). *Large.* Above, tawny-orange with many black marks, including bars near base, zigzag band, row of dots, and several marginal rows of crescents and dashes. *Below,* orange with black and silver spots near FW tip; *HW has rows of silver spots in disk;* disk is greenish-brown in Great Basin but brandy-colored in Rockies and California; *marginal silver spots flattened and capped with broad brown or greenish,* narrow yellow band beyond disk.

Similar Species: Callippe, Zerene, and Egleis fritillaries often not distinguishable from this species except by experts.

Life Cycle: Small, ribbed, tan eggs. Caterpillar overwinters after hatching. Feeds on violets (*Viola*) in spring. Mature caterpillar gray with black patches, gray bands, black spines with orange bristles, and orange side spines.

Flight: 1 brood; June–early September.

Habitat: Chaparral, sagelands, open pine forest,

canyons, and flowery meadows.

Range: Washington and W. South Dakota,
south to S. California, Arizona, and
Colorado.

In Colorado the Coronis Fritillary is a
typical foothill butterfly, although less
common than the Aphrodite. The best
place to find the Coronis is on large
bull thistles where it takes nectar for up
to 10 minutes at a time. Adults fly
swiftly about hillsides and valleys when
not feeding.

607 Zerene Fritillary
(*Speyeria zerene*)

Description: 1⅞–2½" (48–64 mm). Above, orange
or red-orange to yellowish-brown
with complex black pattern of spots,
chevrons, and bars. *Below, FW orange at
base,* with black pattern like upperside
and silver spots near tip; HW has silver
spots (yellowish in California, and S.
Nevada). *HW disk below is usually brown*
(dark brown along Pacific Coast, violet
brown in Sierra Nevada, yellow in
Great Basin, and slightly greenish-
brown in S. Wyoming and Colorado).
Yellow submarginal band beyond disk.

Similar Species: Coronis, Callippe, Egleis, and Atlantis
fritillaries often cannot be distinguished
from Zerene except by experts.

Life Cycle: Host plants are violets (*Viola adunca,
V. cuneata, V. lobata*). Young
caterpillar overwinters. Mature
caterpillar marked with brown and
black, has light back stripe and many
spines.

Flight: 1 brood; mid-June to early September.

Habitat: Coastal dunes, sagelands, meadows
among aspens, open coniferous woods,
mountain roadsides.

Range: SE. Alaska; British Columbia east to
SE. Saskatchewan and South Dakota,
south to central California, N. Arizona,
and central New Mexico.

The Zerene, Coronis, Callippe, Egleis, and Atlantis fritillaries are very similar to each other in some places and also vary within each species from one mountain range to another. Sometimes their variation is parallel from place to place, which further complicates identification. The Zerene Fritillary is usually the most common fritillary in most of its range, but a dark coastal Oregon race, named Hippolyta, has been listed by the Office of Endangered Species; nearly all of its salt-spray meadow habitat has been developed.

610, 621   Callippe Fritillary
(*Speyeria callippe*)

Description: 1⅞–2⅜″ (48–60 mm). Above, usually yellow-brown (orange-brown in N. California and S. Oregon, dark brown with pale yellow spots in California Coast Range) with complex pattern of black spots, sometimes very dark at base. *Silver spots of HW below show through* as pale areas. *Below, FW dull orange* with spots like above; *HW silver spots are elongated ovals and marginal silver spots are triangles capped by thinner triangles of green or brown.* HW disk below green in Plains, Rockies and Great Basin, but brown in California and sometimes in Cascades, with or without yellow submarginal band. HW spots always silvered below except in Sierra Nevada.

Similar Species: Edwards' Fritillary larger, has darker margins. Egleis, Coronis, and Zerene fritillaries often indistinguishable.

Life Cycle: Egg laid haphazardly near host plants, violets (*Viola pedunculata, V. nuttallii*). Caterpillar overwinters after hatching; mature caterpillar gray with black patches, grayish bands, and many orange or black branching spines.

Flight: 1 brood; mid-June to mid-August.

Habitat: Open pine woodlands; sagebrush,

chaparral, and grassland hills; also canyons.

Range: S. British Columbia to SW. Manitoba, south to central and S. California, Nevada, and S. Colorado.

Callippe Fritillaries are among the most beautiful butterflies in the northern grasslands and Great Basin. With their green undersides, they blend almost perfectly among grass and sagebrush. They fly swiftly and stop occasionally to take nectar from thistles and other flowers. Males fly to hilltops in chaparral habitats. Females often lay eggs under shrubs where violets will not appear until the next spring. Populations with green and brown undersides look like different species, but along the eastern edge of the Sierra Nevada and Cascade mountains they mingle and interbreed.

612 Egleis Fritillary
(Speyeria egleis)

Description: 1½–2⅜″ (38–60 mm). *Small.* Above, commonly orange-brown and darker at base with black pattern of spots, bars, and chevrons; brown with yellow-brown outer areas in N. Rockies. *Below, FW dull yellow* with very little orange except in some females and in California and S. Oregon populations; *HW has small silver spots* (lacking in S. California and occasionally elsewhere), marginal spots flattened with narrow brown or greenish caps. *HW disk below usually light brown or dirty tan* but may be reddish-brown, green, or gray-olive.

Similar Species: Mormon Fritillary still smaller, brighter, more rounded; usually more greenish on HW disk. Coronis, Callippe, and Zerene fritillaries difficult to distinguish from Egleis.

Life Cycle: Caterpillar overwinters just after hatching. Older caterpillar gray-brown with black and yellow bands and many yellow or white branching spines. Host plant is violet (*Viola adunca*). Dark brown chrysalis has yellow-brown patches, dark wing cases, and dark cross stripes on abdomen.

Flight: 1 brood; late June to mid-August.

Habitat: Forests including pine, spruce, and aspen; redwood forest clearings.

Range: Washington south to central California, S. Utah, east to central Montana and NW. Colorado.

The Egleis Fritillary is most common in cooler parts of the Great Basin, but also occurs in the mountains surrounding it. Adults gather on hilltops in California, and often feed on flowers. The California Coast Range populations south of San Francisco to Santa Barbara are reddish above and washed out below with a violet tinge and no silver spots. The Adiaste Fritillary (*S. adiaste*) is allied with the Egleis Fritillary by some authors, held distinct by others. It flies in southern California mountains.

618, 624   Atlantis Fritillary
(*Speyeria atlantis*)

Description: 1¾–2⅝" (44–67 mm). Above, orange-brown with complex black pattern of spots, bars, and chevrons, and broadly darkened veins. *Black margins solid and wide in East.* Below, *FW orange on basal* ⅔ with black marks as above and silver spots near paler tip; reddish to chocolate-*brown HW disk has triangular light spots along margin* that are silvered in East, tend to be unsilvered in far West, and may be either color in Rockies. *Narrow yellow submarginal band below.*

Similar Species: In East, Aphrodite has light veins and light margins; Great Spangled Fritillary is larger with broad yellow band below. In West, Hydaspe Fritillary usually has

redder disk, suffused yellow band.
Other western fritillaries, especially
Egleis, Callippe, and Zerene, may be
difficult to distinguish.

Life Cycle: Egg honey-yellow; laid near violets
(*Viola adunca, V. purpurea,* and *V. canadensis*); after hatching, tiny tan
caterpillar overwinters. Mature
caterpillar purplish or blackish with
light brown or gray stripes and orange
spines. Chrysalis brown with black
speckles and light brown mottling.

Flight: 1 brood; mostly July—August.

Habitat: Flowery openings among deciduous and
coniferous forests, frequently along
streams or in moist meadows.

Range: E. Alaska southeast to Nova Scotia,
south to central California, Arizona,
New Mexico, Michigan, and in
Appalachians south to Virginia. Rarely
in N. Iowa and Indiana.

The Atlantis Fritillary spans the
continent from Atlantic to Pacific. In
cool habitats or in wet meadows from
southern Manitoba to the Black Hills
and the Colorado Front Range,
members of this species have silver
spots and chocolate-brown hind wing
disks below. However, in warmer and
hillside habitats they often lack silver
spots and usually have reddish-brown
undersides. Some are in between, with
half-silvered spots. Solar radiation has
been suggested as a major influence on
fritillary variation. Moisture and
temperature are other related factors
which could affect wing patterns from
place to place. The overall result is a
bewildering array of types.

611 Hydaspe Fritillary
(*Speyeria hydaspe*)

Description: 1¾–2⅜" (44–60 mm). Above, orange-
brown with dark wing bases and black
bars, spots, and chevrons. *Below,* FW

orange on disk with yellowish and lavender near tip; *pattern on upperside repeated in very heavy black. HW disk below maroon or reddish-brown, often with lavender tint and large round unsilvered spots* (some Northwest individuals have silver spots); *pale yellow submarginal band suffused with reddish-brown.* HW disk below darkest (deep maroon) in Oregon and N. California and lightest (tan) in Sierra Nevada of California.

Similar Species: Zerene Fritillary in S. Oregon and N. California has dark caps to cell spots in submarginal band, which are only small and vague in Hydaspe.

Life Cycle: Caterpillar feeds on violets (*Viola*); overwinters half grown. Older caterpillar blackish with many branching spines; side spines orange.

Flight: 1 brood; usually July to mid-August.

Habitat: Moist coniferous woodlands, often near aspens; sometimes, especially in California, drier pine woodland, mountain roads, rain forests, openings; alpine meadows in Northwest.

Range: Central British Columbia and S. Alberta south to Sierra Nevada and north coast of California, central Utah, and N. Colorado.

Its lovely maroon disk makes the Hydaspe easier to identify than most other fritillaries. Gathered with Coronis, Zerene, and Aphrodite at thistles and mints, the Hydaspe tends to be the least numerous. Like all fritillaries, this species produces odorous sex attractants, called pheromones—those of the male originate in dark scales along wing veins, while the female's come from the abdomen. It is believed that the pheromones help potential mates to recognize each other, and to distinguish members of nearly identical-looking species.

606, 620 Mormon Fritillary
(*Speyeria mormonia*)

Description: 1⅝–2″ (41–51 mm). *FW rounded.*
Orange-brown above, usually with FW
veins not heavily blackened; complex
pattern of black spots, bars, and
chevrons, and black border (wider in
female). *Below,* FW orange at base with
black pattern of upperside repeated;
silvery or yellow spots at tip. *HW disk
green, golden-tan, olive-brown or reddish-
brown;* spots usually smaller than shown
and heavily silvered (but not always
silver; yellow in Great Basin).

Similar Species: Egleis Fritillary often larger, paler
above, has more pointed wings, darker
FW veins on male, lighter border on
female. Callippe Fritillary larger with
darker veins.

Life Cycle: Egg small, tan; laid near violets (*Viola
adunca*), which caterpillar eats. Mature
caterpillar gray-brown to tan-brown
with back stripe and many pale spines.

Flight: 1 brood; mid-July to early September.

Habitat: High and middle elevation mountain
meadows.

Range: S. Alaska and Yukon southeast to SW.
Manitoba and South Dakota, south to
central Sierra Nevada, Utah, White
Mountains of Arizona, and central New
Mexico.

The Mormon Fritillary swarms in
alpine and subalpine meadows, visiting
asters and tall corn lilies. Usually
confined to mountains, it also descends
to sea level in Alaska and to sagelands
and plains in the Great Basin and Black
Hills, where it takes nectar from
sagebrush and rabbit brush.

638 Napaea Fritillary
(*Boloria napaea*)

Description: 1⅛–1½″ (28–38 mm). Male orange
above with black dots and bars; female

dusky-orange. *Below* both sexes have FW orange with vague light pattern; HW orange to brown with cream-yellow bands and spots, *prominent white bar or crescent inside HW cell* midwing. *HW angled outward* to slight point at middle of margin.

Similar Species: Titania and Kriemhild fritillaries lack white crescent in cream-colored band. Astarte Fritillary larger, more distinctly patterned; lacks white crescent. Freya's Fritillary has white arrowhead on HW below. Chariclea Fritillary has whitish-yellow HW band.

Life Cycle: Eggs laid on alpine bistort (*Polygonum viviparum*).

Flight: 1 brood; July–August.

Habitat: Arctic-alpine and subalpine meadows and tundra.

Range: Alaska southeast to N. British Columbia, Franklin District and Victoria Island in Northwest Territories; also Wind River Mountains of Wyoming.

Males fly fast near the ground, while females tend to fly more slowly; both visit tundra flowers. The isolated colony in western Wyoming indicates a much wider former range. Like other arctic butterflies, the Napaea Fritillary followed the edges of the glaciers southward. When the ice retreated, Napaea was stranded.

633 Bog Fritillary
(*Proclossiana eunomia*)

Description: 1¼–1½″ (32–38 mm). *Small.* Rust-orange above with fine black lines and dots, and darker at base. *Below,* crisply marked: FW like upperside; *HW rich chestnut alternating with cream-white or silver-white bands* across wing, and along margin, *row of submarginal black-rimmed pearly spots* (yellowish in southern Rocky Mountains).

Similar Species: Silver-bordered Fritillary has bolder silver below with dark submarginal spots, darker border above. Titania Fritillary has less crisp markings below, yellower light bands, and dark dots.

Life Cycle: Egg small, cream-colored, with vertical ribs; laid on host plants: alpine bistort (*Polygonum viviparum*), violets (*Viola*), and willows (*Salix*). Reddish-brown caterpillar has many branched spines; overwinters when half grown.

Flight: 1 brood; briefly during June in Northwest, June–August in Rockies.

Habitat: Bogs in spruce and other conifer forests, and arctic-alpine tussock bogs.

Range: Alaska east to Labrador, south to Wisconsin and Maine, and in Rocky Mountains to central Colorado.

Eastern populations were once thought to appear for only 4–5 days, but are now known to fly during several weeks. In cool, cloudy weather, Bog Fritillaries hide among mosses and heaths to await the warming sun.

3, 16, 625 Silver-bordered Fritillary
(*Clossiana selene*)

Description: 1⅜–2″ (35–51 mm). Tawny-orange to ocherous *above*, with black lines and dashes and *broken black rim. Below*, orange FW; orange-brown *HW has 4 rows of metallic silver spots* including a row on margin. Central silver cell spot below is elongated, submarginal dots dark.

Similar Species: Bog Fritillary has silver-white (not metallic) spots below and light submarginal dots.

Life Cycle: Egg, ⁹⁄₂₅₆″ h × ⁷⁄₂₅₆″ w (0.9 × 0.7 mm), cream-colored; laid on or near host plants, violets (*Viola*). Caterpillar, to ⅝″ (16 mm), brownish-black mottled with yellow and gray; has many yellowish branching spines; overwinters half grown. Chrysalis, to

½" (13 mm), tan with brown and green patches.

Flight: Up to 3 broods in East, Plains, and Northwest; May–September. 1 brood in Arctic and Rocky Mountains; June–August.

Habitat: Moist meadows and bogs, often near woodlands or aspen scrub, sometimes wet meadows among plains or sagelands.

Range: Alaska east to Newfoundland, south to Oregon, New Mexico, Illinois, and North Carolina. Holarctic: also Europe and Asia.

The Silver-bordered Fritillary is the only lesser fritillary which has bright silver spots below. It approaches the size of the greater fritillaries (*Speyeria*) in the Midwest but is small elsewhere. In the East, this species is usually abundant in marshy meadows but northwestern colonies are widely scattered. One colony exists in a quaking bog in the arid Columbia Basin desert, and is perhaps an Ice Age relict. Adults seek nectar from red clover, vervain, and other flowers during the morning, and court and mate shortly before sunset.

614   Meadow Fritillary
(*Clossiana bellona*)

Description: 1¼–1⅞" (32–48 mm). Tan to brownish-orange above with black dashes and dots; *FW tip angled outward as if clipped; no heavy black margin. Beneath,* orange FW has black spots and purplish tip; brownish *HW has orange-brown patch band and whitish, keel-shaped patch near base along costa; outer half of wing soft grayish to violet* with row of bluish spots.

Similar Species: Western Meadow, Kriemhild, and Titania fritillaries all lack clipped FW tip and white costal patch on HW below. Frigga's Fritillary also lacks clipped HW tip.

Life Cycle: Egg greenish-yellow. Purplish-black caterpillar has yellow and black mottling and brown branching spines; eats violets (*Viola*); overwinters half grown. Chrysalis yellowish-brown.

Flight: Up to 3 broods; May–September. 1 brood in Rockies and colder parts of Canada; June–July.

Habitat: Moist meadows in wooded areas, hayfields, pastures, and streamsides.

Range: British Columbia and Quebec, south to Washington, Colorado, Missouri, and North Carolina.

Along with the Silver-bordered Fritillary, the Meadow Fritillary is one of the most abundant bog fritillaries, particularly in the East. Rocky Mountain and northwestern populations are fewer and sparser. This species tends to fly low and rapidly in a jerky zigzag.

637 Frigga's Fritillary
(*Clossiana frigga*)

Description: 1¼–1⅝″ (32–41 mm). Orange above with black bars and spots; dark scales at base. *Below,* FW like upperside but paler; HW dark brown at base with golden-brown band and *outer half soft violet-gray; white rectangle containing dark dot* lies near base on costa.

Similar Species: Meadow Fritillary has clipped FW. Western Meadow Fritillary is brownish-violet outwardly on HW below, more rounded, and usually lacks white costal mark. Freya's Fritillary has white arrowhead across HW disk below.

Life Cycle: Egg laid on arctic avens (*Dryas integrifolia*) and possibly raspberry (*Rubus*). Caterpillar also feeds on willows (*Salix*). Chrysalis overwinters.

Flight: 1 brood; June–July.

Habitat: Willow bogs in coniferous forests;

alpine bogs or tundra areas in Arctic.
Range: Alaska southeast to N. Quebec and
Ontario, and south to British
Columbia, Colorado, and Michigan.
Holarctic: Scandinavia across N. Asia.

Frigga's Fritillary slows its activity in
cloudy weather. It prefers to fly or to
bask, with wings spread, on flowers or
grass on sunny days. On Baffin and
Victoria islands the butterfly flies over
the tundra when the brief summer and
intermittent sunshine permit. The
species is named for the wife of the
chief Norse god, Odin.

636 Dingy Arctic Fritillary
(*Clossiana improba*)

Description: 1⅛–1⅝" (28–35 mm). Wings appear
*blurry with indistinct checkering. Above,
dull brown with smeared brown spots* near
edge of rounded FW. Some populations
yellowish-brown with vague pattern.
Below, FW pale tawny-orange, with
even less contrast than above; *HW
bluish-gray or yellowish on outer half, dull
rust-colored orange-brown inwardly,*
separated by narrow darker line. *Keel-
shaped whitish patch or band on base of
HW beneath near costa.*

Similar Species: Other small, northern fritillaries
brighter with more distinct markings.

Life Cycle: Female lays eggs on dwarf prostrate,
arctic willows (*Salix*). Young caterpillar
probably overwinters.

Flight: 1 brood; early July–early August,
depending upon latitude and altitude.
Flies only every other year in some
localities.

Habitat: Poorly drained arctic tundra barrens
and subarctic summits; in Canadian
Rockies, moist, mossy mountaintops
and ridge notches.

Range: Circumpolar: in North America, Alaska
east to Baffin Island and W. Hudson
Bay, south to British Columbia and

central Alberta in Rockies, and Wind River Mountains of Wyoming. Holarctic: also Ural Mountains and Siberia.

A drab, feeble flier, the Dingy Arctic Fritillary spends more time crawling over the tundra than flying. Populations in the Canadian Arctic Archipelago are nearly black, an adaptation to their sun-starved habitat. Others produce yellow individuals.

Uncompaghre Fritillary
(*Clossiana acrocnema*)

Description: 1–1¼" (25–32 mm). *Small with rounded FW. Above,* flat orange-brown, paler toward bases, *large blackish-brown streaks and smeared dots;* 1 prominent black spot along FW costa in cell. Tawny-orange *below; HW ruddy-brown, reddish at base and soft grayish outwardly; whitish band crosses HW* with central tooth jutting outward.

Similar Species: Other small Colorado fritillaries brighter. Frigga's and Titania fritillaries brighter orange above.

Life Cycle: Egg cream-colored, urn-shaped; laid on woody stems of snow willow (*Salix nivalis*). Caterpillar spiny, brown with white lines; overwinters, then resumes feeding in spring.

Flight: 1 brood; mid-July to August.

Habitat: Moist, rocky arctic-alpine tundra meadows, with mats of willow.

Range: San Juan Mountains of SW. Colorado.

Discovered only in 1978, the Uncompaghre Fritillary maintains a precarious existence far to the south of its nearest relatives. As the Pleistocene glaciers retreated northward, this butterfly was able to survive later warming trends at 13,000–14,000' (3965–4270 m) above sea level in Colorado. Apparently all stepping-stone

colonies between the Uncompaghre
Fritillary and the Dingy Arctic
Fritillary perished, leaving the former
to evolve separately from the latter.

616   Kriemhild Fritillary
(*Clossiana kriemhild*)

Description: 1⅜–1¾″ (35–44 mm). Tawny-orange
above with black streaks and dots.
Similar *below, HW has yellow median
band* marked with brown lines, *small
brown chevrons pointing outward near
margin* and small yellow dots near base.

Similar Species: Titania Fritillary has HW marginal
chevrons pointing inward. Western
Meadow Fritillary has rounder wings
with brighter orange and fewer black
spots on margin above.

Life Cycle: Undescribed. Caterpillar feeds on
violets (*Viola*).

Flight: 1 brood; June–July.

Habitat: Moist meadows in mountainous wooded
areas, often near aspen groves.

Range: Rocky Mountains from S. Montana to
N. Utah.

Kriemhild Fritillary sometimes flies
along with the Titania, Napaea, and
Meadow fritillaries. This uncommon
species as well as Hayden's Ringlet and
the Yellowstone Checkerspot is
restricted to the Northern Rockies.

615   Western Meadow Fritillary
(*Clossiana epithore*)

Description: 1⅜–1⅝″ (35–41 mm). Above, orange
with small black dots and bars. Some
populations have heavy markings.
Below, FW paler orange, patterned like
upperside; *HW below violet-brown to dull
brown with median band of yellow patches,
brown line near margin,* and row of
purplish spots in between. *Whitish spot*

along HW costa is anvil-shaped next to, or replaced by, yellow spot.

Similar Species: Meadow Fritillary has clipped FW tip. Frigga's Fritillary has heavy dark markings in outer FW, keel-shaped, dark-spotted white mark on HW costa below. Kriemhild Fritillary duller orange.

Life Cycle: Several violets (*Viola sempervirens, V. glabella, V. ocellata*) are host plants. Caterpillar overwinters half grown.

Flight: 1 brood; mostly June–July, slightly earlier near coast and a little later in higher mountains.

Habitat: Moist, sunny openings in mixed evergreen and deciduous mountain forests, maritime marshes, roadsides, and lush meadows.

Range: Central British Columbia and SW. Alberta south to central California and Idaho.

This species is the most widespread and abundant lesser fritillary in the Northwest. The only member of the group in California, the Western Meadow Fritillary succeeds in a variety of lowland and highland habitats.

634 Polaris Fritillary
(*Clossiana polaris*)

Description: 1¼–1½" (32–38 mm). Above, reddish-orange with black lines and dots and blackish wing bases. *Below,* FW orange, *HW reddish-brown with 4 white dots at base,* many white streaks in middle band, and *white hourglass marks along margin.*

Similar Species: Freya's Fritillary has white diamonds rather than hourglasses around HW base below.

Life Cycle: No record in U.S.; Scandinavian records show eggs laid on mountain avens (*Dryas octopetala, D. integrifolia*).

Flight: 1 brood; June–July, appearing only every other year.

Habitat: Arctic tundra.
Range: Alaska east to Labrador, north to Greenland, south to N. British Columbia and Manitoba.

Because this species flies only every other year, it used to be assumed that it experienced alternating population explosions and extinctions. Now it is known that the life cycles of this and other arctic butterflies require 2 years. The Polaris Fritillary flies in odd-numbered years in Alaska but in even-numbered years in most other places.

Freya's Fritillary
(*Clossiana freija*)

Description: 1⅛–1½″ (28–38 mm). Above, tawny-orange with strong black pattern of spots and bands. Below, tawny-orange FW; reddish-brown or brown HW, sometimes with yellow overlay. *HW below has white bars near margin, sawtooth reddish and white band in middle of wing, with long white tooth in center, large white mark on costa and white arrowhead* at base.

Similar Species: Polaris Fritillary has white hourglasses along HW margin below. Titania's Fritillary has yellowish markings below and lacks white central HW tooth.

Life Cycle: Egg tan, tiny; laid haphazardly near host plants, dwarf and alpine blueberry (*Vaccinium caespitosum, V. uliginosum*), bearberry (*Arctostaphylos uva-ursi*), and black crowberry (*Empetrum nigrum*). Caterpillar brown with cream-colored spots with many branching spines; overwinters nearly grown.

Flight: 1 brood; May–June.

Habitat: Forest clearings, willow bogs; also alpine valleys and arctic tundra.

Range: Alaska east to Baffin Island and Newfoundland, and south to N. Washington, New Mexico, and Wisconsin; also N. Europe and Asia.

Freya's Fritillary flies early in the season, usually before other butterflies in its range. It is frequently found outside of bogs, where dwarf blueberry carpets the forest floor. The butterfly was first named from Lapland for the Norse goddess of love and beauty.

Alberta Fritillary
(*Clossiana alberta*)

Description: 1½–¾" (38–44 mm). FW somewhat rounded. Above, dingy orange with vague pattern of black bars and dots; female suffused with black. *Below*, buff-orange with upperside pattern of black; *HW crossed by paler band of whitish, buff-orange, or tan,* and narrow suffused row of marginal light dots capped with blurred brown chevrons.

Similar Species: Astarte Fritillary larger, wings less round, and bright white and reddish bands on HW below.

Life Cycle: Pale yellow eggs deposited on mountain avens (*Dryas octopetala*).

Flight: 1 brood; July–August; generally in even-numbered years.

Habitat: Alpine tundra hillsides, rockslides, and fell-fields with mountain avens.

Range: S. British Columbia, S. Alberta, and adjacent NW. Montana.

Although they are rather drab, Alberta Fritillaries are interesting to follow as they fly slowly just above the ground, taking nectar at alpine flowers and searching for mates. They can outdistance even a rugged hiker when traveling uphill. Colonies of Alberta Fritillaries may have existed on vegetated refuges that projected above massive ice sheets during the Ice Age.

635 Astarte Fritillary
(*Clossiana astarte*)

Description: 1⅜–1⅞" (35–48 mm). *Large, with
squared-off wings.* Above, light tawny-
orange with black lines and dots.
Beneath, tawny-orange on FW; HW
orange on margin with *rows of white and
black dots,* brick-red toward base crossed
by alternating median and basal bands
of whitish or buff.

Similar Species: Alberta Fritillary has rounder wings,
less distinctly marked below, median
band yellower or browner. Titania's
Fritillary smaller, more heavily marked
above.

Life Cycle: Host plant is spotted saxifrage
(*Saxifraga bronchialis*) in Alberta.

Flight: 1 brood; June–July in Arctic, July–
August southward. Only every other
year in most places.

Habitat: Alpine-arctic ridges, rockslides, and
mountain peaks.

Range: Alaska to N. British Columbia; also
S. British Columbia and adjacent
Washington, N. Montana, and
Alberta.

Some biologists treat the northern and
southern populations of this butterfly as
different species. Both seek ridge tops,
where the males fly swiftly back and
forth just above the ground looking for
females. They are very wary of people's
approach. The Pasayten Wilderness
Area in Washington and Glacier
National Park in Montana are places to
look for this largest of the lesser
fritillaries.

613, 619 Titania's Fritillary
(*Clossiana titania*)

Description: 1¼–1¾" (32–44 mm). Dark orange
above with heavy, variable black dots
and bars. *Beneath,* similar to upperside
on FW; *HW purplish or dull red-brown*

with broken yellowish-white to rust-colored band across midwing and white dashes on margin capped with *brown chevrons which point inward.*

Similar Species: Freya's Fritillary has prominent white tooth in middle of median band. Astarte Fritillary larger and lighter. Kriemhild Fritillary has pale marginal spots with outward-pointing chevrons, or sometimes lacks spots.

Life Cycle: Egg tan, ribbed; laid haphazardly on many plants including host plants. Caterpillar overwinters after hatching. Preferred host plants are willows (*Salix*) and bistort (*Polygonum bistortoides*), possibly also violets (*Viola*) and others.

Flight: 1 brood; June–August.

Habitat: Bogs and moist valley bottoms in coniferous forests, mountainsides, trails, roadsides, alpine meadows, and moist tundra.

Range: Alaska east to Labrador, and south to Washington, Utah, New Mexico, Minnesota, and New Hampshire. Holarctic: W. Europe to Siberia.

Titania's Fritillary has a distribution pattern common to many satyrs, sulphurs, and fritillaries. In the West, they are confined to the moist cool conditions of mountains or arctic meadows and forests. In the East, they cling to the border of boglands and northern woodlands. In the Cascades and Rockies, Titania's Fritillary is usually the most abundant lesser fritillary.

---

639 **Arctic Fritillary**
(*Clossiana chariclea*)

Description: 1¼–1⅜″ (32–35 mm). Above, orange with black spots, bars, and chevrons. *Beneath,* orange on FW; *HW has about 4 silver-white spots* in whitish-yellow middle band, the middle spot elongated, marginal row of white bars capped with brown chevrons, and white spots at base.

Similar Species: Titania's Fritillary lacks silvery spots below, marginal spots yellower. Napaea Fritillary has only 1 silver-white spot on HW below in much yellower band. Freya's Fritillary has longer white arrowhead with white crescents beyond.

Life Cycle: Arctic avens (*Dryas integrifolia*) is suspected host plant, although females have been observed laying eggs on blueberry (*Vaccinium*).

Flight: 1 brood; June–July; only every other year in most places, in odd years in Alaska and Yukon.

Habitat: Arctic tundra and a few alpine peaks.

Range: Alaska and NW. British Columbia east to Labrador and Greenland. Holarctic: N. Europe and Asia.

Some arctic populations cannot be distinguished as either Arctic or Titania fritillaries; they may represent a single species. The Arctic Fritillary is one of only 6 butterflies to occur in Greenland, where it lives on the narrow fringe of greenery rimming the mile-thick layer of ice.

573 Texan Crescentspot
(*Anthanassa texana*)

Description: 1–1½″ (25–38 mm). *FW indented below tip. Above, mostly black with white dots and bars* on margin and center disk, white median band across HW, red-orange areas near wing bases (redder in Southeast). *Below, FW orange at base, black-brown on outer half,* with white marks; HW buff-colored with fine black lines and dots, white midband.

Similar Species: Cuban Crescentspot smaller with yellowish spots, paler below.

Life Cycle: Eggs are laid in clusters on several plants of the acanthus family (*Dicliptera brachiata, Jacobinia carnea, Ruellia*). Caterpillar spiny.

Flight: Several broods; March–November, sometimes year-round in S. Texas.

Habitat: Open areas: scrublands, deserts, and grasslands; mountains in Mexico.

Range: S. California and S. Nevada east to South Carolina and Florida, and south through Mexico to Guatemala; rarely emigrating north from Texas to Colorado, North Dakota, Minnesota, and Illinois.

Male Texan Crescentspots often rest on bushes in gullies, flying out to chase other butterflies. A southeastern population, called the Seminole Crescent, is much more sedentary than the typical species. It flies in Louisiana, Mississippi, Florida, Georgia, and South Carolina, and may represent a separate species. A related species, the Ptolyca Crescentspot (*A. ptolyca*), has been found in the Santa Ana Wildlife Refuge in Texas. This Central American and Mexican species is smaller than the Texan, with yellower spots and less orange below.

629 Cuban Crescentspot
(*Eresia frisia*)

Description: 1⅛–1⅜″ (28–35 mm). *Indented FW tip. Above, black with cream-colored bands and spots* (Texas) *or black with orange areas and spots* (Florida). Below, base of FW dull orange, outer half of wing blackish with cream-colored white patches; HW pale buff with a few black marks near margin.

Similar Species: Texan Crescentspot larger, has white spots on FW margin above, all spots whiter.

Life Cycle: Eggs laid in clusters on shrimp plant (*Beloperone guttata*) and several others of acanthus family (*Dicliptera, Ruellia*). Caterpillar gray mottled with yellow and black; has many grayish spines. Chrysalis brown; horns above eyes.

Flight: Successive broods; most of year.
Habitat: Open areas: scrublands, savannah, and roadsides.
Range: S. Arizona, S. Texas, Mexico, S. Florida and the Antilles, very rarely emigrating north as far as Missouri.

More or less distinctive races of the Cuban Crescentspot occur in the West, but the Arizona population is rare and little known. Latin American members of the genus *Eresia* include mimics of unpalatable longwings, with elongated wings and bright colors.

574 Phaon Crescentspot
(*Phyciodes phaon*)

Description: ⅞–1¼" (22–32 mm). *Small. Above, FW brown with orange-tinged base, followed by black post-basal band, whitish middle-band, and submarginal orange band* and dark margin; brown HW has orange bands. Beneath, FW similar to above; HW chalk-colored to yellowish-cream, with dark veins and crescents (more noticeable in female); white crescents along HW margin outside black dot row in brown area.

Similar Species: Pearly and Field crescentspots yellowish below, lack white band on FW above. Vesta much paler above and on FW below; also lacks white band above.

Life Cycle: Eggs laid in clusters. Caterpillar olive with brown and cream-colored stripes, and branching spines; feeds on fog fruit (*Lippia nodiflora, L. lanceolata*). Chrysalis brown mottled with black and cream-color.

Flight: 2 or more broods; April–September, year-round in Tropics.

Habitat: Fairly moist open areas: desert springs, weedy fields, and marshes.

Range: SE. California and S. Nevada east to North Carolina and Florida, rarely emigrating north to Nebraska and Missouri.

Females often lay clusters of 50–100 eggs. But as with all butterflies, the mortality rate is high; only about 5% survive to the adult stage. The majority are killed by parasitic flies, wasps, and other predators. Adult Phaon Crescentspots often fly to Spanish needles for nectar.

600, 628, 630 Pearly Crescentspot
(*Phyciodes tharos*)

Description: 1–1½" (25–38 mm). *Male has broad, open orange areas above with wide black margin;* female has heavier black markings. Below, orange FW has black patches, especially along margin, and several cream-colored spots; *HW yellowish to cream-colored with fine brown lines and purplish-brown patch containing light crescent* on margin. Spring broods have HW mottled with brown below.

Similar Species: Phaon Crescentspot has cream-colored bands. Painted Crescentspot much paler below. Tawny and Field crescentspots lack purplish-brown patch on HW margin below around crescent and have light bar in FW cell below.

Life Cycle: Eggs laid in clusters on leaves of asters (*Aster*), which caterpillars eat. Caterpillar brown with yellow bands and many branching spines; last brood overwinters when half grown. Chrysalis mottled gray, yellowish, or brown.

Flight: Usually several broods; April–November. 1 brood in Rockies and N. Canada; June–August.

Habitat: Open spaces, moist meadows, fields, roadsides, and streamsides.

Range: Yukon and Newfoundland to S. Mexico, and from E. Washington and SE. California to Atlantic.

One of our most common meadow butterflies, the Pearly Crescentspot flies low over the grasses with alternating flaps and glides. Often described as highly pugnacious, the males dart out from perches, or break their flight pattern to investigate any passing form —butterfly, bird, or human. Adults take nectar from composite flowers, such as asters, fleabanes, and thistles. Recent studies, as yet unpublished, indicate that 2 forms of the Pearly Crescentspot are in fact 2 separate species, now designated as type A and type B. They differ in size, coloring, and chromosomal characteristics, and where their ranges overlap, A and B do not interbreed.

575 Tawny Crescentspot
"Bates' Crescentspot"
(*Phyciodes batesii*)

Description: 1¼–1½" (32–38 mm). Tawny-orange *above with wide black borders,* mottled by dark lines, with paler band in middle of FW; FW blacker in female. *Below, FW orange with large black patch on hind margin,* black patches near costa and outer edge, and yellow patches between; *HW yellowish or orange-tinged with light markings,* seldom with prominent pearly crescent; white, checkered fringes.

Similar Species: Pearly Crescentspot usually has purplish patch with pearly crescent on HW below, and duller checkered fringes. Field Crescentspot normally much less black on FW below.

Life Cycle: Eggs laid in clusters on asters. Brown caterpillar has yellowish stripes, several white patches on head, and many branching spines; eats blue wood aster (*Aster undulatus*) and overwinters half grown. Chrysalis light mottled brown.

Flight: 1 brood; early to mid-July.

Habitat: Open areas in woodlands, preferring moist areas in Michigan and Wisconsin, but dry barren slopes in Northeast.

Range: Northwest Territories south through

SE. Alberta to Nebraska, east to Nova Scotia, south in Appalachians to Georgia.

The Tawny Crescentspot is easily confused with both the Pearly and Field crescentspots. However, the Pearly Crescentspot extends farther south and seems a more versatile butterfly than the rather limited, northern Tawny. The Field Crescentspot essentially tends to replace the Tawny Crescentspot throughout most of the West.

576, 590 Field Crescentspot
(*Phyciodes campestris*)

Description: 1⅛–1⅜" (28–35 mm). Mostly blackish-brown *above, with rows of yellow and orange patches and spots.* Some populations tawny with black and yellow markings above. *Below, orange FW has small black patches, cream-yellow bar across cell,* and several other yellow patches; HW yellowish or orange with vague marginal brown patch containing light crescent. Deeper color at higher altitudes.

Similar Species: Pearly Crescentspot male has much more open orange above; dark form has bold purplish-brown patch below around crescent. Tawny Crescentspot has much more black on FW below. Phaon and Painted crescentspots paler on HW below. Mylitta has more open pattern above. Orseis larger, more angled, lighter below. All but Orseis lack light bar across FW cell below.

Life Cycle: Eggs pale green; laid in clusters on many asters (*Aster, Machaeranthera*). Blackish-brown caterpillar has cream-colored and blackish lines and many branching spines; feeds communally until half grown and then overwinters. Chrysalis light mottled brown with small bumps.

Flight: 1–3 broods; mostly May–September,

April–September in coastal California,
June–August in higher mountains and
Canada.

Habitat:
Open areas such as meadows, forest
clearings, grassland valleys, swamps
and fields; also along canals and
streams. In far West, arctic-alpine
meadows and fell-fields.

Range:
Alaska, Mackenzie District of
Northwest Territories, and
Saskatchewan south to California,
Kansas, and Mexico.

The Field Crescentspot is the most
common crescentspot in western
mountains. Males fly about the
meadows looking for females, while
females slowly flutter through the
vegetation searching for asters upon
which to lay eggs. This species flies at
low to middle altitudes in the Rockies
and in higher mountains of the
Northwest. The Mylitta Crescentspot
seems to replace it in the lowlands of
that region.

Painted Crescentspot
(*Phyciodes picta*)

Description:
⅞–1¼" (22–32 mm). *Small.* Brown
above with cream-colored and orange
bands and spots. Below, FW orange
with buff edges and bars and black
patches and bands; *HW below almost
unmarked yellow in male, pale cream-
colored in female* (both sexes straw-
colored in Arizona). Light crescent on
HW margin below, occasionally with
darker scales around it.

Similar Species:
Phaon Crescentspot has network of
darker lines on HW below. Pearly
Crescentspot lacks whitish bands
above.

Life Cycle:
Eggs tiny, yellow-green; laid in cluster
on host plants. Young caterpillars feed
together and overwinter when half
grown. Mature caterpillar is yellowish-

brown with cream-colored and brown stripes, and branching spines. Chrysalis smooth, light brown. Asters (*Aster*) serve as host plants.

Flight: 2 broods in Colorado; May–June and July–August. 3 broods in South; early spring–late fall.

Habitat: Roadsides, alkaline flats, railroad tracks, alfalfa fields, streamsides, and other open waste places with asters.

Range: W. Nebraska south through Colorado to Arizona, Mexico, and Texas.

At the height of their flight periods, Painted Crescentspots fly by the hundreds among weedy vegetation, often pausing for nectar on alfalfa. However, during the brief summer period between broods not a single butterfly will be seen. This species has replaced the Pearly and Field crescentspots in certain disturbed areas of Colorado.

584 Vesta Crescentspot
(*Phyciodes vesta*)

Description: ¾–1⅛" (19–28 mm). *Small.* Above, light orange *finely marked* with black lines. FW orange *below with black chainlike pattern submarginally;* HW yellow-orange with fine brown-black lines and bands, buff spots, and crescents.

Similar Species: Other crescentspots have heavier dark markings above, lack black chainlike pattern on FW below.

Life Cycle: Eggs laid in clusters on host plant, tube-tongue (*Siphonoglossa pilosella*). Early stages unreported.

Flight: Successive broods; April–October.

Habitat: Open dry areas, including grasslands, deserts, and especially grassy savannahs with mesquite trees.

Range: Texas and New Mexico south to Guatemala, rarely emigrating north to Colorado and Nebraska.

The Vesta Crescentspot is abundant in Mexico and Texas, but the host plant is absent from the central United States, so emigrants perish there without leaving offspring. The slow-flying adults can be observed drinking at mud puddles and flowers. Their small size, light color, and delicate checkering set them apart from other crescentspots.

582 Orseis Crescentspot
(*Phyciodes orseis*)

Description: 1¼–1½" (32–38 mm). *FW slightly indented on margin;* HW somewhat angular. N. *California and Oregon populations* dark brown above with orange bands and spots; below FW orange; *HW below has white median band and marginal brown patch containing white crescent. Sierra Nevada populations* orange above with brown bands and lines; yellowish below with *fine brown lines on HW and marginal white crescent.*

Similar Species: Field Crescentspot less marked and deeper orange below, smaller, less angled. Mylitta Crescentspot smaller, deeper orange above, with more contrasting markings below.

Life Cycle: Eggs small, greenish, ribbed; laid in clusters. Maroon-black caterpillar has 9 rows of branching spines; feeds in group on thistle (*Cirsium*). Chrysalis mottled brown with small bumps.

Flight: 1 brood; May–June.

Habitat: Wooded mountain canyons undisturbed by civilization, near pines and willows.

Range: SW. Oregon (Siskiyous) south into N. California Coast Ranges; NE. Nevada east into Carson Range of Nevada.

Little known and seldom encountered, the Orseis Crescentspot has an aura of mystery. This is due in part to its limited, disjointed range. It may have once occurred in the region north of San Francisco Bay.

Pallid Crescentspot
(*Phyciodes pallida*)

Description: 1¼–1¾" (32–44 mm). Pale orange above with open pattern of fine black lines and spots. *Below,* orange with yellowish patches and marginal lines on FW; *HW buff-yellow with whitish and tan bands, lines, and crescents. FW slightly indented on margin.*

Similar Species: Mylitta Crescentspot smaller with smaller black spot on lower margin of FW below, heavier pattern on HW below.

Life Cycle: Eggs small, cream-colored, and ribbed; laid in clusters. Caterpillar ocher with several brown stripes and many branching spines; feeds in groups when young. Host plant is thistle (*Cirsium*). Chrysalis mottled brown.

Flight: 1 brood; April–June.

Habitat: Foothills and flats, open woodland canyons, sagebrush gullies, and dry streambeds.

Range: S. British Columbia south through Nevada to Arizona, east to South Dakota and central Colorado.

The Pallid and Mylitta crescentspots often fly together, and were once thought to be the same species. But the Pallid Crescentspot is much larger and lighter than any others in its genus. This species remains in its colonies, which are well defined and quite widely separated.

17, 627 Mylitta Crescentspot
(*Phyciodes mylitta*)

Description: 1⅛–1⅜" (28–35 mm). *FW slightly indented along outer margin.* Above, orange with open pattern of fine black lines. *Beneath,* FW orange with dark lines and spots and small black spot on trailing margin; HW has yellowish-brown areas, *white bands, and white marginal crescent in brown patch.*

Similar Species: Pallid Crescentspot larger with lighter color and markings; has prominent black spot on lower margin of FW below.

Life Cycle: Egg cream-colored; laid in clusters. Caterpillar overwinters half grown; mature caterpillar, to ⅞" (22 mm), black with yellowish stripes and spots; many branching spines. Host plants are thistles (*Cirsium*), mild thistle (*Silybum marianum*), and plumeless thistle (*Carduus pycnocephalus*). Chrysalis, to ⅜" (10 mm), mottled brown to gray with golden sheen and tubercles.

Flight: Several overlapping broods; March–October.

Habitat: Agricultural fields, dry canyons, mountains, open woods, shorelines, marshes, vacant lots, meadows, and roadsides.

Range: British Columbia and Montana south to Baja California, New Mexico, and S. Mexico.

In the 1800's weedy thistles spread all over the West and were followed by the Mylitta Crescentspot. Now this species seems ubiquitous and is found in many disturbed as well as natural habitats. Flying from early spring to autumn, it is most abundant in August. Individual butterflies tend to occupy the same territories or stands of nectar flowers for several days at a time.

632 Gorgone Crescentspot
(*Charidryas gorgone*)

Description: 1⅛–1⅜" (28–35 mm). *Above,* orange and black in equal bands, *row of black dots on HW. Below,* FW orange with black bands and whitish tip, brown *HW has band of white arrowheads across disk, ending in deep zigzag near* body; crescents on margin and white bars at base.

Similar Species: Silvery Crescentspot has row of silver-
white ovals on HW below rather than
arrowheads, lacks deep zigzag near
body. Harris' Checkerspot has reddish
spot bands below.

Life Cycle: Eggs cream-colored, ribbed; laid in
clusters. Caterpillars feed together until
overwintering when half grown.
Sunflowers (*Helianthus*), ragweed
(*Ambrosia trifida*), sump-weed (*Iva
xanthifolia*), goldeneye (*Viguiera
multiflora*), and other composites are
host plants. Mature caterpillar orange-
red to orange and black banded to
nearly black. Chrysalis mottled gray to
cream-colored.

Flight: 2 broods in South; May–September.
1 brood northward and in Rockies;
May–July.

Habitat: Grassland roadsides, canals, flats, and
fields; at edge of range, in pine and
open hardwood forests.

Range: S. Alberta, Idaho, and Utah east to
Michigan (rarely New York) and
Georgia, south into Mexico.

Able to exploit many weedy sunflowers,
the Gorgone Crescentspot spreads into
new areas to form large colonies. Yet it
is ecologically limited to the central
part of the continent except for isolated
colonies near the Atlantic Coast. The
butterfly's numbers vary dramatically
from year to year—sometimes scarce,
in other seasons it may be the most
abundant species flying. Adults take
nectar at goldenrod flowers.

---

631 Silvery Crescentspot
(*Charidryas nycteis*)

Description: 1⅜–1¾″ (35–48 mm). *Above,* orange
with black margins and wing bases,
*FW tip black, and row of black dots on
HW. Below,* FW orange with brown
margin and pale crescents; yellowish
HW has marginal band of white crescents

interrupted by brown, plus median band of whitish ovals and whitish spots at base. Often silvery sheen to HW below. Upperside is darkest in Rockies, most orange in Manitoba.

Similar Species: Gorgone Crescentspot has white arrowheads and sharp zigzag below. Harris' Checkerspot has reddish spot band below.

Life Cycle: Eggs greenish; laid in clusters under leaves. Caterpillar black with orange stripes, white and purple specks, and many branching spines; overwinters half grown. Host plants are coneflower (*Rudbeckia laciniata*), wingstem (*Actinomeris alternifolia*), asters (*Aster*), sunflowers (*Helianthus*), and crownbeard (*Verbesina helianthoides* and *V. virginica*). Chrysalis white with brown mottling, pearly-gray, or nearly black.

Flight: 2 broods in South; March—September. 1 brood in North and Rockies; June—July.

Habitat: Grassland and woodland streams, moist meadows, and open deciduous woods.

Range: SE. Saskatchewan east to Quebec and south to Arizona, Texas, and Georgia.

Although the caterpillars employ both sunflowers and wingstems as host plants in the East, they seem to be restricted to coneflowers in the Rocky Mountains. Colonies may be detected easily by the brown blotches the caterpillars make on nibbled leaves. Adults sip nectar from coneflowers along cool mountain streams, where they often fly together with Atlantis Fritillaries. In swampy aspen groves, they can be seen with Pearly Crescentspots. Although this species and the Gorgone Crescentspot are checkerspots by relationship, they are known as crescentspots since they have crescents that resemble those of members of the genus *Phyciodes*.

588 Harris' Checkerspot
(*Charidryas harrisii*)

Description: 1¼–1¾" (32–44 mm). *Above,*
blackish-brown with orange bands
(mostly orange in Manitoba and parts of
Minnesota), *row of black dots on HW.*
Below, orange on FW with rows of
yellow crescents along red margin; *HW*
has alternating bands of brick-red and pale
yellow spots, and row of marginal yellow
crescents.

Similar Species: Gorgone and Silvery crescentspots lack
red below.

Life Cycle: Several hundred eggs laid in cluster on
aster (*Aster umbellatus*) and sometimes
crownbeard (*Verbesina helianthoides*).
Young caterpillars feed together in silk
nest; overwinter half grown. Mature
caterpillar orange with black lines and
black branching spines. Chrysalis white
with black edge around orange, black-
tipped tubercles on back.

Flight: 1 brood; June–July.

Habitat: Moist meadows, edges of bogs and
marshes, openings in forests, and old
fields.

Range: SE. Saskatchewan east to Nova Scotia
and south to North Dakota, Illinois,
and West Virginia.

Although hundreds of thousands of
eggs may be laid in a field, most of the
caterpillars die—they either starve as
soon as the available food is depleted,
or are killed by parasites or predators.
If many survive and adults become
abundant, some fly away to colonize
new areas. The pattern is thus one of
repeated local extinctions and
recolonizations. Colonies are small,
local, and relatively few, especially now
as wetlands diminish.

560, 566   Northern Checkerspot
(*Charidryas palla*)

Description: 1¼–1⅝" (32–41 mm). Above, male orange or brick-red with black lines and patches, female darker, usually cream and black from California to Washington and Idaho. *Below,* FW orange with marginal black, yellowish, and orange rows; reddish *HW has yellowish median bands, yellow crescent row in red margin,* and yellowish basal spots.

Similar Species: Sagebrush Checkerspot paler above and below. Gabb's Checkerspot has very pearly light spots below. Aster Checkerspot has squarer wing shape, often has more contrasting bands on upperside, and darker wing bases.

Life Cycle: Caterpillar black with rows of white dots and orange dashes, and many spines; overwinters half grown. Chrysalis mottled tan, gray, or black, with shiny bumps. Host plants are asters (*Aster conspicuus, A. occidentalis*), rabbit brush (*Chrysothamnus nauseosus, C. viscidiflorus, C. paniculatus*), showy daisy (*Erigeron speciosus*), and goldenrod (*Solidago californica*).

Flight: 1 brood; April–June along Coast and in lowlands, June–August at higher altitudes.

Habitat: Mountain clearings and valleys, open woodlands, and oak and sagebrush canyons; particularly valleys with aspen at middle elevations.

Range: S. British Columbia and Alberta south to N. and central California; south in Rockies to New Mexico. Absent from most of Great Basin.

The Northern Checkerspot and many of its close relatives are often difficult to distinguish when several species fly together. Some populations have striking dark females with little or no orange. Early in the sage-country spring, Northern Checkerspots combat the chill winds and bask on the sun-warmed basalt along with White-lined Green Hairstreaks and Spring Whites. On hot summer days they seek moisture at mud puddles and take nectar from such flowers as mules' ears and lovage.

581, 589 Sagebrush Checkerspot
(*Charidryas acastus*)

Description: 1¼–1¾" (32–44 mm). Male pale orange above with fine to heavy black lines; female is deeper orange but some females are yellowish or blackish. *Below,* pale mottled orange on FW with marginal white crescents; *HW has narrow orange submarginal bands and broad cream-white bands* of black-rimmed spots (submarginal orange band may be vague orange circles on cream-colored background).

Similar Species: Northern Checkerspot is more richly colored above and below. Desert Checkerspot has clear orange FW below, silver-white spot bands on HW below. Gabb's Checkerspot smaller, darker, with narrower cream-colored bands beneath.

Life Cycle: Eggs are clustered. Caterpillar black with cream-colored dots, stripes of orange crescents, and many spines. Host plant rabbit brush (*Chrysothamnus viscidiflorus*); asters (*Machaeranthera canescens, M. viscosa*) suspected.

Flight: 1 brood in Oregon; May–June. Several broods from Utah to New Mexico; May–September.

Habitat: Arid grassland gulches, pinyon-juniper woodlands, sagebrush hills, and canyons.

Range: Great Basin and intermountain areas from SE. Washington and S. Alberta east to North Dakota and south to SE. California and New Mexico.

The Sagebrush Checkerspot generally inhabits pinyon-juniper habitats that are rich in both sagebrush and rabbit brush, but a few colonies have been discovered in open sageland.

602   Desert Checkerspot
(*Charidryas neumoegeni*)

Description: 1¼–1¾" (32–44 mm). *Above, nearly solid orange* with few narrow black lines and spots. *Below*, orange with *pearly tip* on FW; HW has orange bands and *pearl-white bands of spots*. Arizona and New Mexico populations are blacker on upperside.

Similar Species: Sagebrush Checkerspot has cream-colored, not pearly, spot band below. Gabb's Checkerspot has more dark markings above and on FW below.

Life Cycle: Eggs pale green; laid in clusters. Caterpillar black with gray and orange stripes, and branching spines; feeds on goldenhead (*Acamptopappus sphaerocephalus*) and mojave aster (*Machaeranthera tortifolia*); overwinters when half grown. Chrysalis mottled gray and black.

Flight: 1 brood; late March–early May. Occasional 2nd brood in fall if there is heavy rainfall.

Habitat: Rocky desert washes and hills, pinyon-juniper and oak woodlands, and arid mountain canyons.

Range: Desert Southwest: SE California, S. Nevada, and SW. Utah and Arizona south into Baja California and Sonora, Mexico.

After wet winters these checkerspots may be common, along with Leanira Checkerspots, Pima Orangetips, and tiny blues. In April, many butterflies that occur nowhere else can be observed in the desert. For moisture and sugar, they visit desert asters and other brief Sonoran blooms.

585 Gabb's Checkerspot
(*Charidryas gabbii*)

Description: 1¼–1½" (32–38 mm). *Above, orange with black bands* and yellowish median band. *Below,* FW mottled orange with marginal white and red bands; *HW has alternating reddish-brown and lustrous pearly-white bands* of rimmed spots and marginal crescents.

Similar Species: Northern Checkerspot has cream-colored spot band below. Desert Checkerspot has little dark marking on orange upperside and clear orange FW below.

Life Cycle: Clustered eggs are laid on composite family plants, such as telegraph weed (*Heterotheca grandiflora*), golden weed (*Haplopappus squarrosus*), and some others. Caterpillar black with cream-colored dots, orange bars, and many branched spines. White chrysalis heavily mottled with brown and black.

Flight: 1 brood; March–July, according to altitude and weather, mainly April and May.

Habitat: Coastal sand dunes, oak and sometimes pine woodlands, and canyons in lower elevation mountains.

Range: Central California to Baja California along coast, Coast Ranges, and western foothills of Sierra Nevada.

Many colonies of Gabb's Checkerspots on coastal dunes were destroyed by development. However, this species is still common inland in the Laguna Mountains near San Diego and on Santa Cruz Island, largely a preserve of The Nature Conservancy.

---

587 Rockslide Checkerspot
(*Charidryas damoetas*)

Description: 1⅛–1⅝" (28–41 mm). *Dull orange-brown above with greasy sheen marked with blurred brown lines and dusky-brown at base.* (California populations have more distinct markings above.) Below, orange with black lines on FW; HW largely cream-yellow crossed by pale orange spot band.

Similar Species: Northern Checkerspot more brightly marked above and below. Sagebrush Checkerspot less suffused with brown. Both generally occur at lower altitudes. Anicia Checkerspot has longer, redder wings.

Life Cycle: Clustered eggs cream-colored, ribbed; hatch in 1 week. Tiny caterpillars feed together for 1 month, then overwinter half grown. Older caterpillar black with many spines, stripes of orange crescents, and cream-colored dots. Host plants are wild daisy (*Erigeron leiomeris*) and goldenrod (*Solidago multiradiata*). Tan chrysalis has bluish-white tint and orange tubercles. Caterpillars may require 2 or more years to mature.

Flight: 1 brood; July–August.

Habitat: Alpine rockslides and rocky slopes, roads and trails, usually well above treeline.

Range: Sierra Nevada of California and high Rocky Mountains from S. British Columbia and Alberta to Utah and Colorado.

On warm days during midsummer, Rockslide Checkerspots can be found on barren rockslides in the company of Magdalena Alpines and Lustrous Coppers. The checkerspots patrol just above the boulders, often chasing others, and frequently stopping to take nectar. Like other alpine butterflies, these are subject to the vagaries of extreme weather and tend to have good and bad years. Once considered merely a dull, high-altitude form of the Northern Checkerspot, the Rockslide Checkerspot is now seen as a fully distinct species.

586 Aster Checkerspot
(*Charidryas hoffmanni*)

Description: 1¼–1⅝″ (32–41 mm). Orange and brown *above, darkest at base,* banded with lighter orange; *innermost band often yellow or cream-colored. Below, alternating bands of cream-color and brick-red* with black lines and spots.

Similar Species: Northern Checkerspot can be very similar but tends to have lighter base, less contrasting bands, and more rounded wings.

Life Cycle: Eggs light green; laid in clusters. Young caterpillar lives in silk web on showy aster (*Aster conspicuus*) and golden aster (*Chrysopsis breweri*); overwinters half grown. Older caterpillar black with white specks and cream-colored lines. Chrysalis white to mottled brown.

Flight: 1 brood; June–July.

Habitat: Moist mountain valleys and canyons, middle elevation forest glades, meadows, and roadsides.

Range: S. British Columbia south through Washington and Oregon Cascades, and in Siskiyous, Coast Ranges and Sierra Nevada to central California.

Although generally a butterfly of higher, moister habitats, in many places the Aster Checkerspot flies together with the Northern Checkerspot. The 2 species bear many similarities but the different shapes of the male genitalia justify their classification as separate species.

660 California Patch
(*Chlosyne californica*)

Description: 1¼–1⅝″ (32–41 mm). *Brown above with broad orange median bands,* small white dots between bands, and prominent orange spots at base and around rim. *Similar below,* but marginal spots are yellow; *FW base and cell*

orange; HW has yellow bar at base
instead of orange spots and *red spot near
abdomen, blending into yellow median
band.*

Similar Species: Orange-banded forms of Bordered
Patch have brown FW cell and less
prominent orange marginal spots.

Life Cycle: Egg yellow-green. Black or reddish-
orange caterpillar has many white dots
and black spines; feeds in groups on
golden-eye (*Viguiera deltoidea* var.
parishii) and sunflower (*Helianthus
annuus*). Caterpillar overwinters half
grown. Chrysalis variable, black or
white with black patches.

Flight: Several broods; March–November,
more common in spring.

Habitat: Scrubby desert gulches, canyons and
hills.

Range: Colorado and Mojave deserts of SE.
California, S. Nevada, and W. Arizona.

Adults congregate on hilltops and visit
desert flowers, sometimes in the
company of the Bordered Patch.
However, the California Patch is much
more consistent in appearance and more
limited in distribution. It is thought to
have evolved from the Bordered Patch.
Palm Springs and Joshua Tree National
Monument are noted localities of the
California Patch.

659, 675 Bordered Patch
(*Chlosyne lacinia*)

Description: 1⅜–1⅞″ (35–48 mm). FW black *above
with white or orange marginal dots,*
followed by row of white dots, *median
row of narrow to wide white or orange
patches,* and occasionally whitish or
orange spots at base. HW above similar
but middle row usually forms a broad
yellow and orange band or patch.
Below, FW similar to upperside; *HW
black with yellowish basal, median, and
marginal bands and red band or spot* near

corner beyond middle yellowish band.

Similar Species: California Patch has orange FW cell below, brighter orange border spots. Janais Patch has red, not orange, patch in HW disk.

Life Cycle: Eggs light green; laid in clusters on many plants of the composite family, especially sunflower (*Helianthus annuus*), giant ragweed (*Ambrosia trifida*), and cowpen daisy (*Verbesina encelioides*). Young caterpillars feed together, but disperse when overwintering half grown. Older caterpillar black, or orange and black striped, or orange. Chrysalis variable, white with black dots and streaks or nearly solid black.

Flight: Several broods; usually March–November.

Habitat: Subtropical thorn forests, desert hills, weedy edges of agricultural fields, river bottomlands, pinyon pine and oak woodlands, parks and gardens.

Range: SE. California east to Texas and south to Argentina, rarely emigrating north to Utah and Nebraska.

The Bordered Patch has been called our most variable butterfly, as well as the most widespread and abundant checkerspot in the Americas. The enormous range of variation in the appearance of early stages and adults makes this species a good subject for genetic research. This is enhanced by its large broods of up to 500 eggs per female and brief generation time of as little as 30 days.

578 Definite Patch
(*Chlosyne definita*)

Description: 1–1⅜″ (25–35 mm). *Small. Above,* black with *orange spot band* near margin, *cream-yellow median band* of spots, and orange and yellow spots at base. Below, FW orange with darker orange, black, and yellow lines and spots; HW red-orange *below with black-rimmed yellow marginal, median, and basal bands; yellow spot in middle of outermost orange area.*

Similar Species: Theona Checkerspot lacks cream-yellow spot in outermost orange band on HW below. Chinati Checkerspot is larger, with open, uncheckered pattern above, and much less orange on HW below.

Life Cycle: Host plant is acanthus (*Stenardrium barbatum*). Early stages unreported.

Flight: Successive broods; April–October.

Habitat: Desert thorn scrub and subtropical thorn forests.

Range: SE. New Mexico, and Texas; south into Mexico.

The Definite Patch can be quite common in the Gulf region of Texas. Adults fly just above the thorny vegetation, then perch on the ground with wings spread, and visit lantana and other flowers.

---

683 Janais Patch
(*Chlosyne janais*)

Description: 1¾–2″ (44–51 mm). *Above, FW black with white dots; HW black with large red patch* in middle of wing. *Below,* FW black with overlying white dots; *HW yellow with black marks* on basal half, followed by *red band not connected to margin,* black band with white dots, then marginal yellow crescents, and black border.

Similar Species: Bordered Patch has orange, not red, patch that reaches margin.

Life Cycle: Eggs clustered on acanthus shrubs (*Anisacanthus wrightii* and *Odontonema callistachus*). Caterpillar metallic gray-green with black lines, orange patches, and many black spines. Chrysalis gray-green with black lines and points.

Flight: Successive broods; most of year in Mexico, at least July–November in Texas.

Habitat: Subtropical wooded or scrub areas, especially along streams.

Range: Central America north to S. Texas, rarely wandering north to central Texas.

Like many butterflies from the American Tropics, the Janais Patch is occasionally abundant in the lower Rio Grande Valley of southern Texas. But cold winters can kill this species and necessitate periodic recolonization from Mexico. With its brilliant red disks, the Janais Patch clearly illustrates the common name of the "patch" butterflies. The Rosita Patch (*C. rosita*), another Rio Grande resident, is nearly identical but lacks the marginal spots above and below.

559, 572 Leanira Checkerspot
(*Thessalia leanira*)

Description: 1¼–1¾" (32–44 mm). *Above, northern* Coast Ranges and Sierra Nevada *populations black with cream-colored dots* in rows near margin and middle of wings, and in spots near wing bases; often *red-spotted near FW tip. Desert populations mostly orange above with cream-colored spots* and black veins and patches; FW cell orange. *Below,* FW yellow-spotted and orange. *HW cream-colored with or without black blotches near base* (but usually no "Y" or "V" in cell), *black veins, and heavy black chain near margin.*

Similar Species: Cyneas Checkerspot blackish above. Fulvia Checkerspot can usually be told by black "Y" or "V" in HW cell below, and male's dark brown cell.

Life Cycle: Clustered eggs yellow; laid on Indian paintbrush (*Castilleja*) and bird's beak (*Cordylanthus pilosus*). Groups of caterpillars overwinter half grown. Mature caterpillar orange with black bands and 7 rows of spines. White chrysalis has prominent black bands.

Flight: 1 brood; May–June, April–June in
S. California. Reported alternate
year flights may indicate 2 year life
cycle.

Habitat: Open oak woodlands, pine forests,
canyons, streamside hills, and chaparral
near coast; desert hills and basin flats
eastward.

Range: SW. Oregon south in Coast Ranges and
Sierra Nevada to Baja California; SE.
Oregon south and and east in Great
Basin to Arizona, E. Colorado, and
New Mexico.

Along the desert edge north of Los
Angeles, both blackish and orange color
forms occur. Biologists conjecture that
orange may be a more effective
camouflage in lighter desert areas;
numerous checkerspots, fritillaries, and
other butterflies are lighter in the
desert, especially in Nevada, and darker
in moister, more vegetated areas.

558   Fulvia Checkerspot
(*Thessalia fulvia*)

Description: 1⅛–1½" (28–38 mm). Above, veins
black, outer part of wings orange, with
median band of yellow streaks, a row of
yellow spots edged outwardly with
black near margin, orange marginal
spots, and a black line along margin;
*male black in cell and at base with cream-
colored spots, female orange with black
smeared patches. Below,* both sexes orange
with black and cream-colored marginal
bands on FW; *cream-colored HW has
black veins and black chain near margin
and usually a black "Y" or "V" in cell.*

Similar Species: Cyneas Checkerspot is mostly black
above in both sexes. Chinati
Checkerspot has orange spots in HW
chain below, not cream-colored. Leanira
Checkerspot male has orange FW cell,
usually lacks black "V" or "Y" in HW
cell below.

Life Cycle: Eggs cream-colored; laid side by side under leaves; half grown caterpillar overwinters. Mature caterpillar yellow with thick black bands and many spines. White chrysalis has black bars and dots. Host plants are Indian paintbrushes (*Castilleja integra, C. lanata*).

Flight: 2 or more broods; mainly May–September.

Habitat: Grassy hills and hilly open pinyon pine-juniper woodlands, usually on limey soils, scrub oak thickets, and moist canyons.

Range: Arizona east to S. Colorado, W. Kansas, and W. Texas.

Like other checkerspots, adult Fulvias cluster on hilltops. They are usually uncommon, and seldom wander far from Indian paintbrush. The males and females of this species are distinctly different looking, displaying greater sexual dimorphism than other members of this genus.

567 Cyneas Checkerspot
(*Thessalia cyneas*)

Description: 1⅛–1⅜″ (28–35 mm). *Black above with marginal red-orange spot band and 2 rows of cream-colored spots across disk. Below, FW orange with cream-colored spots; HW yellowish with black veins. Black chain enclosing white spots extends submarginally all along both wings below.* Female larger, more rounded, with more reddish.

Similar Species: Leanira and Fulvia checkerspots never as black above, have more and larger red and cream-colored spots, even in dark forms.

Life Cycle: Early stages unreported. Host plants in the figwort family (Scrophulariaceae), *Seymeria tenuisecta* in Mexico.

Flight: Probably 2 or more broods; April–October.

Habitat: Open oak, pinyon, and juniper
woodlands in mountains.

Range: SE. Arizona and Mexico.

The Cyneas Checkerspot just barely
enters our area in the southeastern
Arizona mountains. Sedentary and
never migratory, it flies slowly among
shrubs and low trees and visits flowers.
Some specialists think that the Leanira,
Fulvia, and Cyneas checkerspots are all
the same species, each confined to a
separate range.

583 Theona Checkerspot
(*Thessalia theona*)

Description: 1–1⅝" (25–49 mm). *Above, banded
with light and dark orange* or orange and
cream-color, with black base lines, and
border. Black-veined *below, sharply
banded with bright orange and shiny white
or cream-color.* Innermost orange band
has *small cream-colored spot in cell.*

Similar Species: Most other checkerspots have HW
bands beneath with less contrast
between orange and pale color. Cyneas
Checkerspot black above, lacks orange
bands beneath. Chinati Checkerspot
clearer orange above, mostly cream-
colored on HW below with narrow
orange bands.

Life Cycle: Eggs cream-colored; clustered on Indian
paintbrush (*Castilleja lanata*), cenizo
(*Leucophyllum frutescens*), and vervain
(*Verbena*). Mature caterpillar brownish-
black with cream-colored dots and
bands and many branched spines,
pinkish below; overwinters half grown.
Chrysalis smooth, white, black with
orange stripes.

Flight: Successive broods; April–October in
U.S., most of year in Mexico.

Habitat: Subtropical thorn forests in desert
foothills, open oak or pinyon-pine
woodlands; canyon streamsides, and
plains.

Range: S. Arizona east to central Texas, and
south through Mexico into Central
America.

Two distinct populations of the Theona
Checkerspot enter the Southwest from
Mexico, one via the Gulf Coast and the
other along the Pacific, giving us 2
extremes of a highly variable Mexican
butterfly. Big Bend National Park is a
good place to see this brightly banded
checkerspot.

Chinati Checkerspot
(*Thessalia chinatiensis*)

Description: 1⅛–1⅝″ (28–41 mm). *Clear orange
above* with irregular black patches at
base, black veins, and white-spotted
black border. *Below,* FW orange with
cream-colored marginal band; *cream-
yellow HW has black veins, a few orange
spots at base, and orange-red spots within a
black-rimmed chainlike band* near
margin.

Similar Species: Cyneas Checkerspot blackish above, has
cream-colored spots in chain below.
Theona Checkerspot has much heavier
marking above, more reddish below.
Fulvia lacks orange band on HW
below.

Life Cycle: Host plant is Big Bend silver-leaf
(*Leucophyllum minus*). Mature caterpillar
brown with cream-colored dots and
black branching spines; overwinters half
grown. Chrysalis white with black and
orange stripes.

Flight: Several broods; June–October, possibly
longer depending on rainfall.

Habitat: Desert with many thorny shrubs,
especially lecheguilla and palo verde.

Range: W. Texas.

Adults can be found in the Chinati
Mountains of Texas, but are normally
scarce. They fly just above the ground
or over thorn bushes on desert hills

and washes. While the species may
occur in Mexico, it has not yet been
found there.

677   **Elf**
(*Microtia elva*)

Description: ¾–1⅜″ (19–35 mm). *Small. Coal-black
above and below except for 2 orange bars.*
1 bar crosses FW tip below, the other
begins on FW and continues across
middle of HW. Below, similar to
upperside.

Similar Species: Pixie has orange FW tips rather than
bars, red spots near body and along
HW margin.

Life Cycle: Unknown.

Flight: August in U.S.; most of year in
Mexico.

Habitat: Open tropical woodlands and shrubby
areas.

Range: Central America, most of lowland
Mexico; very rarely wandering north to
S. Texas and S. Arizona.

In Mexico, Elves fly about slowly and
low to the ground in semishaded, dry
woods. It is surprising that any manage
to fly to the United States; perhaps they
enter along with hurricanes or are
accidentally carried north on cars. The
adults take nectar from certain borages
and composites.

Dymas Checkerspot
(*Dymasia dymas*)

Description: ⅞–1⅛″ (22–28 mm). *Small. Orange
above with heavy, suffused, fine black lines
and whitish bar* of tiny white streaks on
FW costa. FW below orange with black
lines and with marginal bands of orange
and white. *HW below has alternating
orange and white bands* and black lines;
outermost spot band is white.

Similar Species: Chara Checkerspot has white bar on FW costa, extending into pale band across FW. Elada Checkerspot has outermost spot band orange on HW below.

Life Cycle: Eggs cream-colored; laid in clusters on tube-tongue (*Siphonoglossa pilosella*). Early stages otherwise unreported.

Flight: Multiple broods; generally February–November, determined by rain.

Habitat: Subtropical thorn forest with mesquite and other trees; also thorny deserts.

Range: S. New Mexico east to SE. Texas.

The smallest checkerspot, smaller even than the crescentspots, the Dymas flies rather weakly near the ground. Although radically different in appearance from the Elf, the genitalic structure denotes a close relationship.

603 Chara Checkerspot
(*Dymasia chara*)

Description: ¾–1⅛" (19–28 mm). *Small.* Above, predominantly orange with several rows of distinct black bars across wings, and *whitish bar on FW costa* leading into paler band. Sexes similar. *Below,* FW orange with black lines and whitish and orange marginal bands; *HW has alternating orange and white spot bands,* the outermost band white.

Similar Species: Dymas Checkerspot female browner; both sexes have blurrier black markings above, and white bar more closely limited to FW costa above. Elada Checkerspot has outermost spot band orange on HW below.

Life Cycle: Eggs cream-colored; laid in clusters on chuparosa (*Beloperone californica*). Caterpillar gray, black, and white, mottled with black spines. Chrysalis whitish-gray with dark specks.

Flight: 2 or more broods; spring and fall following rains, usually early March and late October; varies locally.

Habitat: Deserts with mesquite, saguaro cactus,
and other desert trees, as well as open
deserts and mountain canyons.

Range: SE. California east to SW. New
Mexico, and south into Mexico.

Chara Checkerspots fly slowly and often
visit flowers after rains; males patrol
gulches and hillsides, seeking females.
Arizona populations have slightly
darker borders on the fore wing above,
and the white spots below are more
cream-colored than those of California
populations.

579 Elada Checkerspot
(*Texola elada*)

Description: ⅞–1⅛″ (21–28 mm). *Small. Above,
orange with black borders and 3 bands of
black dashes;* some black dashes on wing
bases, seldom any white on FW costa.
Below, FW orange with fainter black
dashes, orange margin, white band just
inside, and black-edged orange band
further inside. *HW beneath has*
alternating orange and white spot
bands, the *outermost band orange-red.*

Similar Species: Dymas and Chara checkerspots have
white outermost spot bands on HW
below. Vesta Crescentspot larger, lacks
orange and white bands below.

Life Cycle: Host plants are composites (Asteraceae)
and acanthus (Acanthaceae).

Flight: Successive broods; April–October.

Habitat: Desert thorn scrub or shrub savannah
with mesquite, saguaro, and lecheguilla.

Range: Mexico to S. Arizona and central and
S. Texas.

Elada Checkerspots fly together with
the smaller Dymas Checkerspots. Like
other externally similar butterflies,
these look-alikes give off sexual
attractants, called pheromones, which
may help each species to recognize its
own kind.

580 Arachne Checkerspot
(*Poladryas arachne*)

Description: 1¼–1⅝" (32–41 mm). Above, tawny-orange with many black lines along veins and in bands, and black border with white fringe checkered with black; wing bases dark brown. *Below,* FW similar to upperside with cream-white spots at tip; *HW crossed by 3 orange and 3 cream-colored bands, middle cream-colored band has row of black dots* and marginal cream-colored band is edged by *1 black line just inside fringe.*

Similar Species: Dotted Checkerspot has 2 black lines inside fringe on HW below. No others have black dots on cream-colored band below.

Life Cycle: Clustered yellowish eggs laid on underside of penstemon leaves (*Penstemon*). Tiny caterpillars feed together and overwinter half grown. Mature caterpillar white with orange spots near back and many black spines. Chrysalis white with black patches and orange points.

Flight: 2 or more broods; mainly June–September.

Habitat: Foothills and mountains on grassy slopes, usually among open pine or fir forests; oak thicket canyons.

Range: S. Sierra Nevada of California northeast to central Wyoming and NW. Nebraska, south through Arizona and New Mexico well into Mexico.

The pattern on this checkerspot resembles a spider's web, inspiring its scientific name *arachne.* According to Greek mythology, Arachne was a proud weaver whom the jealous Athena turned into a spider. Although the adult butterflies seldom move more than a few hundred yards from where they emerge, they disperse quickly and are seldom seen in large numbers. They take nectar from various yellow flowers, such as ragwort, golden aster, and snakeweed.

Dotted Checkerspot
(*Poladryas minuta*)

Description: 1¼–1⅝" (32–41 mm). Above, orange
with black border and many narrow
lines of black dashes. *Below,* FW
mottled orange with marginal white
crescents; HW has alternating,
somewhat run-together red-orange and
*cream-colored bands, the broad median
cream-colored band bears 1 or 2 rows of
black dots; 2 black lines parallel margin*
just inward from black-checked white
fringe.

Similar Species: Arachne Checkerspot has just 1 black
marginal line on HW below, and
greater contrast in its color pattern.

Life Cycle: Clustered yellowish, ribbed eggs.
Caterpillar overwinters half grown;
mature caterpillar orange with orange
spines on back and many black spines.
Chrysalis white with orange bumps and
black patches. Host plant is
beardtongue (*Penstemon*).

Flight: Several broods; April–September.

Habitat: Usually limestone ridges on grasslands
or open mesquite, oak, and juniper
woodlands.

Range: E. New Mexico, N. and central Texas
south into Mexico.

Both the Dotted and Arachne
checkerspots have black dots on their
hind wings below, suggesting that they
are related. Just how closely, however,
is not certain, but hybrids have been
produced, and a few specialists consider
them to be the same species. These
checkerspots are most often found at
alkaline sites where beardtongue grows.

11, 21, 563   **Baltimore**
(*Euphydryas phaeton*)

Description: 1⅝–2½" (41–64 mm). *Above, black
with numerous cream-colored dots,* several
red-orange spots near base of wings,

and *border of red-orange half moons.*
Similar below with more cream-color
and orange. New England individuals
smallest, with wide red border; Ozark
region individuals largest, with
narrower red border.

Similar Species: Harris' Checkerspot smaller, shorter
wings, with lighter orange.

Life Cycle: Eggs $\frac{1}{32}''$ h $\times$ $\frac{3}{128}''$ w (0.8 $\times$ 0.6 mm)
laid in clusters. Young caterpillar feeds
in silk nests and overwinters half
grown. Mature caterpillar, to 1″ (25
mm), black with orange side stripes
and many black branching spines.
Chrysalis to ¾″ (19 mm), white, black,
and orange; adult emerges after only 10
days. Host plants are turtlehead
(*Chelone glabra*), false foxglove (*Gerardia
grandiflora* and *G. pedicularia*), plantain
(*Plantago lanceolata*), and white ash
(*Fraxinus americana*).

Flight: 1 brood; May–July.

Habitat: Wet meadows in woodlands in the
Northeast; sphagnum bogs in Lake
States; hillsides and drier ridges in open
mixed hardwoods in Ozarks.

Range: SE. Manitoba to Nova Scotia south to
Nebraska, Arkansas, and Georgia.

While often seen in damp turtlehead
stands in the Northeast, the Baltimore
has been found to have many other host
plants and habitats. The caterpillars of
Connecticut colonies, for example, feed
on false foxglove among rocky upland
oak woods. The Baltimore is named for
George Calvert, a 17th-century
American colonist and the first Lord
Baltimore, because its orange and black
colors match those on his heraldic
shield.

22, 557, 571 **Chalcedon Checkerspot**
(*Euphydryas chalcedona*)

Description: 1⅜–2″ (35–51 mm). FW somewhat
drawn out. *Above, black with numerous*

cream-colored dots and red-orange spots on margin. (Some populations mostly red-orange with yellow spots above; *others are black with large yellowish spots* above, appearing more yellow than black or red). *Below,* reddish-orange FW has yellow spots in 1 or 2 submarginal bands; HW has alternating red and *cream-yellow bands,* ending with red border.

Similar Species: Colon Checkerspot tends to have smaller cream-colored spots along margin above. Anicia and Edith's checkerspots usually more orange or cream-colored above.

Life Cycle: Host plants include many members of figwort family (Scrophulariaceae) as well as plantain (*Plantago*), honeysuckle (*Lonicera*), snowberry (*Symphoricarpos albus*), and others. Black caterpillar, to 1″ (25 mm), has speckles or stripes and many spines, those on back and side orange; feeds in clusters in silken nest, overwinters half grown. White chrysalis, to ¾″ (19 mm), has blackish spots and orange tubercles.

Flight: 1 brood; May–July depending on altitude. Several broods in W. Arizona; April–October.

Habitat: Desert hills, chaparral, open oak and pine woodlands, clearings, roadsides, and alpine tundra.

Range: SW. Oregon, California, W. and S. Nevada, and W. Arizona, probably also Baja California.

The Chalcedon, Colon, Anicia and Edith's checkerspots display enormous variation from place to place and often occur together. They are easily confused and sometimes only experts can distinguish the 4 species. The Chalcedon is one of California's most abundant butterflies. This species and related checkerspots were formerly included in the genus *Euphydryas.*

568 Colon Checkerspot
"Snowberry Checkerspot"
(*Euphydryas colon*)

Description: 1⅜–1⅞" (35–48 mm). *Above, black with yellow dots and red margin; HW marginal row of yellow dots smaller* than basal and median spot rows. Below, reddish FW has marginal yellow spots; HW has bands of red and yellow, the outermost red.

Similar Species: Chalcedon Checkerspot has larger cream-colored marginal spots on HW above. Anicia and Edith's checkerspots have more orange or cream-color above.

Life Cycle: Host plants are snowberry (*Symphoricarpos albus, S. vaccinoides*), beardtongue (*Penstemon antirrhinoides* and *P. subserratus*), and common plantain (*Plantago major*). Caterpillar feeds in compact silken nest before overwintering half grown in leaves; mature caterpillar black with white, yellow, or orange areas, long white hair, and many black spines (spines on back and side are orange).

Flight: 1 brood; May–June.

Habitat: Clearings and canyons among evergreen mountains.

Range: British Columbia and W. Montana, south to N. California, NE. Nevada, and NW. Utah.

Where the Colon and Chalcedon checkerspots meet in southern Oregon, it is difficult to tell them apart. For this reason, some specialists consider them the same species. An isolated colony of Colon Checkerspots occupied lush gullies on Mt. Saint Helens in Washington, but was wiped out when the volcano erupted in 1980.

556, 562 **Anicia Checkerspot**
"Paintbrush Checkerspot"
(*Euphydryas anicia*)

Description: 1⅛–1⅞" (28–47 mm). FW usually rather long. *Above,* black with numerous orange or cream-colored spots; *orange* predominates in most places, *black* at high altitudes, and from Nebraska to Great Basin *cream-color and black. Below,* FW orange with cream-colored bands especially toward edge; *HW orange with cream-colored or white bands* of black-rimmed spots.

Similar Species: Chalcedon and Colon checkerspots best separated by locality, both usually blacker. Edith's Checkerspot is never mostly black or cream-colored; has shorter, rounder wings.

Life Cycle: Host plants are Indian paintbrush (*Castilleja*), beardtongue (*Penstemon*), and other plants in several families: figworts (Scrophulariaceae), plantain (Plantaginaceae), and borage (Boraginaceae). Caterpillar feeds in a group in silken nest; overwinters half grown. Mature caterpillar white with black stripes or black with white dots or stripes; has many black branching spines, the back and side spines orange. Chrysalis white with black spots and orange tubercles.

Flight: 1 brood; April–August, latest at high altitudes.

Habitat: Alpine tundra and ridges, pine forests, grasslands with aspen groves, mountain meadows, mountain mahogany chaparral, and sagelands.

Range: W. North America between Cascades, Sierra Nevada, and Great Plains, and from Alaska to Mexico.

Over its vast range, the Anicia Checkerspot is one of the most variable butterflies. At different elevations in the same mountains or even the same locality, these butterflies can be quite dissimilar. Furthermore, specialists disagree about how many species are actually represented by the Chalcedon, Colon, and Anicia. Whatever the name, the caterpillars of all 3 species sometimes defoliate their host plants, leaving large, white silken webs.

561, 565 Edith's Checkerspot
(*Euphydryas editha*)

Description: 1⅛–1⅞" (28–48 mm). Wings somewhat short and rounded. *Above, orange with black and yellowish spot bands, giving checkered appearance;* border orange. Some populations red-orange, others yellowish-orange. Below, FW orange with black bars and cream-colored spots, HW banded with orange and whitish spots.

Similar Species: Chalcedon, Colon, and Anicia checkerspots larger, with more pointed FW; Chalcedon and Colon checkerspots usually more blackish. Edith's Checkerspot does not have mainly black or yellow forms and is not found in desert.

Life Cycle: Host plants include members of Figwort family (Scrophulariaceae), such as Indian paintbrush (*Castilleja*), owl's clover (*Orthocarpus*), and lousewort (*Pedicularis*), as well as plantain (*Plantago*), and others. Clustered eggs laid on leaves or flowers; young caterpillars eat together before overwintering half grown. Mature caterpillar black with orange or white lines and speckles and many black branching bristles, those on back and side often orange at base. White chrysalis has black and orange markings.

Flight: 1 brood; March–April on coast, June in Great Basin, to August above treeline.

Habitat: Coastal bluffs and chaparral, evergreen forests, sagebrush hills, alpine tundra, often on ridge tops; well-drained lowland grasslands.

Range: S. British Columbia and Alberta south to Baja California and SW. Colorado.

Edith's Checkerspot, like other members of its genus, is a highly variable species. Some populations differ from one another in food choice, degree of mobility, and behavior such as where the females lay eggs. The availability of host plants during the hot summer has a great effect on their activity. A population on Jasper Ridge, California, has formed the basis for some of the most extensive butterfly research ever performed in North America; this colony is now threatened by development.

564 Gillette's Checkerspot
"Yellowstone Checkerspot"
(*Euphydryas gillettii*)

Description: 1⅜–1¾" (35–44 mm). FW somewhat short and rounded. Above and below, *brown with bands of yellowish spots* and a few *red-orange spots* in basal ⅔; *broad red-orange submarginal band* edged outwardly with brown.

Similar Species: No other checkerspot has broad red submarginal band; others tend to be mottled with brown, yellow, and red.

Life Cycle: Host plants are honeysuckle (*Lonicera involucrata*) and snowberry (*Symphoricarpos albus*). Eggs laid in clusters under leaves. Caterpillars feed in groups, then overwinter when nearly full grown. Mature caterpillar brown with yellow and white stripes, and many yellow, branching spines.

Flight: 1 brood; June–July.

Habitat: Moist clearings among lodgepole pine forests and mountain meadows.

Range: Rocky Mountains in S. Alberta, Montana, Central Idaho, and W. Wyoming.

With its striking red bands, this checkerspot is one of the only western species that is easy to recognize.

Gillette's Checkerspot, in common
with Hayden's Ringlet and the
Kriemhild Fritillary, is found only in
the northern Rockies, and has closer
relatives in Asia and Europe than in
America. Formerly included in the
genus *Euphydryas.*

9, 15, 373, 374 **Question Mark**
(*Polygonia interrogationis*)

Description: 2⅜–2⅝″ (60–67 mm). Wing margins
ragged, *HW distinctly tailed.* Above,
bright rust-orange blotched with black;
*in fall and spring; margin lined with
violet;* summer brood black over HW
above. Below, violet- or red-brown, in
fall and spring; in summer brown
mottled with maroon and blue. *Silvery
comma in cell of HW beneath has an offset
dot,* forming a question mark.

Similar Species: No other anglewing has pronounced
tails or dot opposite silver comma.

Life Cycle: Egg, $\frac{5}{128}$″ h × $\frac{7}{256}$″ w (1 × 0.7 mm),
light green, keg-shaped, and ribbed;
may be laid atop other eggs in vertical
columns, in horizontal strings, or
singly on lower side of leaves of host
plants, which include nettles
(Urticaceae), hops (*Humulus*), elms
(*Ulmus*), hackberries (*Celtis*), and related
trees. Caterpillar, to 1⅝″ (41 mm),
buff to rust-colored; when mature,
bears pair of black, branched spines on
head and several on each segment.
Chrysalis, to ⅞″ (22 mm), gray-brown
with olive mottling; hangs down like a
shriveled leaf. Adults overwinter and
reappear in April.

Flight: 2 broods in North, 5 in South; spring–
fall.

Habitat: Woodland glades, roads, and other
sunny openings; also orchards and
streamsides.

Range: East of Rockies from Saskatchewan to
Texas and Mexico and east to Maritimes
and Florida.

Confusingly similar, all of the anglewing butterflies of the genus *Polygonia* have ragged wing borders. None extends as far south as the Question Mark, which is absent from the West. Like the other anglewings, the adult Question Mark loves sap and rotting fruit. Normally highly alert, anglewings can actually become intoxicated if the fruit they are drinking has fermented in the sun.

371, 375   **Comma**
"Hop Merchant"
(*Polygonia comma*)

Description: 1¾–2" (44–51 mm). Wing margins ragged, *HW has short tail.* Rust-brown above, with black blotches. HW above has broad dark margin, spotted with yellow in fall and spring broods; summer brood HW is suffused with blackish-brown above. *Borders above may be violet in fall forms.* Underside of fall brood male patterned brown, female dull brown; summer brood golden-brown. *All forms below have silver comma mark usually clubbed or hooked* at both ends.

Similar Species: Other anglewings extremely similar; see individual descriptions. Question Mark has silver dot opposite comma and pronounced HW tails.

Life Cycle: Egg pale green, keg-shaped, ribbed; 2–9 deposited in vertical columns. Caterpillar, to 1" (25 mm), light green to brown with complex spines along length. Chrysalis brown with a few gold or silver spots along sides; curved, irregular shape resembles a bit of twisted wood. Hops (*Humulus lupulus*) and nettles (*Urtica dioica, Boehmeria cylindrica*) are preferred host plants; also elms (*Ulmus*). Adults overwinter, emerging in spring.

Flight: 2 broods in North, 3 farther south; March or April until cold nights.

Habitat: Clearings in edges of thickets, groves, and forests; also watercourses and moist, open woods.

Range: Saskatchewan to E. Colorado, east to Maritimes and North Carolina, and south to Mississippi and central Georgia.

Like the other anglewings, the Comma is wary. It darts rapidly and frequently at other butterflies, birds, and people, but when challenged retreats to the woods and perches upside down, camouflaged on a tree trunk. In the sun, it alights on sand, gravel, or mud to bask or drink, usually with its wings closed above its back, but sometimes spreading or opening and closing its wings with a regular rhythm.

372 Satyr Anglewing
(*Polygonia satyrus*)

Description: 1¾–2" (44–51 mm). Wing margins ragged. *Bright tawny-golden above* with black blotches. *HW above lacks strong dark margin,* having only band of golden spots narrowly lined with brown. *Beneath, male bright yellow-tan marked with tiny dots and striations; female brown, slightly violet-tinged,* darker on basal half. *Silver comma mark in HW below clubbed or hooked.*

Similar Species: Comma has dark, spotted margin above; Satyr paler and tannish below. Other western anglewings are gray beneath.

Life Cycle: Caterpillar has greenish-white stripe and chevrons on back, branched spines along blackish segments; feeds on stinging nettles (*Urtica*), drawing leaf edges around itself for shelter. Chrysalis tan, angled; hangs from tree trunks or stones. Adult overwinters.

Flight: 2 or more broods; early spring–late autumn.

Habitat: Wooded canyons, streamsides, and

canals; also foothills, northern forest edges, and wooded city parks.

Range: British Columbia to Newfoundland in S. Canada; in U.S. from the Pacific to eastern edge of Rockies, south to Mexico.

Most anglewings look rust-orange as they flash by; the Satyr is golden. Its broods are not so strikingly different from each other as are those of the Question Mark and the Comma. The Satyr roughly replaces the Comma as the common anglewing of the West; it is not entirely clear just how the 2 species relate in eastern Canada. The Satyr Anglewing is not to be confused with the satyrs of the family Satyridae; fanciful lepidopterists dubbed them both satyrs because they inhabit woodland glades. Other mythological rustic deities are commemorated in this group of anglewings, including Faunus, Zephyr, and Oreas, as well as the nymphs, for whom the family Nymphalidae is named.

378 Faunus Anglewing
"Green Comma"
(*Polygonia faunus*)

Description: 1⅞–2" (48–51 mm). Wing margins irregular and ragged. Rich dark russet above. *HW above has broad, dark margin, crossed by a number of small yellow-orange spots,* and edge may be narrowly purplish-gray. *Fresh individuals have greenish sheen mostly on margin above.* 2-toned and striated brownish-gray beneath, *with 2 rows of blue-green bars and chevrons along outer part* usually more pronounced than in other species. *Silver-blue comma mark on HW below.*

Similar Species: Other anglewings generally lighter above. Zephyr lacks dark margins; others lack green sheen and/or blue-green marks.

Life Cycle: Egg pale green. Solitary caterpillar, to
1¼" (32 mm), tan with whitish patches
and spines; feeds on birch (*Betula*),
alder (*Alnus*), willow (*Salix*), and
currant (*Ribes*), perhaps wild
rhododendron and azalea (*Rhododendron*)
in West. Chrysalis tan with dusky
green streaks and metallic spots. Adult
overwinters, sometimes emerging on
warm midwinter days.

Flight: 1 brood; March–April or May at higher
altitudes well into autumn.

Habitat: Sunny glades, roadsides, and especially
coniferous woods.

Range: British Columbia south to central
California, east through Northwest
Territories and Montana to Hudson
Bay, Iowa, and Adirondacks, south in
Appalachians to N. Georgia.

More confined to forests than the
Comma, Satyr, or Question Mark, the
Faunus seeks out sunny glades for
basking and feeding. This species varies
considerably within its range—smaller
and grayer in the Northwest
Territories, medium-sized and brighter
in the Pacific Northwest, it is large and
dark in the Appalachians.

Colorado Anglewing
(*Polygonia hylas*)

Description: 1¾–2" (44–51 mm). *Wing margins very
ragged.* Bright rust-colored above with
heavy black spots and broad dark
margin spotted with yellow on HW.
*Underside mottled and striated with various
tones of gray,* sometimes with vague
blue-green spotting near edge; female
usually poorly marked gray. *Silver
comma mark on HW clubbed or hooked.*

Similar Species: Faunus is darker with green sheen
above; ranges may not overlap. Zephyr
lacks dark margin above.

Life Cycle: Unreported.

Flight: April–September, where snow permits.

Habitat: At 6500–9000' (1983–2750 m) beside aspen-lined or brushy streams, in canyons, and montane valleys.

Range: W. Nebraska, SE. Wyoming, and montane Colorado into New Mexico and NE. Arizona.

One of the most restricted and least known anglewings, the Colorado Anglewing is heavily centered in its namesake state, and may be a southern Rockies subspecies of the Faunus. Most of the anglewings occupy relatively northern latitudes, while this species exploits boreal habitats in mountains farther south.

14, 379 Sylvan Anglewing
(*Polygonia silvius*)

Description: 2–2⅜" (51–60 mm). Wing margins ragged. Rust-orange with light-dotted, *dark margin above. Purplish-brown below; browner toward base, grayer toward margin.* Pale steel blue-gray marks along margin *below, silver comma knobbed or barbed.*

Similar Species: Faunus has green sheen above, is brownish-gray below with stronger blue markings.

Life Cycle: Egg light green, keg-shaped. Caterpillar, to 1¼" (32 mm), blackish-brown, has rust-colored and white bands and many spines. Chrysalis, to ⅞" (22 mm), tan mottled greenish-brown; has 2 head-horns and other hornlike protuberances along body. Host plant is western azalea (*Rhododendron occidentale*).

Flight: Late spring–early autumn.

Habitat: Streamsides and azalea glades in forests and foothills.

Range: N. central California.

The identity of this butterfly has puzzled lepidopterists. Although it is allied to the Faunus, the Sylvan's purplish-brown underside as well as its life cycle is quite distinctive. The species name refers to the wooded settings in which the butterfly dwells.

370 Zephyr Anglewing
(*Polygonia zephyrus*)

Description: 1¾–2″ (44–51 mm). Wing margins ragged. *Above,* light russet to golden-orange with fairly light black spotting and *border tawny or narrowly brown interrupted by large tawny crescents. Cool gray beneath,* 2-toned with darker patch down middle, lightly striated, with faint light blue submarginal markings. *Silver comma tapered at both ends* (not clubbed or hooked).

Similar Species: Satyr is brown or light tan beneath. Faunus and Colorado anglewings have broad dark borders above and clubbed commas below.

Life Cycle: Mature caterpillar, to 1½″ (38 mm), variable, black with reddish spines in front, white spines behind, projecting out of light patches with black chevrons. Host plants are squaw currant (*Ribes cereum*) and other currant species (*Ribes*), elms (*Ulmus*), and rhododendrons (*Rhododendron*). Chrysalis slender, marbled with salmon and olive and studded with small silver dots; bears beaklike projection from underside of head and thorny projections from top of head and along thorax. Adult develops in about 2 weeks and overwinters.

Flight: Number of broods uncertain; March well into autumn.

Habitat: In and around mountain forests along streams, roads, and trails; also in meadows, ranging from above timberline to plains along wooded watercourses.

Range: British Columbia to S. California in the Cascades and Sierra Nevada, east to Manitoba and New Mexico along E. Rockies.

The commonest anglewing of the western mountains, the Zephyr is well named, for it darts about with great speed. When frightened, it heads for the forest but usually returns again and again to the same spot after a moment or two. Unlike most anglewings, the Zephyr avidly visits asters, rabbit brush, and other meadow and streamside flowers. Such visits are among the few times it sits still for long.

368 Hoary Comma
(*Polygonia gracilis*)

Description: 1⅜–1⅝" (35–41 mm). *Relatively small. Wing margins very ragged; tails quite prominent.* Dark rust-colored above with black spots and very heavy dark border; lined with light spots on inner edge. *Below strikingly 2-toned: inner half dark gray-brown and striated; outer half frosty-white or hoary,* darkening toward edges. *Silver comma on HW below tapered at both ends.*

Similar Species: No other anglewing has hoary outer part below.

Life Cycle: Unreported. Adults overwinter.

Flight: 1 brood; July well into autumn, emerging again in spring.

Habitat: Forests, clearings, and rivers; also granite balds and mountain summits in Northeast.

Range: Alaska and Yukon east across N. Canada, south to Great Lakes, N. New England, and Adirondacks.

The relatively diminutive Hoary Comma has a vast northern range but it appears rarely and in small numbers. Only in the northeastern states does it occur somewhat regularly yet, even in Maine, it can be entirely absent for years. This little species is less fleet than other anglewings except for the Gray Comma.

377 Oreas Anglewing
(*Polygonia oreas*)

Description: 1⅝–2″ (41–51 mm). Wing margins
very ragged, tails fairly prominent.
Above, dark reddish-tawny and black-
blotched, *with broad dark borders* lined
by golden spots. *Beneath, California race
is 2-toned brown and striated; northwestern
race usually nearly uniform black or may be
2-toned* dark brown and gray-brown, or
black and brown. *Silver comma on HW
below is right-angled and tapered to point*
rather than clubbed or hooked.

Similar Species: Faunus lacks green sheen. Oreas has
dark upperside and brown underside
with distinctive, tapered comma.

Life Cycle: Unreported. Straggly gooseberry (*Ribes
divaricatum*) is host plant in California,
perhaps other currants (*Ribes*) and
rhododendrons (*Rhododendron*) in the
Northwest. Adults probably
overwinter.

Flight: Early spring–autumn.

Habitat: Canyons and streams in coastal redwood
forests in California; coastal and
mountain forests up to alpine meadows
in Northwest.

Range: British Columbia south through Coast
Ranges and Cascades of Washington
and Oregon to California, where
confined to redwood belt south to Santa
Cruz County; also east into Idaho,
Montana and NE. Wyoming.

The 2 populations of this unusual
butterfly—California coastal and
Northwest forest and montane—differ
greatly in appearance and habitat.
However, intermediate forms occur
through much of the range and the
habitats blend in southern Oregon. The
northwestern variety seldom appears;
perhaps its present-day scarcity reflects
the decades of intensive logging in its
coniferous forest habitats, or Oreas may
simply occur only sparsely in nature.
One of the few butterflies resident in
the western Washington rain forest, its

darkness in the Northwest has been attributed by some to the effects of moisture. The California race frequents coastal redwood forests.

---

376 Gray Comma
(*Polygonia progne*)

Description: 1⅝–1⅞″ (41–48 mm). Wing margins ragged. *2 seasonal forms:* fall and spring brood tawny-orange above with yellow-spotted dark borders and relatively little black spotting. Summer butterflies have very dark chocolate-brown HW, above with tawny base, yellow spots, and heavier black spotting. *Gray-brown beneath,* more uniform but heavily striated in summer, more 2-toned in fall–spring. *Silver comma on HW below L-shaped, 1 arm shorter in summer form; both arms tapering to points.*

Similar Species: Roughly similar to all other anglewings. Faunus has green sheen and clubbed comma. Question Mark has purple tails, which dark form Gray Comma lacks. Dark form Comma is brown beneath.

Life Cycle: Egg, to ³⁄₆₄″ (1.2 mm), green and ribbed; laid singly. Caterpillar, to 1⅛″ (28 mm), of variable color: tan or rust, marbled with dull green and bearing short, branched spines along back and on head. Chrysalis tan to brown with dark streaks. Currants (*Ribes*) are host plants. Fall and spring adults overwinter.

Flight: 2 broods; April–October.

Habitat: Woods, especially deciduous; also along roads and trails and in clearings, such as campsites and homesteads.

Range: British Columbia, Wyoming, and Kansas east to Nova Scotia, Missouri, and North Carolina; possibly Alaska.

Like its relatives, the Gray Comma is fond of sap, rotting fruit, carrion, and scat, but it also takes nectar from flowers somewhat more often than most anglewings, except for the Zephyr. When the Gray Comma perches on tree trunks with its wings folded, its ashen, finely lined underside blends superbly with the weathered wood. This species flies more slowly than the other anglewings.

369 Compton Tortoiseshell
(*Nymphalis vau-album*)

Description: 2½–2⅞" (64–73 mm). Broad, ragged wings. *Above, rich rust-brown at base, blending into yellow-gold crescents* near margin, with heavy black spotting on FW; *HW costa has 1 big black spot bordered on outside edge by frost-white patch.* Beneath, various shades of gray, brown, and tan, striated and dotted, usually darker on inner half. Pale blue-gray crescents often line outer margin; small silver *"V" or comma mark in HW cell below.* FW trailing edge straight.

Similar Species: Anglewings are smaller and narrow-winged, with more ragged borders, and lack white spot above; trailing FW margin concave (not straight). California Tortoiseshell is smaller, lighter, and less spotted; lacks silver "V".

Life Cycle: Eggs laid in clusters. Caterpillar pale green, chartreuse-speckled with branched black spines; feeds communally on birches (*Betula*), willows (*Salix*), and perhaps poplars (*Populus*). Brown chrysalis angled, with horned head; often hangs from wood. Overwintering adults lay eggs in spring for new brood in June or July.

Flight: 1 brood; any month.

Habitat: Clearings and tracks in dense deciduous woodlands; also canyons and rivercourses.

Range: Subarctic Alaska and Canada south to Oregon, Colorado, Minnesota, and mountains of Missouri and North Carolina.

The *"vau-album"* of the scientific name refers to the whitish-silver "V" below, which places this butterfly close to the anglewings, although in every other respect it is clearly a tortoiseshell. It is occasionally numerous when conditions are right, clustering around fallen fruit, sap, or wet earth. But this big butterfly is notorious for its unpredictability; for years at a time it may be absent from a given district.

367   California Tortoiseshell
(*Nymphalis californica*)

Description: 1⅞–2⅜" (48–60 mm). Wing margins ragged. *Above, rich russet with black blotches on FW,* small white spots on FW costa, *black and white pair of spots on HW costa,* margin dark often with hint of blue. *Below, dull barklike brown,* lighter outwardly, contrasting with darker base; margin lined with deep blue marks. FW trailing edge straight.

Similar Species: Compton Tortoiseshell larger, has silver "V" on HW beneath, and heavier black spotting above. Anglewings have narrower, more ragged wings and concave trailing FW edge.

Life Cycle: Caterpillar velvety black with yellow and blue-black spines and white specks. Chrysalis gray and tan, may have blue cast; horned head. Various buckthorns (*Ceanothus*) are host plants; probably also other plants.

Flight: 1 brood in North, 2 or 3 farther south; adults may be encountered year-round.

Habitat: Mountainous terrain below subalpine zone, especially canyons; also lowland forest edges, glades, and parklands.

Range: British Columbia south to S. California, and east in mountains to Saskatchewan, Nebraska, and New

Mexico; very rarely wandering or introduced into Midwest and Northeast.

Some biologists consider this the same species as the European Large Tortoiseshell (*N. polychloros*), which feeds on elms (*Ulmus*). Both appear irregularly. Although much more common on the whole than the Compton, the California Tortoiseshell may be rare or absent from large parts of its range for several years. These dearths are followed by periods of enormous abundance, involving emigrations over immense areas, which seem to be related to population pressures, host plant availability, and climate conditions.

10, 24, 681 Mourning Cloak
(*Nymphalis antiopa*)

Description: 2⅞–3⅜″ (73–86 mm). *Large.* Wing margins ragged. *Dark with pale margins.* Above, rich brownish-maroon, iridescent at close range, with ragged, *cream-yellow band, bordered inwardly by brilliant blue spots* all along both wings. Below, striated, ash-black with row of blue-green to blue-gray chevrons just inside dirty yellow border.

Life Cycle: Egg, to ⁹⁄₂₅₆″ h × ⁷⁄₂₅₆″ w (0.9 × 0.7 mm), pale, becoming black before hatching; laid in groups on or around a twig. Caterpillar, to 2″ (51 mm), velvety black with white speckles, a row of red spots on back, and several rows of branched black bristles; has rust-colored legs. Feeds in groups on many broadleaf deciduous plants: willow (*Salix*), elm (*Ulmus*), hackberry (*Celtis*), and cottonwood (*Populus*). Chrysalis, to ⅞″ (28 mm), tan to gray, with 2 head horns, a "beak," and several thorny tubercles down the body; hangs upside down.

Flight: Number of broods varies with latitude
and altitude; year-round, most common
in spring, late summer, and early
autumn.

Habitat: Watercourses, sunny glades, forest
borders, parks, gardens, open
woodlands, and groves.

Range: Much of Northern Hemisphere south to
N. South America; virtually all of
North America where sufficient
moisture occurs, except for high Arctic
and subtropical regions.

Absolutely unique, the Mourning
Cloak camouflages itself perfectly
against dark bark at rest, then flaps
instantly into flight at the approach of
any predator, emitting an audible
"click." Few butterflies show such a
great contrast between the drab
underside and colorful upperside. In
summer, adults may be attracted with
fruit for closer observation. While less
common in some areas than in others,
Mourning Cloaks do not seem to
undergo the massive fluctuations in
abundance typical of the Compton and
California tortoiseshells.

658 Milbert's Tortoiseshell
(*Aglais milberti*)

Description: 1¾–2" (44–51 mm). *Above and below
2-toned; inner dark and outer light,
separated by a sharp border. Above,
inner dark area chocolate-brown* with 2
red-orange patches along FW costa,
*outer part has yellow band blending into
bright orange band. Below, inner area
purplish-brown; outer band tan.* Dark
margin above and below punctuated by
faint blue bars, has irregular but not
really ragged outline.

Life Cycle: Egg pale green; deposited in clusters,
often several hundred together.
Caterpillars at first live in colonies in
silken nests, later become solitary leaf-

folders. Mature caterpillar black with narrow yellow band above and green side stripes, white speckled with 7 rows of short spines. Host plants are nettles (*Urtica*). Chrysalis grayish or greenish-tan, thorny. Adults overwinter.

Flight: 2 or 3 broods where season permits; spring, summer, and fall.

Habitat: Dry stream beds and canals, riversides, beaches, meadows, alpine rockslides, roads, and trails.

Range: Far North except Alaska, south to S. California, Oklahoma, and West Virginia.

This unmistakable butterfly prefers northern latitudes and higher altitudes, although it occupies lowlands if they are cool enough. Extremely versatile, it inhabits every kind of place within its range, from cold desert to rain forest and city lot to alpine summit. Milbert's Tortoiseshell may sometimes be seen even in midwinter on a warmish day in many temperate areas. As with the Compton Tortoiseshell, the abundance of Milbert's fluctuates radically, but unlike the California Tortoiseshell, Milbert's never emigrates in impressive masses. Some specialists consider this butterfly to be the same species as the European Small Tortoiseshell (*A. urticae*). These 2 small, bright nettle-feeders may have evolved from common stock and, having been successful in their respective hemispheres, now differ significantly in appearance.

13, 669 **American Painted Lady**
"Hunter's Butterfly" "Virginia Lady"
(*Vanessa virginiensis*)

Description: 1¾–2⅛" (44–54 mm). FW tip extended, rounded. Above, pinkish-orange with black marks across FW; margins black-spotted and FW tip has white spots; row of black-rimmed blue spots crosses outer HW. *Underside has complex pattern* of olive, black, and white, *dominated by* large bright pink area on FW and *2 large blue eyespots in olive field on HW.*

Similar Species: Other 2 painted ladies have only small blue eyespots below.

Life Cycle: Egg yellowish-green, barrel-shaped; laid singly. Caterpillar, to 1⅜" (35 mm), black with yellow crossbands and white to rust-colored spots; makes solitary nest of silk and leaves on species of everlastings (*Gnaphalium, Antennaria, Anaphalis*) or other composites. Gold-spotted brown chrysalis, to ⅞" (22 mm), is often formed in nest. Reported to overwinter as adult or chrysalis.

Flight: 2 or 3 broods; summer–fall.

Habitat: Sunny, flowery open spots, sandy wastes, gardens, streambeds, riversides, and canyons.

Range: Subarctic North America south to Mexico; quite common in East, rarer in West; naturalized in Hawaii.

Of the painted ladies and their close relative the Red Admiral, this species seems to be the most tolerant of cold, and is quite likely the only one able to overwinter in the North. Its numbers fluctuate, although the American Painted Lady does not emigrate in the impressive numbers of the other 2 painted ladies. While common in the East, its numbers nonetheless never seem very large in any area; in much of the West its appearances are quite unpredictable. The American Painted Lady was formerly called *V. huntera;* this species and the other painted ladies are sometimes grouped in their own genus, *Cynthia.* Here *Cynthia* is regarded as a subgenus of *Vanessa.*

18, 670 Painted Lady
"Thistle Butterfly" "Cosmopolite"
(*Vanessa cardui*)

Description: 2–2¼" (51–57 mm). FW tip extended
slightly, rounded. *Above, salmon-orange
with black blotches,* black-patterned
margins, and broadly black FW tips
with clear white spots; outer HW
crossed by small black-rimmed blue
spots. *Below,* FW dominantly rose-pink
with olive, black, and white pattern;
HW has small blue spots on olive
background with white webwork. *FW
above and below has white bar* running
from costa across black patch near tip.

Similar Species: American Painted Lady has large
eyespots below. West Coast Lady has
orange bar across black patch near
FW tip.

Life Cycle: Egg pale green, barrel-shaped; laid
singly. Caterpillar, to 1¼" (32 mm),
varies from chartreuse with black
marbling to purplish with yellow back
stripe; has short spines. Chrysalis, to
⅞" (22 mm), lavender-brown, bumpy,
bluntly beaked; hangs upside down.
Preferred host plant is thistle (*Cirsium*)
but also feeds on great array of other
composites (Asteraceae) and mallows
(Malvaceae).

Flight: 2 or more broods; all year in southern
deserts, April–June until hard frosts in
North.

Habitat: Anywhere, especially flowery meadows,
parks, and mountaintops.

Range: All of North America well into sub-
Arctic, and south to Panama;
naturalized in Hawaii.

This species deserves its alternate name,
"Cosmopolite." Despite its inability to
overwinter in any stage above a certain
undetermined latitude, the Painted
Lady is perhaps the most widespread
butterfly in the world, found
throughout Africa, Europe, Asia and
many islands, as well as in North
America. Most of North America is

devoid of Painted Ladies between the
first heavy frosts and the onset of
spring, although they occur year-round
in the Sonoran deserts and perhaps
other warm regions. In February and
March, Painted Ladies begin
infiltrating the North and East from the
Southwest, and by late spring, they
have recolonized the continent. The
number of immigrants fluctuates
greatly from one year to the next.
Several mechanisms have been proposed
to explain the variance: cycles of
parasite attack, host plant defoliation,
and superabundance of nectar following
heavy winter rains. Unlike the
Monarch's annual round-trip outings,
the movements of the Painted Lady are
essentially one-way.

668 West Coast Lady
(*Vanessa annabella*)

Description: 1¾–2" (44–51 mm). FW tip extended,
clipped. *Orange to salmon-orange above
with black patterning;* small white spots
near FW tip and blue to purplish spots
along outer HW. *Beneath, complex
blurred marbling* of olive, tan, and white
webbing on HW, *with small, indistinct,
unrimmed blue spots outwardly;* FW pink
with blue, black, and buff spotting
below. *Orange to pale yellow bar outside
FW cell both above and below,* runs from
costa across black area near tip.

Similar Species: American Painted Lady has 2 big blue
eyespots below. Painted Lady has white
FW bar beyond cell in black area; HW
pattern less blurred beneath.

Life Cycle: Egg greenish, barrel-shaped; laid
singly. Caterpillar, to 1–1¼" (25–32
mm), variable, from light brown to
black, with yellow or orange blotches,
lines, and spines. Chrysalis olive-straw,
rounded, bluntly beaked, with whitish
tubercles. Host plants are mallows
(Malvaceae), including cheeseweed

(*Malva parviflora*), sidalceas (*Sidalcea*), hollyhocks (*Althaea*), and globemallows (*Sphaeralcea*); also occasionally nettles (*Urtica*).

Flight: Year-round in warmer parts of California; elsewhere spottily from early spring—late fall; autumn in eastern part of range.

Habitat: Vacant lots, flowerbeds, canals, fields, mountain canyons, and slopes.

Range: Pacific Slope from British Columbia to Baja California, east as a transient to western edge of Great Plains.

The West Coast Lady's numbers vary dramatically from year to year, but it does not undergo the massive emigrations of the Painted Lady. Its cold tolerance may be in between that of the other ladies: it can withstand moderate winters but not frigid conditions. The West Coast Lady probably travels eastward from the coast in late summer, appearing in the Rockies in fall. Formerly *V. carye*.

672 Red Admiral
"Alderman"
(*Vanessa atalanta*)

Description: 1¾–2¼″ (44–57 mm). FW tip extended, clipped. *Above, black with orange-red to vermilion bars across FW and on HW border. Below,* mottled black, brown, and blue with *pink bar on FW.* White spots at FW tip above and below, bright blue patch on lower HW angle above and below.

Life Cycle: Egg greenish, barrel-shaped. Caterpillar, to 1¼″ (32 mm), patterned light and dark from shiny black and yellow to brown and tan; warty and spiny. Chrysalis brown, gold-flecked; has dull short tubercles on thorax and curved abdomen. Nettles (*Urtica*) are best-known host plants, but other species in family Urticaceae used, such

as pellitories (*Parietaria*), false nettles
(*Boehmeria*), and hops (*Humulus*). Adults
and chrysalises overwinter in mild
areas.

Flight: 2 broods in most of range; generally
April or May–October, year-round in
far South.

Habitat: Forest margins and glades, rivers,
shorelines; also barnyards, gardens,
parks, roads; meadows, fields, and
savannahs; open woods and clearings.

Range: Holarctic: in New World from
subarctic Canada to Central America;
naturalized in Hawaii.

Unmistakable and unforgettable, the
Red Admiral will alight on a person's
shoulder day after day in a garden. This
species emigrates north in the spring,
and there is some evidence of a
dispersed return flight in the fall. If the
season is mild, occasional individuals
may pass a winter in the North;
however, Red Admirals are not usually
year-round residents in freezing
climates. In midsummer it is not
unusual to see them chasing each other
or Painted Ladies just before a
thunderstorm or at dusk. During the
full sunshine hours, they are more
likely to be found quietly drinking
from flowers or fruit.

671 Kamehameha
(*Vanessa tameamea*)

Description: 2⅜–2¾" (60–70 mm). *Large.* FW
tip extended; *wings appear scalloped.*
Bright salmon-orange above, rust-colored
toward base, black margins and black
FW tip with white spots. *2 distinct
black spots on salmon part of FW,* 1 in cell
and 1 above lower margin; angled black
line runs across HW above. Below,
very variable in tans and buffs with
vague blue eyespots and chevrons.

Similar Species: Red Admiral has red-orange confined to

narrow bars. American Painted Lady has 2 large blue eyespots below. Painted Lady's salmon area much more broken up with black. All 3 are significantly smaller.

Life Cycle: Caterpillar varies from green to purplish to multicolored, always with yellow stripe down each side; has small red, green, and black branched spines down back and sides, very large ones on rear, and blunt, white-tipped horns all over head. Host plant is mamake (*Pipturus albidus*), but also feeds upon other plants in nettle family (Urticaceae), including olana (*Touchardia*) and opahe (*Urera*).

Flight: 2 or 3 broods; all year.

Habitat: Koa forests, edges of woods, clearings, and kipukas (islands of vegetation in volcanic landscapes); also mountaintops and high canyons.

Range: Major islands of Hawaiian Archipelago.

Named for Kamehameha, the celebrated Hawaiian king, this large, brilliant butterfly probably derived from the Asian Admiral (*V. indica*). Now this is a uniquely Hawaiian species, one of only 2 native butterflies on the islands, and a common one. It is able to fly powerfully but spends much of its time resting on the trunks and limbs of old koa trees and basking on bracken. The nature trail through Kipuka Puaulu in Hawaiian Volcanoes National Park, Hawaii, is a good place to see Kamehamehas courting, jousting, and feeding at sap flows.

673, 674 Mimic
(*Hypolimnas misippus*)

Description: 2–2⅞" (51–73 mm). *Wings large, broad; FW tip extends somewhat. Male black above with large white oval on HW. 1 small and 1 large white bar on FW,* white spots usually edged with iridescent blue. *Female bright orange with black veins* and margins; *tip half of FW black with large white spots and bars.* Below, similar but paler.

Similar Species: Monarch FW has more orange, less white, than female Mimic.

Life Cycle: Caterpillar has spines on head and body; feeds on various plants of the mallow (Malvaceae) and purslane (Portulacaceae) families; figs (*Ficus*) also reported. Chrysalis thick, with 2 rows of tubercles.

Flight: Any time.

Habitat: Woods and brushlands where native; gardens and waste ground in Florida.

Range: Limited to Florida and Caribbean in North America; also Africa and Australasia.

Very common throughout much of the Old World, the Mimic is said to have come to the Caribbean with the slave ships. Now established in the Guianas and Antilles, it shows up in Florida periodically and may breed there. Sexual dimorphism is very striking. The female replicates the wing pattern of a particular distasteful butterfly species in each part of its range, which explains the butterfly's common name, the Mimic.

23, 688, 691 Buckeye
(*Junonia coenia*)

Description: 2–2½″ (51–63 mm). Wings scalloped and rounded except at drawn-out FW tip. Highly variable. *Above, tawny-brown to dark brown;* 2 orange bars in FW cell, orange submarginal band on HW, white band diagonally crossing FW. *2 bright eyespots on each wing above:* on FW, 1 very small near tip and 1 large eyespot in white FW bar; on HW, 1 large eyespot near upper margin and 1 small eyespot below it. *Eyespots black, yellow-rimmed, with iridescent blue and lilac irises.* Beneath, FW resembles above in lighter shades; HW eyespots tiny or absent, rose-brown to tan, with vague crescent-shaped markings.

Similar Species: West Indian Buckeye lighter and redder, has smaller eyespots of nearly equal size on HW above.

Life Cycle: Egg dark green, stubby, ribbed, flat-topped. Caterpillar, to 1¼" (32 mm), dark or greenish to blackish-gray with orange and yellowish markings. Wide variety of host plants include plantain (Plantaginaceae), figwort (Schrophulariaceae), stonecrop (Crassulaceae), and vervain (Verbenaceae) families. Chrysalis, to 1" (25 mm), mottled pale brown.

Flight: 2–4 broods; year-round in Deep South, elsewhere March–October.

Habitat: Shorelines, roadsides, railroad embankments, fields and meadows, swamp edges, and other open places.

Range: Resident throughout South, in North to east and west of Rockies to Oregon, Ontario, and New England.

Although the Buckeye flies in summer throughout much of North America south of the Canadian taiga, it is not able to overwinter very far north. In the autumn along the East Coast, there are impressive southward emigrations. In places such as Cape May, New Jersey, the October hordes of Buckeyes drifting southward rival those of Monarchs in number and spectacle. The classification of Buckeyes has puzzled generations of lepidopterists. They are sometimes listed under the genus *Precis* (which includes the Old World species) and under the old species name *lavinia.* The Dark Buckeye (*J. nigrosuffusa*) is nearly black above, with buff wing tips, orange fore wing cell bars, and smallish, blue-centered eyespots. The underside looks very different: the fore wing is orange, black, and buff with a prominent, blue-centered spot, while the hind wing is clear sandy-buff, crossed by a vague brown line or band, and has minute eyespots near the margin. This species dwells in the canyons of the Southwest from southeastern California and Arizona across southern Texas into Mexico. Its caterpillars feed on *Stemodia*, a member of the figwort family. Adults fly in the fall.

689, 692 **West Indian Buckeye**
"Florida Buckeye"
(*Junonia evarete*)

Description: 2–2½" (51–64 mm). Wings scalloped and rounded except at drawn-out FW tip. *Very variable.* Usually warm brown above with light orange FW band and submarginal HW band, 2 red-orange bars in FW cell, and *4 eyespots* (1 small and 1 large on FW, *2 nearly equal in size on HW*), *containing little if any reddish-purple* in blue iris. Below, similar but lighter on FW, HW brownish to reddish with vague lines and minute eyespots.

Similar Species: Buckeye has white FW band, upper eyespot on HW larger than lower one, usually darker brown with some purplish in eyespots.

Life Cycle: Soft black caterpillar has small yellow and blue dots and 6 rows of branching bristles; light rings on body. Host plant in Florida and Antilles is black mangrove (*Lippia*).

Flight: Successive broods; year-round.

Habitat: Subtropical shorelines and open scrub.

Range: S. Texas, S. Florida, south through Mexico and West Indies to South America.

For a long time the West Indian Buckeye was confused with the Buckeye, and some still consider them the same species. But the 2 butterflies mingle in southern Florida and the

West Indian species seems largely to replace the Buckeye in the Florida Keys, at least during the summer.

690, 693 White Peacock
(*Anartia jatrophae*)

Description: 2–2⅜″ (51–60 mm). FW tip extends slightly, rounded; HW bluntly tailed. *Above and below, white,* blending to buff and orange in margins, *overlaid with complex pattern of brown and orange scrawls and small black eyespots.*

Life Cycle: Egg pale yellow. Caterpillar black and spiny with silver spots. Chrysalis green, darkening with age; smooth. Host plants are ruellia (*Ruellia occidentalis*) and water hyssop (*Bacopa monniera*).

Flight: Year-round except in cold weather.

Habitat: Swampy places, watersides, shorelines, and disturbed ground.

Range: S. Florida and S. Texas, straying occasionally as far north as Kansas and Massachusetts; also much of the American Tropics.

Much more limited to the Tropics than its relative the Buckeye, the White Peacock also invades the North. However, it is neither as strong a flier nor as hardy as the Buckeye, and remains a rarity outside its southern strongholds.

682 Fatima
(*Anartia fatima*)

Description: 2–2⅛″ (51–54 mm). FW tip extended; HW has small, blunt tails; both wings scalloped. *Above, dark brown, banded and spotted across all wings with white or cream-color;* white spots near FW tips. 3–6 offset ruby spots in middle of HW.

Life Cycle: Unreported.

Flight: 2 broods; March—May and October—December.

Habitat: Sunny, flowery open spaces and watercourses.

Range: S. Texas south to Mexico, rarely straying as far as Kansas.

The bands and bright red spots distinguish Fatima from any other medium-sized brown butterfly in southern Texas. Fatima was the daughter of Mohammed; Fabricius gave this name to the butterfly in 1793.

648 Malachite
(*Siproeta stelenes*)

Description: 2½–3" (64–76 mm). *Large.* Wings scalloped; HW tailed. *Black above, diagonally banded and spotted with green*—from pale jade (sometimes white) to deep emerald. Beneath, marbled green, pearl, black, and tawny, with tawny borders and a pair of tawny and white stripes crossing HW. Female paler above and below than male.

Life Cycle: Egg laid singly. Caterpillar, to 2" (51 mm), velvety black with red bristles and 2 long, red horns curving back from head. Lime-green chrysalis may have pink spots; hangs from vegetation. Yerba papagayo (*Blechum brownei*) and other related species are recorded as host plants.

Flight: Year-round in Tropics.

Habitat: Shrubby forest edges, disturbed secondary growth scrub, and woodland pathways and glades.

Range: Texas and S. Florida, occasionally straying farther north; south well into Latin America; also Antilles.

A fresh Malachite is one of the most spectacular North American butterflies, but the marbled green fades upon exposure to sunlight. Florida Malachites may be strays rather than residents; more sightings have occurred in recent years but this probably reflects more searchers rather than more butterflies.

650 White Admiral
"Banded Purple"
(*Basilarchia arthemis*)

Description: 2⅞–3⅛" (73–79 mm). *Fairly large. Coal-black with milk-white band* across middle of wings above and below. Often has *blue or blue-green iridescence above*, especially on HW. *Above, brick-red spots may occur in row between HW band and marginal row of blue crescent-shaped marks. Below, red spot row is expanded;* prominent brick-red spots alternate with blue spots at wing bases. All red markings more pronounced in western populations.

Similar Species: Weidemeyer's Admiral lacks blue reflections above and red below; its range rarely overlaps.

Life Cycle: Egg compressed, oval. Caterpillar mottled off-white, olive, and greenish-yellow, with enlarged, light hump behind head; hump has long, dark bristles. Chrysalis cream-colored with enlarged wing cases and a darker, projecting mid-back "saddle horn." Host plants include birches (*Betula*), willows (*Salix*), poplars (*Populus*), occasionally hawthorns (*Crataegus*), and some other hardwood trees and shrubs.

Flight: 1 or 2 broods; June–August.

Habitat: Deciduous forest borders and glades.

Range: Northeast from Manitoba and Minnesota to New York and Maine; also from Alaska southeast through British Columbia.

Before winter, when admiral caterpillars finish feeding for the year, they secure themselves within the rolled-up, silk-tied bases of leaves. In spring, ravenous caterpillars emerge to complete their development. All adult admirals alternately sail and flap, darting out at insects or other interlopers in their territories. They also perch on leaves, twigs, or other prominences, taking aphid honeydew or liquids from flowers and carrion. This and related species were formerly included in the Old World genus *Limenitis.*

317, 320 **Red-spotted Purple**
(*Basilarchia astyanax*)

Description: 3–3⅜" (76–86 mm). *Large.* FW long, HW very squared, sometimes quite scalloped. *Above, coal-black with brilliant blue to blue-green iridescence,* especially over HW. *Below, brick-red spots line borders and cluster around wing bases.* Eastern populations have some red in FW tips above.

Similar Species: Female Diana Fritillary has much rounder wings, lacks red spotting. Pipevine Swallowtail is tailed.

Life Cycle: Caterpillar humped, dark-saddled and mottled; basically cream-colored with 2 prominent brushlike bristles behind head. Host plants include willows (*Salix*), poplars and aspens (*Populus*), cherries (*Prunus*), hawthorns (*Crataegus*), apples (*Malus*), hornbeams (*Carpinus*), and others.

Flight: Up to 3 broods in South; mid-spring through summer.

Habitat: Open woodlands, forest edges, nearby meadows, watercourses, shorelines, and roads and paths; arroyos and canyons in Southwest.

Range: E. Dakotas and NE. Colorado east to S. New England, south to central Florida, and west to Arizona and Mexico.

Along the northern edge of its range, the Red-spotted Purple hybridizes with the White Admiral to produce partially banded offspring. Some lepidopterists consider them one species, but genetic evidence suggests that they have come together relatively recently. The Red-spotted Purple mimics the toxic, bright blue Pipevine Swallowtail, thus gaining protection from birds.

26, 597 **Viceroy**
(*Basilarchia archippus*)

Description: 2⅝–3″ (67–76 mm). Above and below, *rich, russet-orange with black veins, a black line usually curving across HW,* white-spotted black borders, and *white spots surrounded by black in diagonal band across FW tip.* Color ranges from pale tawny in Great Basin to deep, mahogany-brown in Florida.

Similar Species: Monarch and female Mimic, Queen, and Tropic Queen lack black line across HW.

Life Cycle: Egg compressed oval. Caterpillar, 1–1¼″ (25–32 mm), mottled brown or olive with saddle-shaped patch on back; fore parts humped; 2 bristles behind head. Chrysalis, to ⅞″ (22 mm), also brown and cream-colored with brown, rounded disk projecting from back. Willows (*Salix*) are preferred host plants but also poplars and aspens (*Populus*), apples (*Malus*), and cherries and plums (*Prunus*).

Flight: 2, 3, or more broods depending upon latitude; April–September in middle latitudes, later in South. Sometimes a distinct gap between broods, with no adults for some weeks in mid- to late summer.

Habitat: Canals, riversides, marshes, meadows, wood edges, roadsides, lakeshores, and deltas.

Range: North America south of Hudson Bay, from Great Basin eastward, and west to eastern parts of Pacific States.

In each life stage, the Viceroy seeks protection through a different ruse. The

egg blends with the numerous galls
that afflict the willow leaves upon
which it is laid. Hibernating
caterpillars hide themselves in bits of
leaves they have attached to a twig. The
mature caterpillar looks mildly fearsome
with its hunched and horned foreparts.
Even most birds pass over the chrysalis,
thinking it is a bird dropping. The
adult, famed as a paramount mimic,
resembles the distasteful Monarch.
Since birds learn to eschew Monarchs,
they also avoid the look-alike Viceroy.
Southern populations of Viceroys mimic
the much deeper chestnut-colored
Queen instead. In flight, the Viceroy
flaps frenetically in between brief
glides.

649 Weidemeyer's Admiral
(*Basilarchia weidemeyerii*)

Description: 2¾–3⅜" (70–86 mm). *Large. Above,
coal-black without significant blue
reflections,* both wings *crossed by broad
white bands.* Small white spots line
black border and cross FW tip. Below,
FW similar with 2 reddish bars in cell;
HW bluish-gray with black cross lines
inside white band; outside band is row
of reddish spots followed by another
row of blue-gray crescents. Amount of
reddish and width of bands vary among
populations.

Similar Species: White Admiral has outstanding blue
reflections and more reddish below.
Lorquin's Admiral has orange to buff
wing tips and off-white bands.

Life Cycle: Caterpillar mottled gray and white,
humped; feeds upon willows (*Salix*) and
cottonwoods (*Populus*), including
aspens. Chrysalis resembles bird
dropping, with contrasting light and
dark, and rounded protrusion on back.

Flight: 1 brood; late May or June–August.

Habitat: Chiefly watercourses lined with willows
or cottonwoods; also sand hills, sage

flats, parklands, gardens, washes, and mountainsides with nearby streams, canals, or lakesides.

Range: SE. Alberta, E. Oregon and Nevada east to Dakotas, Nebraska, and New Mexico.

The several species of American admirals neatly divide up the continent. Only the Viceroy occupies nearly the whole. The White Admiral is basically northeastern, the Red-spotted Purple southeastern, and each extends westerly to the north and south respectively. Lorquin's Admiral occupies the West Coast, and Weidemeyer's Admiral the Rocky Mountains and their adjacent lowlands. Where the species do meet, a measure of hybridism often takes place. Throughout most of the Rockies, Weidemeyer's is the only banded admiral encountered. Territorial battles often take place between Weidemeyer's Admirals and other waterside denizens, such as Mourning Cloaks, Tiger Swallowtails, small skippers, and crescentspots, as well as dragonflies.

651 Lorquin's Admiral
(*Basilarchia lorquini*)

Description: 2¼–2¾" (57–70 mm). *Smallest admiral. Above, brownish-black with cream-colored, off-white bands across all wings and buff to bright or dark rust-orange FW tips,* sometimes extending along much of FW margin. FW cell has white spot. Below, a complex pattern of alternating brick-red and blue-gray bands beyond white band, reddish and gray spots and black lines inside band. Submarginal reddish spot bands may also show on HW above.

Similar Species: All other banded admirals are larger and have pure white bands; all lack orange wing tips, except much larger

California Sister, which has a more crisply defined and rounded orange patch.

Life Cycle: Egg globelike, pale green; deposited near leaf tip. Young caterpillar dark with white saddle. Mature caterpillar mottled olive and yellow-brown with light side band and white back patch; plumelike bristles behind head are smaller than those on other admiral caterpillars, as are humped segments behind them. Chrysalis irregular, lilac-brown in front and whitish behind; has greenish wing cases and dark, raised disk extending above back. Host plants include willows (*Salix*), poplars (*Populus*), and chokecherry (*Prunus virginiana*).

Flight: 2 broods in California; April–September. 1 brood in Northwest; June–September.

Habitat: River bottomlands, canals, and lakeshores; also parks, forest margins, and elsewhere near host plants.

Range: Central British Columbia and SW. Alberta south to Baja California, NE. Nevada, and SE. Idaho.

Lorquin's Admiral is an abundant West Coast species, avoiding only the wettest coastlines and higher alpine reaches. When basking, it holds its wings open at a 45° angle, periodically opening and closing them altogether. Without warning it will burst into flight, flapping and gliding in an alternating rhythm, often returning to its original perch. These admirals inspect or attack almost anything that passes by, and have been seen to lunge at gulls 20′ (6 m) overhead, harass the birds until they fly away, then resume their station. The butterfly bears the name of the mid-19th century French collector Pierre Lorquin, who sent it back to Europe from California.

654 Mexican Sister
(*Adelpha fessonia*)

Description: 2–2½" (51–64 mm). FW tip drawn out to a blunt point. Grayish-brown above with chestnut and chocolate-colored bars, *large orange patch on FW costa near tip* and small patch at HW outer angle (tornus), distinct *complete white band crosses both wings above.* Orange and white repeated below, along with blue-gray and russet patches and bars.

Similar Species: California Sister larger, has broken white bar, lacks drawn-out wing tip. Laure male has blue reflections, much more orange on FW. Laure and Pavon females have clipped-looking FW tips and paler orange, and lack blue-gray or russet beneath.

Life Cycle: Unreported. Caterpillar found on *Rondia,* a member of the bedstraw family (Rubiaceae) in Costa Rica.

Flight: Late summer and autumn in U.S.; probably year-round in Tropics.

Habitat: Subtropical woodlands, river bottomlands, and tropical dry forests.

Range: Southernmost Texas south to Costa Rica.

The Mexican Sister is a member of a large Latin American group of butterflies, which are mostly rather similar in appearance. Vegetated irrigation ditches in the Rio Grande Valley near Pharr, Texas, are usually good spots to search for this species in the United States. Males perch and patrol in forests or along river edges. They take nectar from small white or greenish flowers of tropical shrubs, such as cordias and caseanas, and probably also drink tree sap and rotting fruit.

31, 652   California Sister
(*Adelpha bredowii*)

Description: 2⅞–3⅜″ (73–86 mm). *Large. Above,
dark brown, narrowly banded with
white, the band broken into spots on FW;
bright orange patch lies on FW tip,* neither
reaching nor extending down along
brown margin. Underside complexly
marked with auburn, pale blue, orange,
and white bands and spots. Red-brown
bars on FW cell and HW inside angle.

Similar Species: Lorquin's Admiral smaller, lacks blue
bands below, and orange tip extends to
and runs down along margin.

Life Cycle: Egg spherical. Caterpillar, to 1¼″ (32
mm), dark green on back, olive-brown
beneath, with 6 pairs of brushy
tubercles front to back. Both caterpillar
and chrysalis are slightly humpbacked.
Chrysalis light brown, with 2 head
horns and metallic marks. Host plants
are canyon live oak (*Quercus chrysolepis*)
and coast live oak (*Q. agrifolia*), and
perhaps giant chinkapin (*Chrysolepis
chrysophylla*).

Flight: 2 broods; April–October. 1 brood in
desert mountains; May–June.

Habitat: Oak groves in middle and low elevation
mountains; on coast and on offshore
islands.

Range: SW. Washington, rarely south to Baja
California, east through Nevada and
Arizona into Colorado, New Mexico,
and (rarely) Kansas.

The coloring of this butterfly's wings is
like that of a nun's habit, inspiring its
common name. California Sisters visit
buckeye flowers, but take their
moisture more often at damp mud,
riverside sand, and fallen fruit. Spilt
grape juice attracts them to wineries.

645 Blue Wing
(*Myscelia ethusa*)

Description: 2⅞–3¼" (73–83 mm). *Large.* FW tip extended; HW rounded. *Iridescent blue above crossed by bands of black; more white spots on female; FW tip dotted with white.* Below, HW heavily mottled with brown and gray.

Life Cycle: Unknown.

Flight: May–November in Texas.

Habitat: Tropical and subtropical forests.

Range: Much of American Tropics, extending into S. Texas around Pharr.

The Blue Wing may be rather numerous at times, although it is probably an irregular member of the Texas butterfly fauna. No other large bluish North American butterfly has the Blue Wing's intense sapphire coloring. The Cyananthe Blue Wing (*M. cyananthe streckeri*) is the only other member of the genus to occur in this country straying to New Mexico. It lacks white spots and some of the stripes on the fore wing, and its wings are largely blotted with black; the blue is much reduced.

684 Florida Purplewing
(*Eunica tatila*)

Description: 1⅝"–2" (41–51 mm). FW tip concavely indented at tip. Dark brown *above, inner ⅔ heavily overlaid with iridescent purple. FW tips have 7 white spots and bars.* Mottled dark and dull brown below. HW has eyespots above and below.

Life Cycle: Unknown.

Flight: Several broods; any month.

Habitat: Dense coastal hardwood hammocks.

Range: Central America and Antilles north to S. Florida.

Although many purplewings fly farther south, few reach North America. The

Florida Purplewing may be fairly common in southern Florida. The related Dingy Purplewing (*E. monima*) infrequently wanders to Florida and southern Texas; it feeds on prickly ash (*Zanthoxylum*) in Mexico. It is smaller, has fewer white spots in the more rounded fore wing tip, and its purple shines with less brilliance. Purplewings haunt the undeveloped Florida hardwood hammocks. Their undersides are camouflaged against the tree bark on which they perch. The purple iridescence shows only in direct sunshine; they look brown in the shade. Sometimes purplewings take to the ground to drink from fruit or mud, or wander out to the edges of hammocks or along roadsides.

647 Mylitta Greenwing
(*Dynamine mylitta*)

Description: 1½–1⅞" (38–48 mm). *Male above bright green* with broad black border, projecting toward base in middle of FW. Female above dark brown with broken white median band and 5 smaller white spots nearer FW margin. *Below, cocoa-brown with white patches and white or orange basal bands on FW, HW patch near margin containing 2 yellow and brown-rimmed blue eyespots.*

Life Cycle: Unrecorded. Host plants may be dalechampia (*Dalechampia*) and tragia (*Tragia ramosa*) in Texas.

Flight: Probably any month.

Habitat: Quite local in mixed scrub and forests.

Range: South America through Mexico, very rarely in S. Texas.

All greenwings are rare strays in the United States. The Dyonis Greenwing (*D. dyonis*) has been recorded near San Antonio. The male is lighter green above and a bit paler below. Numbers of greenwings sometimes band together and may roost communally at night. Many other greenwing species dwell in Mexico and southward.

---

687 **Eighty-eight Butterfly**
"Leopard-spot"
(*Diaethria anna*)

Description: 1⅝–1⅞" (41–48 mm). *Black above with pale green bands* across FW and around HW margins. *Below, FW rose-red near base, black and white banded near tip; HW white with 2 black "8" shaped marks in center,* ringed by 3 concentric, thin, black lines. Innermost "8" may coalesce, resembling a "9."

Life Cycle: Unreported. Feeds on trema (*Trema*) in Brazil.

Flight: 2 or more broods; year-round in Tropics.

Habitat: Tropical and subtropical forest glades.

Range: South America and Mexico, very rarely S. Texas.

The Eighty-eight Butterfly is one of a large number of Central and South American butterflies known collectively as "eighty-eights" because of their hind wing markings. *Diaethria anna* sometimes enters the Rio Grande Valley, but its appearance has not been officially recorded yet. The South American Eighty-eight Butterfly (*D. clymena*), seen in Florida, resembles the Eighty-eight Butterfly but has more red in the fore wing cell below, and thicker hind wing markings.

73, 679 **Amymone**
(*Mestra amymone*)

Description: 1⅜–1⅝" (35–41 mm). *Small.* Wings rounded and slightly scalloped. *Grayish* above and below near bases and on FW tips and margins; *broad golden-yellow HW borders,* the rest of HW broadly dusted with pearl-white.

Life Cycle: Host plant may be tragia (*Tragia*).
Flight: Successive broods; any month.
Habitat: Open brush country.
Range: Louisiana and S. Texas south through Central America, emigrating as far north as Nebraska.

A rather subtly marked, small brush-foot, the Amymone becomes quickly worn and may seem rather unimpressive in flight. It occurs regularly in the southern part of Texas and it sometimes flies considerably farther north into the plains states. In certain years, huge emigrations of Amymones leave the South, traveling to places hundreds of miles beyond their normal range.

680 Crimson-banded Black
(*Biblis hyperia*)

Description: 2–2⅛" (51–54 mm). *All black above except for crimson bands around HW,* which are bordered outwardly by black, scalloped margin. Bands pink below.
Similar Species: Ruby-spotted Swallowtail larger; longer wings have crimson patches, not bands.
Life Cycle: Egg off-white, grooved and tufted on top. Caterpillar thickened in middle, green with red and brown spots, many spiny red knobs on body, longer horns on head. Chrysalis irregular, mottled brown. Host plant is tragia (*Tragia volubilis*) in Brazil.
Flight: Year-round in Tropics, in greater abundance following rainy season.
Habitat: Subtropical thorn scrub and roadsides.
Range: Mexico to Panama; occasionally to S. Texas.

The Crimson-banded Black probably does not breed north of Mexico. It is one of many species of tropical and subtropical butterflies that can be considered part-time members of the Texas butterfly fauna.

759 Guatemalan Calico
(Hamadryas guatemalena)

Description: 3⅛–3¼″ (80–84 mm). *Above,* very
complicated calico pattern of dark gray
and black markings against gray
background; *prominent HW black circles.*
Below, FW with black markings and
with large white patches; HW tan or
buff brown with submarginal black
circles.

Similar Species: Ferentina Calico is less boldly marked
above and has some red in black or
brown semi-circles.

Life Cycle: Egg white, spherical, grooved.
Caterpillar and chrysalis: light, dark,
or mottled but usually with some
greenish stripes. Caterpillar has black
head horns. Chrysalis bears 2 very long
earlike projections. Host plants are
dalechampia *(Dalechampia)* and tragia
(Tragia).

Flight: 2 or more broods, any month.

Habitat: Tropical and subtropical forests,
especially disturbed primary forests.

Range: N. Mexico to Chiapas, Mexico,
straying into Texas.

All calicoes blend in with their
environment when perched upside
down on gray bark with their wings
spread. When they stir from their tree
trunk positions, they produce sharp
"clicks" that may be heard 30 yards
(27 m) away.

758 Ferentina Calico
(Hamadryas februa)

Description: 2–2⅝″ (60–67 mm). Mottled calico
pattern above, mainly brown or blue-
gray with fine dirty white and black
spotting. *Eyespots on HW above mostly
gray with some red in black or brown semi-
circles.* Below, FW gray with black and
white spots; HW largely gray-white
with circular black and white eyespots.

Similar Species: Guatemalan Calico tan or buff brown on HW below, and HW eyespots above are heavily circled with black.

Life Cycle: Caterpillar and chrysalis mottled, vary in color; both possess prominent head horns. Host plants are dalechampia (*Dalechampia*) and tragia (*Tragia*).

Flight: Uncommon in autumn in Texas.

Habitat: Small patches of subtropical woodlands.

Range: S. Texas and Mexico.

Of the 7 species and subspecies of *Hamadryas* that have been recorded in the United States, only the Ferentina Calico (*H. februa ferentina*) probably breeds here. All the calicoes in the United States are rare, casual strays except the Ferentina Calico.

686 Blomfild's Beauty
(Smyrna blomfildia)

Description: Blomfild's Beauty was only recently recorded in southern Texas. Both this species and the very similar Karwinski's Beauty (*S. karwinskii*) are rarely encountered in the United States. Karwinski's Beauty has a less bold pattern below and the female is not as dark brown above. In dry season, both of these Central American species roost in the high mountains. Adults perch head down on tree trunks. They take moisture from sap, mud, and rotting fruits but seem to ignore flowers.

Similar Species: Monarch has longer wings, black veins and margins, less of FW tip black, and lacks pattern below.

Life Cycle: Egg green, spherical. Mature caterpillar, to 1⅝–1¾" (41–44 mm), round with black marbling and short, branched spines. Chrysalis rounded, light to dark brown, with rows of lighter bumps and small black markings. Host plants are members of nettle family (Urticaceae); *Urticastrum* and *Urera* in El Salvador.

Flight: 2 or more broods; year-round in
Tropics, more common in wet season.

Habitat: Moist tropical and subtropical forests
and seasonal dry forests.

Range: Mexico and most of Central America,
straying very rarely into S. Texas.

It has been many years since
Karwinski's Beauty has been seen near
Brownsville, Texas. In the dry season,
both of these Central American
butterflies roost in the high mountains.
Adults perch head down on tree trunks.
They take moisture from sap, mud, and
rotting fruits but seem to ignore
flowers.

381 Waiter
(*Marpesia coresia*)

Description: 2–2⅝″ (51–67 mm). Strikingly *long,
white-tipped tails. Chocolate-brown above
and contrasting white and brown below.*
Basal half of wing below is white, outer
portion cocoa-brown, separated by dark
brown line.

Life Cycle: Unrecorded.

Flight: 2 or more broods; any month.

Habitat: Tropical forest clearings and edges, and
a wide array of tropical and subtropical
habitats.

Range: SE. Texas south to Brazil and Peru.

Named for its crisp, tuxedo-like
coloring, the Waiter is an abundant
butterfly throughout much of tropical
America. Its presence north of the
Mexican border consists of infrequent
visits to the southeastern tip of Texas.

364 Banded Daggerwing
(*Marpesia chiron*)

Description: 2–2⅜" (51–60 mm). *Long-tailed. Tiger-striped light and dark brown above,* with white spots in FW tip. Beneath, similarly banded with browns, tans, and white.

Life Cycle: Egg green, broadly conical. Yellow caterpillar has black stripes, patches, and horns. Angular chrysalis mottled gray, black, mahogany, and cream-colored. Host plant is breadfruit (*Artocarpus integrifolia*).

Flight: 2 or more broods; year-round in Tropics.

Habitat: Fields and forest borders.

Range: Southernmost U.S. south into South America.

Although both daggerwings and swallowtails have tails, they are very different butterflies in flight, behavior, and structure. Both share a liking for damp earth. The Banded Daggerwing occurs sometimes in Florida and Texas, and rarely wanders farther north.

363 Ruddy Daggerwing
(*Marpesia petreus*)

Description: 2⅝–2⅞" (67–73 mm). *FW drawn out into squared-off, curving, hooked tips; HW long-tailed;* 2nd, stubby tail arises from inside angle of HW. *Above, clear orange crossed by weak, thin, brown lines; tails brown.* Below, pale orange-tan with slight purplish cast.

Similar Species: Julia is similar in color and size but has even outline and lacks tails.

Life Cycle: Caterpillar rust-colored and yellow, spotted and lined with black; hair extends down back; prominent bristly head horns. Host plants in Florida are figs (*Ficus*) and cashews (*Anacardium occidentale*).

Flight: 2 or more broods; all months. In

S. Florida, most common May–July.
Habitat: Hardwood hammocks and thickets.
Range: E. Texas and Florida Panhandle (occasionally wandering farther north), and south through much of Latin America and Antilles.

Such a bright, large butterfly draws the attention of birds; perhaps the hooked fore wing tips and hind wing tails divert attacks from vital parts. While the caterpillars thrive on fig leaves, adults love rotting figs, other fruits, and giant milkweed. Nature trails through hardwood hammocks in Everglades National Park are good places to watch for the Ruddy Daggerwing.

361 **Florida Leafwing**
"Florida Goatweed Butterfly"
(*Anaea floridalis*)

Description: 2¾–3" (70–76 mm). FW tip extended, hooked; *small tail on HW. Outline finely scalloped. Above,* bright red-orange with dark, crescent-shaped markings along margin and across wings. *Below,* FW yellowish, HW *brown striated with black.*
Similar Species: Goatweed Butterfly overlaps only narrowly in S. central Florida; is not so brightly or heavily marked, and wing outline is smooth, not scalloped.
Life Cycle: Caterpillar green, white-dotted and yellow-striped. Chrysalis green, stubby, and rounded. Woolly croton (*Croton linearis*) is host plant.
Flight: 2 or more broods; any month.
Habitat: Hammocks, woodland edges, limestone bluffs, and keys.
Range: S. Florida including Keys; occasionally as far north as Gainesville.

Older books call this butterfly Portia (*A. portia*), now recognized as an Antillean species. Leafwings differ in

shape from brood to brood; summer
individuals have wings that are less
hooked. When at rest, they can be
mistaken for leaves. All leafwings are
extremely wary, rapid fliers.

362 Goatweed Butterfly
(*Anaea andria*)

Description: 2⅜–3″ (60–76 mm). *Outline of wings
smooth* (not sawtoothed or scalloped),
*HW has blunt, narrow tail. Male bright,
fiery orange-red above,* with dusky
margins; *female duller orange* with broad
dusky margins and black-edged
submarginal yellow band. Both sexes
below, purplish brown or gray, mottled
and *striated with fine brown marks.* Fall
and spring adults have very sharp,
drawn-out hooks on FW tips; summer
brood individuals have FW less so.

Similar Species: Florida Leafwing barely overlaps in
range; has finely scalloped wing
outline.

Life Cycle: Caterpillar green-gray, narrowed
toward rear; has many minute warts
and small orange horns on narrow-
necked head; folds host plant leaves
into protective sheaths and ties them
with silk. Chrysalis blunt, thick; looks
as if it has lid near rear, from which it
hangs by a silk button. Host plants are
goatweeds (*Croton capitatum, C. linearis,
C. monanthogynus*). Adult overwinters.

Flight: Overlapping broods; early spring,
summer, and fall.

Habitat: Farmyards, canals, country roads,
fallow fields, swamps, pine barrens,
wood edges, and prairie groves.

Range: Michigan south through Nebraska and
E. Colorado, West Virginia, and
Georgia, to Gulf and into Mexico;
absent from most of Florida; rarely west
to Wyoming and Arizona.

The only semi-northern representative
of a large tropical group of leafwings,

the Goatweed Butterfly not only breeds throughout the central plains states but, like the anglewings and tortoiseshells, also apparently survives their harsh winters as an adult. Its pointed wing tips, stemlike tails, and mottled brown underside make the Goatweed an excellent leaf mimic. The 2 seasonal forms seem to be moisture-related; individuals with longer tips and tails follow periods of greater rainfall. Three other species of leafwings occur occasionally in southern Texas. The Tropical Leafwing (*A. aidea*), is quite regular there and ranges north to Kansas and Oklahoma in some years. The male is even brighter orange-red than the Goatweed but with darker markings; the female has separate bands of yellow dots crossing the upperside. This species has a scalloped outline rather than the smooth outline of the Goatweed. The Crinkled Leafwing (*A. glycerium*) resembles the Tropical Leafwing but is paler, approaching a bright tan, and has stronger dark markings both above and below; its fore wing outline has a notable in-and-out curvature. The Blue Leafwing (*A. pithyusa*) retains the leaflike shape and tails of the others but is black with blue spots on the wing tips, and has blue-green reflections above and mottled gray beneath. It is the only North American blue leafwing; most leafwings dwell in the Tropics. Both the Blue and Crinkled leafwings find their way to Texas only rarely.

45, 664 Hackberry Butterfly
"Hackberry Emperor"
(*Asterocampa celtis*)

Description: 1¾–2¼″ (41–57 mm). Complexly and variably marked above with brown, black, and purplish-gray. Typical male brown to grayish-brown above; female lighter, tawnier. *Above,* both have dark FW tips with white spots, *1 black, broad eyespot, usually without pupil on FW,* and a submarginal HW row of black eyespots, which sometimes have white pupils; *FW cell has 1 dark bar and* 2 dark spots. Below, eyespots repeated, amid purplish gray-brown bars, spots, and chevrons and whiter patches. Male's wings narrow, concave and elongated at tip; female's broader, rounder. Green sheen to wings when fresh.

Similar Species: Empresses Antonia and Leilia have 2 black eyespots on FW. Empress Alicia is larger. Tawny Emperor, Empresses Flora and Louisa, and Pale and Texas emperors lack black eyespots on FW and together with Leilia have 2 bars in FW cell.

Life Cycle: Egg pale yellow to white. Caterpillar, to 1¼″ (32 mm), rather sluglike, bright grass-green with yellow and chartreuse lengthwise stripes, 2 tails projecting from rear and small, branched horns on head. Chrysalis, to ⅞″ (22 mm), bluish-green, sharply horned and razor-backed. Hackberry trees (*Celtis*) are sole host plants.

Flight: 1 brood in North; June–August. 3 broods southward; March–October.

Habitat: In vicinity of hackberry trees in deciduous woodlands along roads, trails, and margins; also suburbs, city parks, and streets.

Range: S. Ontario and North and South Dakota east to Massachusetts, south to N. Florida and E. Texas.

Populations of hackberry butterflies in the United States have recently been divided into 3 main species groups; those related to the Hackberry Butterfly, to the Empress Leilia, and to the Tawny Emperor. The Hackberry Butterfly's closest relatives are the Empress Antonia and the Empress Alicia (*A. celtis* form "alicia"), but the latter may actually be a population of the Hackberry Butterfly that occurs along the Gulf. To confuse matters further, the name Empress Alicia has also been applied to a different eastern Florida population of the Hackberry Butterfly.

663 Empress Antonia
(*Asterocampa antonia*)

Description: 1¾–2⅛" (41–54 mm). Warm olive-brown, tawny orange, or gray *above*, marked with black and white spots and eyespots; FW above with whitish spots; *2 black eyespots on FW*, usually with pupils; row of dark eyespots around HW; *FW cell has 1 dark bar* and 2 dark spots. Below, eyespots repeated often with blue pupils on background of gray and purplish-tan with complex darker markings. Female broader and lighter than male.

Similar Species: Hackberry Butterfly and Empress Alicia with only 1 black eyespot on FW. All others with 2 bars in FW cell.

Life Cycle: Unreported. Host plants are hackberries (*Celtis*).

Flight: 2 or more broods; spring through fall.

Habitat: Canyons, foothills, and woodlands with hackberries.

Range: Colorado east to W. Nebraska and W. Kansas, south through Arizona, New Mexico, W. Oklahoma, and S. and W. Texas to N. Mexico.

The Empress Antonia and close relatives may be subspecies of the Hackberry Butterfly. Populations of Empress Antonia in the mountains of western New Mexico, Arizona, and northwestern Mexico were formerly thought to be a separate species, the Mountain Emperor (*A. montis*). Empress Antonia often flies in canyons and arroyos. Like other hackberry butterflies, adults drink from scat, mud, tree sap, or fruit, such as overripe wild plums or persimmons.

662 Empress Leilia
(*Asterocampa leilia*)

Description: 1¾–2" (44–51 mm). *Above, fawn-brown with dark FW tips spotted with white; several eyespots on HW, 2 on FW; 2 dark bars in FW cell.* Below, purplish-gray; FW brown near base, eyespots have blue pupils. Male has narrower wings than female.

Similar Species: Empress Antonia has only 1 dark bar and 2 spots in FW cell. Tawny, Texas, and Pale emperors and Empress Louisa lack black eyespots on FW.

Life Cycle: Egg light yellow. Caterpillar sluglike, bright green with 2 pairs of narrow, longitudinal yellow stripes; head green with long brown antlers; tail typically forked. Chrysalis yellow-green, with dorsal crest; head with blunt projections. Host plant is spiny hackberry (*Celtis pallida*).

Flight: 2 broods; spring-fall, year-round in Mexico.

Habitat: Hackberry groves in chaparral and surrounding desert canyons and arroyos.

Range: SE Arizona to W. Texas, and southward into Mexico.

Although closely related to the several varieties or species of the Hackberry Butterfly, the Empress Leilia is quite different biologically. In arid Sonoran canyons, it seeks moisture near streams and perches upside down on shaded trunks. The Empress Alicia (*A. celtis* form "alicia") occurs to the east from the Texas Gulf to northwestern Florida along the coastal lowlands. It differs from Leilia in having only 1 bar in the fore wing cell, and is bigger and brighter than the Hackberry Butterfly.

12, 666 Tawny Emperor
(*Asterocampa clyton*)

Description: 1⅞–2⅜″ (48–60 mm). *Above,* rich *tawny orange-brown* with darker bars and patches; yellow (not white) *spots in usually darker FW tip;* submarginal row of *black pupilless eyespots along HW but absent from FW.* HW may be very dark above. Below, whitish with violet-brown areas and black markings; eyespots on HW often have blue pupils. Female broader, larger, more rounded and paler than male, which has wings concavely curved and drawn out.

Similar Species: Hackberry Butterfly, close relatives, and the Empress Leilia have black FW eyespots. Empress Flora is larger and brighter.

Life Cycle: Light yellow egg, ⅟₂₈″ h × ⅟₂₅″ w (0.9 × 1.0 mm); laid in tightly packed, large clusters. Caterpillar sluglike, bright green with parallel rows of yellow, buff, and bluish-green stripes; 2 taillike projections at rear; head bears large, antlerlike, brushy protuberances. Chrysalis bluish-green; has raised, sawtoothed ridge along back, otherwise quite rounded, with head horns. Hackberry trees (*Celtis*) are host plants.

Flight: 1–2 broods; mid-June to mid-August; earlier and later southward.

Habitat: Deciduous woodlands containing hackberry trees, windbreaks in agricultural areas, and hackberry thickets on rich bottomland soils.

Range: S. Ontario, Nebraska, and Wisconsin; Massachusetts south to E. Texas and S. Georgia.

The Tawny Emperor often occurs with the Hackberry Butterfly, but is usually much less common, particularly northward and westward. Less of a colonist than the Hackberry Butterfly, it does not adapt to city habitats as successfully as its near relative. The Empress Flora replaces the Tawny

Emperor in southern Georgia and
Florida. The Pale Emperor (*A.
subpallida*) occurs in Arizona south into
Mexico in canyons and along bases of
mountains. The male has a clear
purplish-gray hind wing below, while
the female is tan; most lack dark
eyespots on the hind wing below. The
Texas Emperor (*A. texana*) in western
Texas typically has more pronounced
hind wing markings below.

665   Empress Flora
(*Asterocampa flora*)

Description: 2–2¾ " (51–70 mm). *Larger than most
hackberry butterflies.* Tawny reddish-tan
above with bands of brown bars and
*yellow spots across FW and a row of brown
spots around HW.* Female has less
brown, more orange above than male.
Below, complex blend of pale orange,
white, and rusty-tan, with black lines
and spots and a submarginal row of
blue eyespots around HW. Female has
broader, rounder wings than male.

Similar Species: Tawny Emperor smaller, not so richly
colored. Hackberry Butterfly and
Empress Alicia with a black eyespot on
FW.

Life Cycle: Light yellow egg; laid in clusters.
Caterpillar sluglike, green with rows of
buff, yellow, and bluish stripes;
projections at rear and on head.
Rounded chrysalis bluish-green with
sawtoothed ridge on back and head
horns. Hackberry trees (*Celtis laerigata*)
are host plants.

Flight: 2 broods; March–September.

Habitat: Mixed woodlands on coastal lowlands.

Range: S. Georgia and Florida.

The Empress Flora replaces the Tawny
Emperor in the Florida peninsula.
W. J. Holland first popularized the
fanciful names of these hackberry
butterflies in *The Butterfly Book* in

1898. The regal connotation comes from the subfamily to which these butterflies belong, the Apaturinae, known in Europe as the emperors.

661 Empress Louisa
(*Asterocampa louisa*)

Description: 1⅞–2¼″ (48–57 mm). Above, male oak-brown; female yellower; both have black eyespots in row along HW margin; *FW has no eyespots but dark FW tips with white spots.* Beneath, pale violet-tan and white with eyespots and other markings modified, especially on female. Male's wings concavely curved, female's rounder and fuller.

Similar Species: No other hackberry butterfly has combination of white spots in dark FW tip, black antennae and lacks FW eyespots.

Life Cycle: Unreported. Hackberry (*Celtis*) is host plant.

Flight: Long season, varies with rainfall and temperature.

Habitat: Rio Grande Valley hackberry stands.

Range: Southernmost Texas and NE. Mexico.

Some lepidopterists now believe that Empress Louisa is a form of the Texas Emperor. The Empress Louisa was only discovered and named in 1947, but like all the Hackberry butterflies has been undergoing revision since the early 1950s. Hackberry butterflies share a strong fondness for the juices of rotting fruit, and can be lured out of their trees to overripe figs, pears, peaches, and persimmons.

655 Pavon
(*Doxocopa pavon*)

Description: 1¾–2⅜″ (51–60 mm). FW tips somewhat extended and clipped. *Above, male brown with suggestion of whitish* or lighter brown *bands across wings,* vague orange patches near FW tips, and *strong purplish reflections. Female lighter brown above* with white bands and large orange patches on FW tips. Below, both sexes tan with white bands.

Similar Species: Mexican Sister resembles female Pavon but has brighter orange patch and orange HW spot; chestnut and gray below. Laure male has more orange and blue reflections above; female larger, has orange patch on FW base below, small tail.

Life Cycle: Egg round, green. Caterpillar has light green stripes; branched head horns. Chrysalis angular, warty. Caterpillars feed on hackberry (*Celtis pallida*).

Flight: 2 or more broods; year-round.

Habitat: Forest openings.

Range: Mexico to Panama, straying into SE. Texas.

Pavon looks rather dull in the forest shade but the male can be brilliant in the full sun. Only the subspecies "theodora" enters Texas, appearing there rarely.

380, 653 Laure
(*Doxocopa laure*)

Description: 2–2¼" (51–63 mm). FW tips extended, curved, clipped; short HW tail. *Above, brown and banded. Bands on male mostly orange on FW,* white at bottom, continuing white across HW; *bright blue iridescent reflections surround white bands* on fresh males. *Female larger, lighter, mostly white-banded with orange or yellow FW tip patch above.* Both sexes below are tan to brown with broad whitish band, metallic silvery or pearly sheen, and sometimes orange or pinkish patch near FW base.

Similar Species: Mexican Sister has clipped FW with blunt tip, chestnut below, lacks blue reflections. Pavon female lacks blue, has less orange than Laure male; is smaller, lacks orange below.

Life Cycle: Hackberry (*Celtis pallida*) is host plant.
Flight: All months in favorable conditions.
Habitat: Riverbanks and roadsides in tropical forests.
Range: Mexico to Panama, straying into Texas.

Fresh males have bright blue reflections around their bands, and their margins shimmer purple in the sunshine. Laure and Pavon are hackberry butterfly relatives. The genus *Doxocopa* commonly ranges through much of the American Tropics, where members fly in sunny jungle openings, settling to drink on rotting fruit or wet sand. Laure has appeared north of the border only a few times, but on one occasion many invaded Brownsville, Texas.

SATYRS OR BROWNS
(Satyridae)

3000 species worldwide; about 50 in
North America. Most satyrs of
temperate climates are dingy, dull
brown or gray, with subtle yet complex
tones and textures. Their wings span
$1-2\frac{7}{8}''$ (25–73 mm), and usually have
eyespots, which may serve as targets to
divert attacks of predators from their
bodies. Like the brush-foots, with
which they are sometimes combined in
a single family, satyrs have reduced fore
legs, but differ in having the bases of
the fore wing veins conspicuously
swollen.

Satyrs usually inhabit places where their
mythical namesakes might reside: wood
nymphs live in forest glades, while grass
satyrs fly around meadow tussocks;
arctics and alpines haunt mountain
heights and treeless tundra. The North
American species span the range of
available grassy habitats. Their green-
striped caterpillars feed on grasses,
sedges, and canes, after emerging from
dome-shaped eggs. Most of the tapered
and twin-tailed caterpillars overwinter;
the arctic species sometimes take 2
summers to reach maturity. The brown
or green chrysalises are rounded and
smooth; they hang from grass stems or
among leaf litter.

Adult satyrs usually do not wander far;
their erratic, dancing flight carries
them briskly between grass blades and
tree trunks. Most species do not
emigrate, nor do they feed a great deal.
However, a few satyrs do wander,
taking nectar, sipping at puddles, and
drinking sap. Alpines cease activity and
hide at the slightest hint of clouds, but
pearly eyes thrive under shady
conditions. Some satyrs even fly at
dusk, when they may easily be
mistaken for moths.

694, 695 Pearly Eye
(*Enodia portlandia*)

Description: 1¾–2" (44–51 mm). Wings slightly
scalloped. *Above, male cocoa-brown to tan,
female brighter;* some populations have
yellowish areas; on both, zigzag pattern
of darker brown crosses FW and is
suggested on HW; outer half of HW
has row of black spots on lighter brown
field. Below, inner part of both wings
violet-brown to greenish-gray crossed
by darker brown lines toward base; tip
half whitish or yellowish with brown,
yellow-rimmed eyespots that have blue
or pearly pupil. Pearly sheen below.
*Female FW below has upper part of brown
line in from eyespots bowed out toward tip;
female usually has 4 eyespots on FW.*
Antennal club has orange tip.

Similar Species: Creole Pearly Eye male has dark patches
above; female above usually has more
complete FW spots and wavier convex
line inside eyespots on FW below; both
have more HW eyespots. Northern
Pearly Eye smaller, paler, less yellowish
below; antennal club has orange, black-
ringed tip; range more northerly.

Life Cycle: Egg greenish-white. Caterpillar
yellowish-green with matching red-
tipped horns at both ends; feeds on
giant cane (*Arundinaria gigantea*) and
perhaps maiden cane (*A. tecta*).
Chrysalis green, sometimes has bluish
tint.

Flight: Up to 3 broods; May–September.

Habitat: Moist, shady woodland spots near
stands of cane.

Range: S. Illinois and S. Virginia south to
Florida, and west to Texas, mostly on
coastal plain and in Mississippi Valley.

Formerly, the Pearly Eye and Northern
Pearly Eye were considered a single
species. However, their mating
behavior and other aspects of their
biology, structure, and appearance
clearly separate them. Pearly Eyes
alight on tree trunks, which males

energetically defend. Courtship takes place during the twilight hours. The 3 North American pearly eyes were previously included in the genus *Lethe*.

700, 703 Northern Pearly Eye
(*Enodia anthedon*)

Description: 1⅝–2″ (41–51 mm). Wings slightly scalloped. *Above, light brown* with zigzag dark brown lines across wings and *prominent row of 5 HW submarginal brown-black spots. Below,* opalescent gray, lilac, and light brown; *upper branch of main brown line on FW straight or slightly concave.* 4 FW *eyespots,* usually 7 on HW; *all have pupils.* Antennal club has orange, black-ringed tip.

Similar Species: Pearly Eye larger with richer coloring; ranges narrowly overlap. Creole Pearly Eye male has dark patches above, FW below of female has brown line with uppermost portion convexly curved.

Life Cycle: Caterpillar green with red-tipped, forked protuberances at both ends; feeds on grasses, including dropseed (*Muhlenbergia*) and long-awned wood grass (*Brachyelytrum erectum*).

Flight: 1 brood; June–August.

Habitat: Deciduous forest glades and margins.

Range: Manitoba and N. Arkansas east to Maine and Virginia.

Only recently distinguished from the Pearly Eye as a separate species, the Northern Pearly Eye feeds on forest grasses instead of the cane favored by the more southern Pearly Eye. Farther north, the Northern Pearly Eye is smaller and paler. It is local, confined to woods and their borders, and can tolerate more shade than most butterflies. It is not attracted to flowers, preferring willow or poplar sap, carrion, and scat; it often perches on tree trunks. In the much cooler

northwesternmost portion of its range,
the Northern Pearly Eye shifts its
habitat to more open woodlands with
sedge marshes; the butterfly changes its
behavior as well, gathering in groups
and perching on bushes.

696 Creole Pearly Eye
(*Enodia creola*)

Description: 2–2¼" (54–57 mm). Wings slightly
scalloped. *Above, olive-brown;* below,
lavender to bluish-gray. Wings above
and below have submarginal rows of
prominent dark eyespots (without
pupils above) and darker brown
irregular lines. Underside highlighted
by pearly reflections. *Male has
conspicuous patches of dark, raised scales
between veins of FW above. Female
normally has 5 full FW eyespots; portion
of brown line nearest eyespots irregularly
convex.*

Similar Species: Pearly Eye and Northern Pearly Eye
males both lack raised patches of dark
scales on FW above; both females
usually have 4 FW eyespots, with
straight brown line near uppermost
eyespot.

Life Cycle: Unrecorded. Host plant is maiden cane
(*Arundinaria tecta*).

Flight: 2 broods; April–October.

Habitat: Shaded hardwoods, especially among
swampy canebrakes and near marshy
streams.

Range: S. Illinois and SE. Virginia south to
E. Texas and Georgia.

The noticeable dark patches on the
wings of Creole males are composed of
raised sex scales, which produce sex
attractants called pheromones. The
Pearly Eye lacks these patches and
mates at dusk, while the Creole Pearly
Eye mates during the daytime.
Nonetheless, the Creole Pearly Eye flies
well into dusk and perhaps sometimes

at night. While these 2 shade-tolerant satyrs sometimes share habitats, they use different niches within them.

701, 704 **Eyed Brown**
(*Satyrodes eurydice*)

Description: 1⅝–2″ (41–51 mm). Wings rounded. *Above, warm tan to olive-brown,* often but not always with light patch on outer third of FW. *Variable dark eyespots near margins of both wings above and below;* eyespots have small white pupils below. *Below, light brown crossed by darker, deeply zigzagged lines near yellow-rimmed eyespots.*

Similar Species: Appalachian Brown darker above, violet-brown or gray-brown below, with gently wavy darker lines inward from eyespots.

Life Cycle: Caterpillar slender, light green, with lengthwise yellow and dark green stripes and red-tipped horns extending from head and tail. Host plants are sedges (*Carex*). Chrysalis green, with small, blunt hook on head.

Flight: 1 staggered brood; June–September.

Habitat: Open, damp meadows, sedge marshes, and wetter parts of prairies.

Range: S. central Northwest Territories, south through Dakotas to NE. Colorado, east across Canada and NE. United States south to N. Illinois and Delaware.

This locally abundant species occupies a very broad range and is familiar throughout much of the Northeast; its colonies are small, separate, and local. Some lepidopterists consider a dark, large race that is indigenous to the prairie states to be a separate species, called the Smoky Eyed Brown (*S. fumosus*). It has disappeared from much of its former territory as the moist grasslands have been drained, plowed, or inundated by reservoirs. For many years, the Eyed and

Appalachian browns were thought to be a single species. The browns were previously included in the genus *Lethe*.

702, 705 **Appalachian Brown**
(*Satyrodes appalachia*)

Description: 1⅝–2″ (41–51 mm). Wings rounded. *Above, olive-brown, paler toward tip of FW.* Below, gray-tan or violet-brown. *Black submarginal eyespots above and below,* those below ringed with white and yellow and with white pupils. *Darker brown lines cross underside; line nearest eyespots is mildly wavy, not radically zigzagged.*

Similar Species: Eyed Brown paler above; below, brown line near eyespots sharply zigzagged.

Life Cycle: Caterpillar slender, green and yellow striped; has 2 horns with red tips in front and rear; feeds on sedges (*Carex*). Chrysalis green, streamlined.

Flight: 1 brood in northern part of range, more farther south; June–August in Maine and June–October in N. Florida.

Habitat: Wooded areas near standing or slow-moving water: swamps, streamsides, bogs, and springs; also edges and roadsides along woods.

Range: E. South Dakota, Quebec, and Maine, south in Appalachians to Mississippi and W. central Florida.

The Appalachian Brown and the Eyed Brown are excellent examples of sibling species—close relatives, similar in appearance, that may occur in many of the same areas, and yet possess quite different physical and ecological characteristics. The Appalachian Brown dwells in dispersed, damp, brushy marshes and is never plentiful. In contrast, the Eyed Brown lives in wet, open meadows where it reaches much greater abundance. Both species are vulnerable to development of inland wetlands.

Nabokov's Satyr
(*Cyllopsis pyracmon*)

Description: 1½–1¾" (38–44 mm). *FW tip extends slightly.* Above, dull brown with reddish cast, especially along veins; small dark marks on outer HW margin. Grayish-tan below; HW has black marginal marks with silver centers, more silver spots, and a silver line around HW margin, lined with reddish-brown. *Faint brown lines crossing underside are continuous, not offset between FW and HW; their coloring runs out along foremost veins on HW in form of rays* from 2nd line in from margin.

Similar Species: Sonoran Satyr has HW lines below that do not form rays along veins. Canyonland Satyr's lines below are offset between FW and HW; lacks HW rays.

Life Cycle: Unknown. Host plants probably grasses (Poaceae).

Flight: Probably 2 broods; mostly August–September.

Habitat: Sunny openings in arid woodlands.

Range: S. Arizona, S. New Mexico south through Mexico to Guatemala.

The North American race of this species was named in honor of the novelist Vladimir Nabokov, who studied the satyrs of the genus *Cyllopsis*. The Nabokov's, Sonoran, and Canyonland satyrs are subtly-colored butterflies and impossible to tell apart in flight. The genus *Cyllopsis* was previously included in the genus *Euptychia*.

Sonoran Satyr
(*Cyllopsis henshawi*)

Description: 1½–1¾" (38–44 mm). *Above,* male rust-brown with *small dark marks* on outer margin of HW; female has reddish cast and similar markings. Both sexes *below* have tan FW; HW overcast with grayish-purplish, crossed by weak brown lines; the *2nd line in from margin straight and continuous, not offset between FW and HW;* marginal marks are rimmed with silver, which extends around HW margin.

Similar Species: Canyonland Satyr has brown line just in from margin offset between FW and HW. Nabokov's Satyr has brown lines below running out along veins in rays.

Life Cycle: Egg green, smooth, rounded. Otherwise unknown. Host plants probably grasses (Poaceae).

Flight: 2 broods; June—July and September—October.

Habitat: Arid country in canyons with moisture and vegetation; also open woods.

Range: E. Arizona, W. New Mexico, and W. Texas into Sonora, Mexico.

Both Sonoran and Canyonland satyrs inhabit much of the same southwestern range, but the former does not venture as far northward as the latter. In the heat of the desert mountain day, these satyrs seek shade in open woods.

717 Canyonland Satyr
(*Cyllopsis pertepida*)

Description: 1½–1¾" (38–44 mm). Above, reddish-brown. *Below,* FW reddish-brown; HW purplish-brown (female) or rust-colored (male) *with silvery markings along HW margin and blue-black spots in bluish-silver field near middle of outer margin.* Light brown lines cross underside, *2nd line in from edge* is not continuous between FW and HW, but *begins on HW somewhat offset toward edge* from where it ends on FW.

Similar Species: Sonoran and Nabokov's satyrs differ in shape of brown line along disk below, which is continuous from FW to HW.

Life Cycle: Unknown. Probably feeds on grasses (Poaceae) or sedges (Cyperaceae).

Flight: 2 broods southward; May—June and

August–October, a briefer period
farther north.

Habitat: Foothills, low mountains, and plateau
canyons among arid lands.

Range: Colorado, S. Utah, S. Nevada south
through Arizona, New Mexico, and
Texas into Mexico.

The Canyonland Satyr haunts cooler
parts of arid lands in the spring and
again during the hottest part of the
summer. It can be found dodging in
and out among scrub oaks and junipers
along the rims of the Black Canyon of
the Gunnison River and among the
ruins of Mesa Verde. Like most satyrs,
it settles among the duff and vanishes
from sight before flying again. This
species was previously known as
Euptychia dorothea, which is now
considered a subspecies (*C. p. dorothea*)
of the Canyonland Satyr.

716 Gemmed Satyr
(*Cyllopsis gemma*)

Description: 1¼–1⅜″ (32–35 mm). Wings
rounded. *Above, soft brown with tiny
black eyespots on HW margin.* Below,
gray-brown or warm brown, crossed by
darker brown wavy lines; *purplish-gray
patch on HW margin contains small
metallic-blue and silvery eyespots.*

Similar Species: Carolina Satyr has eyespots below
instead of metallic patches, and has
stronger dark lines.

Life Cycle: Egg globelike, covered with network of
lines. Caterpillar light green in early
summer, light brown toward fall; both
forms are striped with darker lines and
have 2 head horns and 2 tails. Chrysalis
also green or brown. Host plants are
grasses (Poaceae), including Bermuda
grass (*Cynodon dactylon*) in Texas.

Flight: 2 broods; March–October in Georgia,
briefer period farther north.

Habitat: Shaded, moist, and grassy areas: along

streams and ponds in open woods, meadows with water and long grasses; also pine flats.

Range: S. Illinois and S. Virginia south to Texas and Mexico and east to central Florida.

Gemmed Satyrs from Texas and Mexico are redder than those from farther north and east. Although widespread in the sun belt, this species occurs locally and is rarely common. It has been suggested that the caterpillar's change in color, from green in summer to brown in the fall, allows it to blend with the seasonally changing colors of the grasses.

736, 739 **Hermes Satyr**
(*Hermeuptychia hermes*)

Description: 1⅛–1½″ (28–38 mm). *Small.* Wings rounded. *Above, dark gray-brown with eyespots tiny or absent. Below,* brown with whitish overscaling, crossed by darker lines and *submarginal HW row of 6 eyespots* with light rims and bluish pupils, *2nd and 5th being largest.*

Similar Species: In Texas, where ranges overlap, Carolina and Hermes satyrs cannot be distinguished in the field.

Life Cycle: Unreported. Host plants probably grasses (Poaceae).

Flight: Year-round in overlapping broods.

Habitat: Moist subtropical woodlands.

Range: S. Texas south into Mexico.

The Hermes and Carolina satyrs were formerly considered subspecies of a single species in the genus *Euptychia.* Although they are very similar, they are now known to be different species and have been assigned a new generic name. The Carolina replaces the Hermes in most of the southeastern United States.

712 Carolina Satyr
(*Hermeuptychia sosybius*)

Description: 1⅛–1⅝″ (28–41 mm). Wings rounded. *Above, dark brown with eyespots minute or absent. Below,* brown frosted with white scales, crossed by darker brown lines and *submarginal HW row of 6 small eyespots with light rims and bluish pupils, 2nd and 5th largest.*

Similar Species: Little Wood Satyr larger, tanner beneath, with conspicuous eyespots above and below. Mitchell's Marsh Satyr and Georgia Satyr larger, with more prominent eyespots surrounded by reddish lines, the 3rd and 4th being largest. Hermes and Carolina satyrs cannot be distinguished in the field where ranges overlap.

Life Cycle: Egg green, rounded. Caterpillar light green with dark green stripes and fine, yellowish pile. Host plants are various grasses (Poaceae). Chrysalis curved, olive.

Flight: Successive broods in Florida; year-round. 2 broods farther north; spring–late summer.

Habitat: Deciduous woodlands with standing water, pinelands, and shady meadows; more common at lower altitudes.

Range: Southeast, from New Jersey to Florida and around Gulf to Texas, north in Mississippi Valley at least to Kentucky.

The Carolina Satyr is one of the smallest satyrs in North America; it is abundant and widespread in the Southeast. The region's luxuriant growth suits this satyr's liking for moisture, shade, and grasses. Unlike most members of the family, the Carolina Satyr visits flowers frequently. Some specialists consider the Carolina and Hermes satyrs the same species.

714 Georgia Satyr
(*Neonympha areolatus*)

Description: 1½–1¾″ (38–44 mm). Wings rounded. Above, dull brown. *Below, dull brown crossed by prominent brick-red lines;* HW lines enclose *row of long and narrow oval eyespots* with yellow rims (red farther south) and bluish-silver centers.

Similar Species: Other satyrs have round eyespots, not elongated ovals.

Life Cycle: Egg pale yellow-green. Caterpillar chartreuse with dark stripes, brownish pile, and 2 small horns on head. Host plants are grasses (Poaceae), including Indian grass (*Sorghastrum nutans*). Chrysalis green with whitish markings.

Flight: 1 or 2 broods in North; April–September in Virginia, a briefer period beginning in June farther north. 3 or more broods in Florida; year-round.

Habitat: Open, grassy wetlands, especially among taller grasses; drier areas in South, including pine flats and rural roadsides.

Range: Coastal New Jersey south along seaboard to S. Florida, around Gulf to Texas, and north to mid-Louisiana, Mississippi, Tennessee, and Kentucky.

The Georgia Satyr was studied extensively and painted in all its stages by John Abbot, the pioneer Georgia naturalist of the late 18th and early 19th centuries. Southern populations tend to have longer and narrower eyespots. This species often lurks in high grass, weaving and wending quite close to the ground among the stems and blades. This and related species were formerly included in the genus *Euptychia*, the grass satyrs.

713 Mitchell's Marsh Satyr
(*Neonympha mitchellii*)

Description: 1½–1¾" (38–44 mm). Wings rounded. Above, mahogany-brown, sometimes with eyespots from below vaguely showing through. *Below,* dull brown crossed by dark orange-brown lines; *4 submarginal eyespots on FW and 5–6 on HW have yellow rims and bluish-silver centers; eyespots are round or slightly oval* (but not narrow and elongated); *largest HW eyespots usually 3rd and 4th.* Beyond eyespot row is reddish line, followed by gray line near margin.

Similar Species: Little Wood Satyr has eyespots above; below, 2nd and 5th HW eyespots are largest. Georgia Satyr has long, narrow, oval eyespots.

Life Cycle: Egg pale green, globe-shaped. Caterpillar lime-green with contrasting stripes, fine, white, raised stippling all over, and 2 fleshy horns extending at rear. Host plants probably sedges (*Carex*). Chrysalis lime-green, rounded over back with large bump protruding from back of head, small horns at front.

Flight: 1 brood during 2 week period; normally first 2 weeks of July, may be advanced or delayed depending upon weather.

Habitat: Tamarack bogs with poison sumac, adjacent wet meadows, and slightly drier meadows.

Range: S. Michigan, N. Indiana, N. Ohio, and N. New Jersey.

Discovered in Michigan in the 1880's, this little brown butterfly has one of the most restricted ranges in North America. The special kinds of bogs it requires have largely been eliminated by agriculture and urban development. Although there are populations in both New Jersey and Ohio, the species has never been found in intervening Pennsylvania. Males fly for a week or 10 days before the females make their appearance. Fragile and lightly scaled, these butterflies quickly lose their rich brown coloring. Mitchell's Marsh Satyrs are characterized by their bobbing flight pattern, unlike the stronger flight of the Little Wood Satyrs and Eyed Browns with which they may occur.

706, 709 Little Wood Satyr
(*Megisto cymela*)

Description: 1¾–1⅞" (44–48 mm). Wings rounded. *Above, dull brown; below, dull brown to tan, crossed by darker brown lines. Each wing, both above and below, has 2 prominent black eyespots* with yellow rims and 2 light pupils; smaller eyespots may be clustered around large ones; margins are rimmed with brown lines. Upperside violet-gray in S. Florida.

Similar Species: Eyed Brown larger, has more eyespots of even size. No other small satyr has prominent eyespots in pairs both above and below.

Life Cycle: Egg pale yellowish-green. Caterpillar brown, stippled with minute white tubercles; overwinters partially grown. Host plants are grasses (Poaceae) and possibly sedges (Cyperaceae). Chrysalis rounded and curved at rear toward point of attachment, usually on a sedge stem.

Flight: 1 brood in North; April until midsummer, beginning as late as June in Wisconsin. 2 broods in far southern part of range; March–October.

Habitat: Deciduous woods with glades and ponds, pinelands, nearby thickets and meadows, prairie reservoir margins and groves, salt bays and brackish streamsides, hammocks, plantations, and clearings.

Range: Saskatchewan, Dakotas, NE. Colorado, Texas, and NE. Mexico east throughout S. Canada and E. United States.

Despite its name, the Little Wood Satyr is larger than most of the other small satyrs formerly grouped together in the genus *Euptychia*. It is highly adaptable to moderate environmental change, requiring only that some woods, brush, grass, and moisture remain to provide shelter and food. With a fairly long flight period and prolific reproduction, the Little Wood Satyr can be enormously abundant under the right conditions. The adult expertly negotiates tall grass and thick shrubbery with its dancing, slow-motion flight.

707, 710   Red Satyr
(Megisto rubricata)

Description: 1¾–1⅞″ (44–48 mm). Wings rounded. *Above, rich brown with strong copper-colored highlights on middle of each wing. Below, FW has large copper-orange patch,* otherwise both wings frosty gray-brown crossed by dark lines. 1 prominent black eyespot near tip of FW, 1 black eyespot near outer edge of HW both above and below; HW below also has 1 black eyespot near HW costa and a few silvery eyespots near larger ones.

Similar Species: Red-eyed Wood Nymph has reddish patches, but is larger, darker below, and has 2 large FW eyespots above.

Life Cycle: Unreported. Caterpillar has been reared on St. Augustine grass (*Stenotaphrum secundatum*) and Bermuda grass (*Cynodon dactylon*) in Texas.

Flight: 2 or more broods; in Texas, March–May and August–September; in montane Arizona, July and August.

Habitat: Hill country, live oak groves, and oak-lined canyons and waterholes.

Range: Arizona, New Mexico, Oklahoma, and Texas south to Mexico.

Most North American satyrs are subtly shaded, but with its bright copper-colored patches, the Red Satyr is one of the most colorful members of the group. It frequents hot and arid terrain.

697 Pine Satyr
(*Paramacera allyni*)

Description: 1⅜–1¾" (35–44 mm). *Olive-brown above; male has darker sex patches on FW;* female has some reddish shading, especially on FW. *Broad tan submarginal bands enclose a number of prominent black eyespots (2 or 3 on FW, 4 or 5 on HW). Below, reddish-brown FW has darker rust-colored lines and near tip 1 large, yellow-rimmed eyespot with white pupil, 2 or 3* smaller eyespots around it; *HW violet-gray, with brown median band between 2 rust-colored lines;* around margin row of *rimmed eyespots with pupils.*

Similar Species: Red-eyed Wood Nymph has large rust-colored patches around eyespots.

Life Cycle: Unrecorded.

Flight: 2 broods likely; June–September, most common July–August.

Habitat: Mountain pine forests.

Range: SE. Arizona, perhaps W. New Mexico, Texas, and N. Mexico.

Until recently, the Pine Satyr was thought to be the same species as the Xicaque Satyr (*P. xicaque*) of Mexico. It has now been named a fully separate species. The Pine Satyr and the Chiricahua Pine White can be found in the same mountain ranges in southern Arizona. They are among our most unusual butterflies.

715 Hayden's Ringlet
"Wyoming Ringlet"
(*Coenonympha haydenii*)

Description: 1½–1⅞" (38–48 mm). Wings rounded. *Above, rich brown* aging to a

dull gray-brown, *without markings.*
Below, grainy, tan-gray; *HW margin*
rimmed by row of small, distinct black
eyespots with pale pupils and saffron rings.
Between eyespots and HW fringe is
thin silver line; silver is repeated
minutely within eyespots.

Life Cycle: Unknown. Caterpillar feeds on grasses
(Poaceae) or sedges (Cyperaceae).

Flight: 1 brood; late June–August.

Habitat: Mountain streamside meadows, open
coniferous woods with grassy floors,
grassy slopes and trailsides in middle
elevation mountains.

Range: SW. Montana, SE. Idaho, and W.
Wyoming.

Hayden's Ringlet is a member of a
small guild of Northern Rockies
butterflies living nowhere else, and
perhaps left behind from a larger,
earlier distribution. In spite of its
narrow range, it is common within its
local colonies. Yellowstone, the Tetons,
and the Absaroka Wilderness Area
nurture large populations. Males spend
time on drier hillsides, scudding along
mountain trails, while females tend to
prefer moister brookside meadows.

722 Kodiak Ringlet
(*Coenonympha kodiak*)

Description: 1–1⅜" (25–35 mm). *Above,* uniform
dark olive-gray or deep, dusky ocher,
unmarked. Below, overall dusky olive or
ocherous at FW base, pale olive-gray at
tip, and HW dark olive at base, lighter
at margin with a vague whitish line
between color fields on both wings;
may have 1 or 2 faint eyespots.

Similar Species: Ocher Ringlet is larger, brighter ocher
above, and usually has some eyespots
above.

Life Cycle: Undescribed. Probably uses grasses
(Poaceae) as host plants.

Flight: 1 brood; briefly June–July.

Habitat: Arctic tundra, boreal forest openings.
Range: Alaska, Yukon and Mackenzie District of Northwest Territories; also Siberia.

All of the North American ringlets in the genus *Coenonympha* except Hayden's Ringlet may actually belong to the Eurasian species known as the Large Heath (*C. tullia*). The Kodiak Ringlet on Kodiak Island is one of the most distinct races, lacking any ocher scales whatever. Elsewhere the Kodiak Ringlet appears smaller and duller than the Northwest and Ocher ringlets, which fly farther to the south.

721 Prairie Ringlet
(*Coenonympha inornata*)

Description: 1–1⅞" (25–48 mm). Pale orange-brown to rich ocher above, sometimes dusky, without markings. *Below, FW mostly ocher, paler at tip, sometimes with 1 small black, yellow-rimmed eyespot;* HW below pale dusky olive with fragmentary, curved white band across disk and whitish veins; HW often darker at base and sometimes has a partial row of minute eyespots around margin.

Similar Species: Best separated from Northwest and Ocher ringlets by locality; where they overlap, those species usually have more developed eyespots or more white below, but they may not be distinguishable.

Life Cycle: Egg yellowish. Caterpillar tan or olive with 2 tails. Various grasses (Poaceae) are host plants. Chrysalis rounded, greenish-brown.

Flight: 2 broods; June–September.

Habitat: Prairies, meadows, pastures, and grassy woodland glades and embankments.

Range: Western Ontario east to Labrador, south to E. British Columbia, Dakotas, Iowa, Wisconsin, New England, and N. New York.

Along with its similar relatives, the Prairie Ringlet may belong to the *C. tullia* complex. The Nipisquit Ringlet (*C. nipisquit*), another dark, mustard-colored ringlet of the Northeast, may also be a separate species. It occurs only in northeastern New Brunswick and on a few of the Thousand Islands of New York and Canada in the St. Lawrence River. A single brood of these dark ringlets flies in late summer. The relatively recent colonization of northern New York by the Prairie Ringlet has raised some concern that it might swamp the local, endemic Nipisquit Ringlet.

719, 724 Ocher Ringlet
(*Coenonympha ochracea*)

Description: 1–1⅞″ (30–48 mm). Above, bright orange-ocher, sometimes with 1 small eyespot near FW tip, otherwise unmarked. *Below,* FW ocher, crossed by long or short cream-colored streak; *olive-brown tip has 1 eyespot;* HW grayish-olive with a streaky cream-colored median band that may be complete or fragmented, irregular and broader in middle; HW margin usually has at least 1, often a full row of 6 or more black eyespots with yellow rims and sometimes yellow pupils.

Similar Species: Northwest and Prairie ringlets difficult to separate from this species, but normally have fewer or no eyespots, paler or duskier ocher. Kodiak Ringlet much less ocherous than Mackenzie District race of Ocher Ringlet.

Life Cycle: Unreported. Host plants probably grasses (Poaceae).

Flight: 1 brood; May–July.

Habitat: Foothills, sagelands, grassy glades and meadows in ponderosa pine forests, alpine tundra, and mountain canyons.

Range: Rocky Mountain States from Montana south to Arizona and New Mexico, probably also Black Hills; an isolated population in Mackenzie District, Northwest Territories.

Some lepidopterists refer to the orange-tinged ringlets of North America as Ocher Ringlets, under the scientific name *C. tullia*. Others separate the ringlets into several species and give the name *C. ochracea* to the brightest of the ocher ringlets. Despite its abundance in early summer, the Ocher Ringlet's small size, soft colors, and weak, low flight make it inconspicuous among the grasses.

723 Northwest Ringlet
(*Coenonympha ampelos*)

Description: 1–1⅞″ (25–48 mm). Yellow-buff to deep ocher *above, clear and unmarked* (Great Basin race may be very pale, with whitish males). *Below,* FW ocher, crossed by partial whitish band, frosty olive-gray at tip; *HW* greenish-gray or brownish-olive *crossed by broad, zigzag, and broken cream-colored streak* and often light-scaled veins. Usually without eyespots, although sometimes has 1 FW eyespot below and small eyespots in partial row on HW margin below, especially in eastern part of range.

Similar Species: Ocher and Prairie ringlets are difficult to distinguish from this species. Prairie usually is less bright, with smaller cream-colored band below. Ocher normally has more eyespots and richer color.

Life Cycle: Probably resembles that of California Ringlet. Feeds on grasses (Poaceae). Caterpillars of 2nd brood overwinter.

Flight: 2 overlapping broods; April–September.

Habitat: Many grassy places, including marshes, mountain forests and meadows, sagelands, maritime bluffs, disturbed areas, pastures, and vacant lots.

Range: S. British Columbia, Washington, Idaho, Oregon, and N. Nevada.

The Northwest Ringlet seems to dislike too much moisture: it thrives in the arid Columbia Basin, can tolerate the Puget Sound area, but is absent from the very wet Olympic Peninsula and the coast of Washington. In the spring, meadows around Seattle and Portland abound with Northwest Ringlets and Sara Orangetips. How the Northwest Ringlet overlaps with the Prairie and Ocher ringlets on the north and east, and the California Ringlets on the south, is not clear. Some lepidopterists believe all are part of the superspecies known as *C. tullia*.

720 California Ringlet
(*Coenonympha california*)

Description: 1–1¾" (25–44 mm). Clear, *dull white above* and dust-gray below. *Underside is crossed by a pale, irregular, and usually intermittent median band*, often with small eyespots on both wings. Southern summer brood is slightly ocher-washed above and tan below, darker inside pale band.

Similar Species: Northwest Ringlet ocher-colored above.

Life Cycle: Egg yellowish, ridged, and pitted, rounded at bottom, flattened on top. Mature caterpillar brownish or olive-green with alternating light and dark lengthwise stripes and 2 tails. Chrysalis brown or green with brown streaks. Host plants are grasses (Poaceae).

Flight: 3 broods in S. California coastal hills; February–October. 2 broods northward; spring and late summer.

Habitat: Grassy hillsides, chaparral, foothills, and lower mountains, coastal lowland grasslands, and oak savannah.

Range: S. Cascade and Siskiyou mountains of SW. Oregon and California. Absent from higher peaks and Mojave Desert.

Of all the closely related North American ringlets, the California is perhaps the most distinctive with its whitish wings. Abundant in the grassy hills around San Francisco and Los Angeles, this ringlet is often mistaken for a white moth. California Ringlets, unlike those of the Northwest and Rockies, fail to reach the alpine heights. Fragile and weak fliers, their wings become tattered and lose scales after only a few days. Like all ringlets, they virtually never perch with their wings spread but close them tightly as soon as they alight.

698, 725 **Large Wood Nymph**
"Blue-eyed Grayling"
(*Cercyonis pegala*)

Description: 2–2⅞" (52–73 mm). *Large.* Highly variable. *Above, light cocoa-brown to deep chocolate-brown* (very pale in N. Great Basin). *Below, paler and heavily striated* with darker scales. *Normally FW above and below has 1 or 2 small to very large black eyespots,* often yellow-rimmed, with small white or large blue pupil; *eyespots may lie in a vague or discrete broad band of bright or dark yellow.* HW above may have small eyespots; HW below may have 1 or 2 small eyespots or a full row of 6 eyespots. *HW below usually divided* into darker inner and lighter outer portion *by single zigzagged, dark line.* Female normally larger, paler, with bigger eyespots.

Similar Species: No other large eastern satyr has 2 large FW eyespots. In West, Great Basin Wood Nymph is smaller and often has darker band within 2 dark lines across HW below.

Life Cycle: Egg lemon-yellow, keg-shaped, and ribbed. Caterpillar grass-green, with 4 lengthwise yellow lines, fine, fuzzy pile, and 2 reddish tails; overwinters shortly after hatching; feeds on various grasses (Poaceae). Chrysalis green, rather plump.

Flight: 1 brood; generally June–August or September, varying with locality.

Habitat: Open oak, pine, and other woodlands; meadows, fields, and along slow watercourses with long, overhanging grasses; marshes, prairie groves, thickets, and roadsides.

Range: Central Canada to central California, Texas and central Florida. Absent from Pacific Northwest Coast and much of Gulf region.

The Large Wood Nymph occupies much of North America; it is the largest wood nymph and the only one east of the Mississippi. Extremely variable, this butterfly has been given dozens of names. Today, all are considered a single species. As they perch on tree trunks or boughs to bask or drink sap, Large Wood Nymphs blend beautifully with the bark. When disturbed or seeking mates, they fly erratically through tall grasses, with little speed but great skill and endurance. Western wood nymphs visit such flowers as alfalfa and spiraea, while eastern populations seem to favor rotting fruit.

---

726 Red-eyed Wood Nymph
"Mead's Wood Nymph"
(*Cercyonis meadii*)

Description: 1⅝–2″ (41–56 mm). *Above, dark brown with bright rust-red flush* over most of FW or in patch surrounding *2 prominent FW eyespots* (larger on female). *Below, striated; FW red-flushed, HW 2-toned* brown with dark dividing line; lighter marginal area has small eyespots. HW below frosty-whitish in San Luis Valley, Colorado.

Similar Species: Other wood nymphs lack reddish flush, except Large Wood Nymph in Grand

Canyon, which is larger. Red Satyr has only 1 eyespot on FW outside reddish area. Pine Satyr is light tan above, has more eyespots on lighter HW below.

Life Cycle: Egg pale yellow-orange. Caterpillar green with lighter stripes and spots on sides; feeds on grasses (Poaceae). Chrysalis light green; hangs in tuft of grass.

Flight: 1 brood likely; mainly August, July at lower altitudes and latitudes, into September in higher ones.

Habitat: Clearings, roadsides, and dry meadows in mid-elevation mountain pine forests.

Range: W. Dakotas and E. Montana, south discontinuously through E. Wyoming, Colorado, Utah, Arizona, New Mexico, W. Texas, and N. Mexico.

The Red-eyed Wood Nymph has a patchy widespread distribution with relatively little variation between populations. It may have more specific ecological requirements than other wood nymphs or may have once occurred more continuously. It generally lives in higher, moister, cooler, and more forested places than other western wood nymphs. The species bears the name of Theodore Mead, a pioneer lepidopterist who came West in 1871 with the Wheeler Expedition.

708, 711 Great Basin Wood Nymph
(Cercyonis sthenele)

Description: 1⅜–2" (35–51 mm). Above, light to dark brown. *Below, pale brown to silvery gray-brown striated,* with outer half of HW paler than inner and often bearing 2 small but prominent eyespots; *HW also has band composed of darker scaling between 2 gently wavy or mildly zigzagged dark lines. 2 large eyespots usually appear on outer FW,* equidistant from margin above and below; sometimes yellow around eyespots but never in strong patches.

Similar Species: Large Wood Nymph larger and HW below usually lacks whole band. Red-eyed Wood Nymph has rust-colored flush. Dark Wood Nymph usually smaller, lower FW eyespot closer to margin than upper and HW band below radically zigzagged, especially on the innermost dark edge of band.

Life Cycle: Egg whitish, keg-shaped. Caterpillar light green with dark back stripes and yellowish side stripes; feeds on grasses (Poaceae). Chrysalis light green.

Flight: 1 brood; June–July.

Habitat: Basin sagelands, dry shrub steppes, oak-lined arid canyons, and pinyon-juniper woodlands.

Range: E. British Columbia and Washington south and west through Great Basin to Four Corners area of Utah, Colorado, Arizona, and New Mexico; also central and S. California and Baja California.

One widespread population of the Great Basin Wood Nymph has a dappled and frosty-whitish underside, but populations from outside or along the edges of the Great Basin retain the chocolate-brown hue characteristic of wood nymphs. The name *C. sthenele* used to apply only to a San Francisco sand dune subspecies which is now extinct; the Great Basin Wood Nymph then went by the names *C. paula, C. silvestris,* and *C. behrii.*

699 Dark Wood Nymph
(*Cercyonis oetus*)

Description: 1½–1⅞" (38–48 mm). *Above,* black-brown or dark-brown, *usually with 1 or 2 black FW eyespots.* Male has prominent, dark sex patch on FW above. *Below, FW similar to upperside but lighter, with 2 eyespots with pupils and usually yellow rims,* the lower one noticeably smaller and nearer the wing margin than the upper (female usually has larger eyespots); *HW ranges from nearly black to mottled* gray-brown and black (frosty-whitish in parts of central Nevada), crossed by darker band defined by dark brown lines, which are sharply zigzagged (especially inner one).

Similar Species: Other wood nymphs generally larger. Great Basin Wood Nymph larger, paler, with FW eyespots below equidistant from wing edge and HW band below only gently zigzagged.

Life Cycle: Egg cream-colored, keg-shaped. Caterpillar green with dark green and white back stripes, yellow side stripes. Host plants are grasses (Poaceae). Chrysalis may be green, brown, or black with light stripes.

Flight: 1 brood; July–August.

Habitat: Dry, open grasslands including valleys, sage flats, ponderosa pine forest openings, grassy mountain roadsides, and lakesides.

Range: Central British Columbia south in Cascades and Sierra Nevada to central California, and east to eastern edge of Rockies and Black Hills.

Strictly limited to the mountainous West, the Dark Wood Nymph eschews the coast, the Arctic, and the arctic-alpine, but is otherwise widespread and tremendously abundant throughout late summer. It dodges in and out of sagebrush clumps, skillfully avoiding predators. The Dark Wood Nymph and the other *Cercyonis* species visit flowers more than most satyrs, choosing the nectar of mock orange, mint, alfalfa, rabbit brush, and others.

757 Northwest Alpine
"Vidler's Alpine"
(*Erebia vidleri*)

Description: 1¾–2" (44–51 mm). *Above, dark chocolate-brown crossed by rich rust-orange or pale orange irregular bands,* usually reduced on HW and not reaching lower margin. *Below, gray-brown with jagged yellow-orange band on FW and ash-gray band across entire width of HW; band has many small eyespots with white pupils.* Conspicuously checkered fringes give appearance of scalloped outline.

Similar Species: Common Alpine has smooth, russet (not orange) patches around eyespots; may be grayish on HW beneath but lacks distinct gray band across middle, and fringes are not checkered.

Life Cycle: Caterpillar feeds on grasses (Poaceae).

Flight: 1 brood; July–August.

Habitat: Moist, flowery alpine and subalpine meadows and slopes; steep, wet, open ravines in high forests.

Range: N. central British Columbia, south through Coast Ranges (but not on Vancouver Island) to Washington in Olympic Mountains, North Cascades, and Okanogan Highlands; absent from Oregon.

Seven of the 10 North American species of alpines (*Erebia*) also occur in Asia. Even the 3 strictly North American species have close Old World relatives. The Northwest Alpine is extremely similar to the Japanese Alpine (*E. nipponica*). If they are indeed separate species, they must have evolved fairly recently from common stock. Diverse and widespread in the Alps and across Asia, alpines dispersed into the New World across ancient Bering Sea land bridges, and became distinct species in the New World. The Northwest Alpine lives nowhere else but the Pacific Northwest; its exact southern limits remain unknown, lying probably within the Alpine Lakes

Wilderness Area of Washington. It is
the most colorful North American
Erebia.

Arctic Alpine
"Ross's Alpine"
(*Erebia rossii*)

Description: 1½–1⅞″ (38–48 mm). *Above, dark
brown, usually with 2 small eyespots near
tip of FW, orange-circled* and without
pupils, sometimes absent. *Below,* FW
tinged reddish and eyespots repeated;
*HW banded gray-brown and dark brown,
lacking eyespots, white spots, or checkered
fringes.*

Similar Species: Spruce Bog Alpine has brown-
checkered fringes, white spots on HW
below, and usually more FW eyespots.
Yukon Alpine usually has 4 FW
eyespots.

Life Cycle: Unknown. Probably feeds on grasses
(Poaceae) or sedges (Cyperaceae).

Flight: 1 brood; late June–July.

Habitat: In Alaska, usually below timberline in
wooded valleys; arctic barrens to east.

Range: Alaska south in mountains to N.
British Columbia and east along
Arctic Ocean to W. Hudson Bay,
Southampton Island, and Baffin Island
(southern limits across most of Canada
uncertain); also Siberia.

The Arctic Alpine exists across a vast
expanse of the Northland, in an array of
upland and high latitude habitats.
There is a good chance of finding it in
Mount McKinley National Park in
July.

749, 752 Spruce Bog Alpine
(*Erebia disa*)

Description: 1¾–2″ (44–51 mm). *Above, blackish-
brown, normally FW has 4 eyespots,*

orange-rimmed and often with minute white pupils; orange rings may be broad and nearly contiguous or become a diffuse, orange patch. *Below, FW chestnut-brown with 4 eyespots repeated; HW frosted hoary-brown usually with broad, darker band across middle and 2 white spots*—1 along costa, 1 at end of cell; spots sometimes join, becoming white patch bordering dark band. *Fringes checkered tan and gray.*

Similar Species: Arctic Alpine usually has 2 eyespots. Yukon Alpine lacks white spots below. Common Alpine usually has HW eyespots.

Life Cycle: Unreported. Host plants are probably grasses (Poaceae).

Flight: 1 brood; June–July.

Habitat: Spruce bogs, taiga forests, and damp arctic slopes.

Range: Alaskan North Slope east to Hudson Bay, south to S. British Columbia and N. Minnesota; also arctic Eurasia.

As its name implies, the Spruce Bog Alpine is particularly fond of moist locations. Unlike most alpines, it even takes moisture at mud puddles. When it settles in its characteristic posture with its wings closed, the white spots are clearly evident.

756 Magdalena Alpine
"Rockslide Alpine"
(*Erebia magdalena*)

Description: 2–2½" (51–64 mm). *All black or dark brown with no markings whatever.* Some Colorado populations have vague whitish band across HW below; Alaskan-Asian race has more or less auburn cast to wings, and sometimes almost red patches.

Similar Species: Arctic Alpine has red-rimmed eyespots on FW. Banded Alpine prominently marked below with light and dark bands.

Life Cycle: Egg yellow-brown. Caterpillar sluglike, bright green, with contrasting stripes, 2 tails at rear; feeds on alpine grasses (Poaceae); may take 2 years to mature. Chrysalis blackish, compactly rounded.

Flight: 1 brood; late June–early August, depending upon snowmelt, usually peaking mid-July.

Habitat: Usually confined to rockslides above timberline in higher Rockies and to arctic screes and fell-fields.

Range: Alaska and Yukon, S. Montana, W. Wyoming, NE. Utah, Colorado, and N. New Mexico; also Siberia.

When very fresh, the Magdalena's plush black wings shine with green and mauve iridescence, but the loose dusting of prismatic scales soon brushes off and its wings weather from deep black to brown. Agile and fleet, male Magdalenas course up and down rockslides, sometimes stopping to take nectar from pink clumps of campion. The females are seldom seen at all, although they may be found basking on bare granite. Trail Ridge Road in Rocky Mountain National Park, Colorado, is one of the few places where this impressive ebony alpine can be readily watched.

718 Banded Alpine
(*Erebia fasciata*)

Description: 2–2¼" (51–57 mm). *Above,* male all black, female dark reddish-brown, both *unmarked. Below, alternating broad bands of pale gray and brown cross wings* (more contrasting in male).

Similar Species: Magdalena Alpine lacks bands below. White-veined Arctic gray-brown above.

Life Cycle: Undescribed. Host plants probably grasses (Poaceae).

Flight: 1 brood; late June, for 2 weeks or less.

Habitat: Wet tundra tussocks, lee sides of ridges, and gullies.

Range: High Arctic Alaska north of treeline, east to Hudson Bay; also Asian High Arctic.

When its wings are closed, displaying its distinctively banded underside, this species looks more like an arctic (*Oeneis*) than any other alpine (*Erebia*). The bands break up the wing pattern, perhaps making the insect less readily apparent to predators. The actual tones of the bands and the reddishness of the upperside vary dramatically from place to place. While arctics settle among lichen-covered stones, the Banded Alpine tends to land in tundra tussocks or on boggy vegetation.

748, 751 Red-disked Alpine
(*Erebia discoidalis*)

Description: 1¾–2″ (44–51 mm). *Above,* blackish-brown, *FW dominated by diffuse rust-red patch covering much of cell. Below, mottled brown and gray,* becoming frosty purplish-gray toward outer edge; *FW patch repeated beneath.*

Similar Species: All other brown and rust-colored alpines have eyespots and/or bands of light and dark, except Alaskan Magdalena Alpine, which lacks gray scaling beneath and FW patch.

Life Cycle: Unknown. Host plant probably grasses (Poaceae) or sedges (Cyperaceae).

Flight: 1 brood; May–July, depending upon latitude and weather.

Habitat: Woodland meadows; dry open northern forests, swampy clearings among willows; spruce bogs.

Range: Alaska south and east to Banff, North Dakota, Minnesota, Wisconsin, N. Michigan, W. central Hudson Bay, and Laurentides Provincial Park north of Quebec City; also Asia.

This species and the Arctic Alpine extend farther east in North America than any other alpines. The increasing number of records in recent years for the Red-disked Alpine around the northern Lake States, Ontario, and Quebec probably reflects an expansion of knowledge rather than an increase in the butterfly's range. The Red-disked Alpine flies earlier than other alpines and in a wider variety of places than most. Nonetheless, it is chiefly a butterfly of the sub-Arctic, flying neither as far north or south in the Rockies as other alpines.

754 Theano Alpine
(*Erebia theano*)

Description: 1¼–1½" (32–38 mm). *Small.* Wings rounded. *Above,* dark brown, *each wing crossed by light russet spot bands;* spots may merge into almost solid bands, and are reduced or nearly absent on HW in some populations. Below, similar to upperside, but *spots on HW underside are cream-colored or pale yellow.* FW cell above and below may be reddish; HW cell below cream-spotted. *No eyespots.*

Similar Species: Colorado Alpine lacks bright spot rows, is gray on HW below.

Life Cycle: Unknown. Caterpillar probably feeds on grasses (Poaceae).

Flight: 1 brood; July, into August in some years and areas.

Habitat: Alpine meadows, bogs, and lakeshores above and below timberline; grassy glades among arctic scrub birch woods; dry, long grass in pine forest openings; taiga.

Range: Alaska, Yukon Territory, N. Manitoba on Hudson Bay, Montana, Wyoming, and Colorado; also Siberia.

This brightly ornamented alpine flies in small, tightly circumscribed colonies within regional populations which may be great distances apart. Isolated races, such as the one restricted to the San

Juan Mountains of Colorado, tend to
develop distinctive appearances. Their
habitat preferences may also differ but
only 1 or 2 types of sites will be used
by a single population. Like many
alpine butterflies, Theano Alpines seek
shelter when clouds obscure the sun,
disappearing deep into grass or moss.
With the first direct rays of sun, they
spring back into flight. Although
Theano Alpines are approachable and
may occur in great abundance, few
colonies are easily accessible. One of the
easier ones to reach is at Togwotee Pass
east of Grand Teton National Park in
Wyoming.

Yukon Alpine
"Young's Alpine"
(*Erebia youngi*)

Description: 1⅜–1⅞" (35–41 mm). *Above,* dark
brown, *usually with 4 equally small
eyespots ringed in russet along FW margin;*
sometimes similar eyespots along HW
margin. *Below,* FW brown; eyespots
may lie in broad rust-colored patch;
*HW gray-brown crossed by jagged, darker
median band, often without eyespots.*

Similar Species: Common Alpine usually has eyespots
on HW below; upper 2 eyespots on FW
are larger than lower ones. Spruce Bog
Alpine has white spots on HW cell and
costa below. Arctic Alpine normally has
only 2 FW eyespots.

Life Cycle: Undescribed. Host plants probably
grasses (Poaceae) or sedges (Cyperaceae).

Flight: 1 brood; late June–early July.

Habitat: Bogs, damp forest edges, tundra, and
shale ridges.

Range: Upper drainage of Yukon River in
Alaska and Yukon, east into Northwest
Territories on both sides of Arctic
Circle, and N. British Columbia.

Small and subtly marked, the Yukon
Alpine is one of the few butterflies

strictly limited to the far North. To see it, an observer must contend with hordes of mosquitoes; the short flight period on the few warm days is shared by the numerous insects of the muskeg. Eagle Summit, north of Fairbanks, is an accessible and favorite spot for finding the Yukon Alpine and other butterflies of the North American Arctic.

750, 753 **Common Alpine**
"Butler's Alpine"
(*Erebia epipsodea*)

Description: 1¾–2" (44–51 mm). Wings rounded. *Above,* dark brown, crossed by *reddish-orange patches or bands usually containing eyespots with white pupils. 2 eyespots uppermost on FW are larger than others. Below,* FW similar to upperside but paler, or sometimes reddish-orange; *HW usually frosted with gray over outer third,* and sometimes also at base; *darker brown median band usually not very outstanding.* HW eyespots often ringed with yellow or pale orange. Rare individuals lack any eyespots above in rust-colored bands, which are then broader and more suffused.

Similar Species: Yukon Arctic has FW eyespots of about equal size. Northwest Alpine has paler, broader orange, more jagged bands; grayish median band on HW beneath, and checkered fringes.

Life Cycle: Egg pale yellowish-white, nearly spherical; laid singly or in small groups on grasses (Poaceae). Caterpillar striped green; overwinters when young and resumes feeding in spring, pupating in loose shelter of silk and grass blades.

Flight: 1 brood; early June–late August for about 3 weeks, earlier in lowlands and southern part of range and later at higher elevations and latitudes.

Habitat: Mountain meadows, bogs, clearings, and lower arctic-alpine tundra; sage flats; northern prairie parklands; often in association with aspens.

Range: Central Alaska south in Coast Ranges and Rockies to E. and central Oregon and New Mexico, and east to central Montana and W. Manitoba.

Strictly a North American species, the Common Alpine occurs farther south than any other New World alpine. While the other Rocky Mountain alpines are confined to the high country, this species flies down almost to the foothills, where moisture permits. Its adaptability and broad taste in habitat have allowed it to spread more widely than most satyrs.

Colorado Alpine
(*Erebia callias*)

Description: 1⅜–1⅝″ (35–41 mm). Wings rounded. *Above, dark brown,* with srong greenish reflections when fresh; *2 merged, black-ringed eyespots with white pupils in rust-colored patch near tip of FW;* often a ring of smaller eyespots around HW. *Below,* FW bright copper-colored with eyespots repeated and gray margin; *HW soft, woolly gray* with vague lines but no strong marks.

Similar Species: Theano Alpine has bright yellow spots on HW below. Common Alpine larger, has more eyespots, less uniformly gray below.

Life Cycle: Undescribed. Caterpillar feeds on grasses (Poaceae) or sedges (Cyperaceae); may take 2 years to grow from egg to adult.

Flight: 1 brood; mid-July to late August.

Habitat: Mountain summits, slopes, and passes in arctic-alpine meadows; tundra, fellfields.

Range: S. central Montana, W. Wyoming, NE. Utah, and much of high altitude Colorado.

Neither as continuous and widespread as the Common Alpine, nor as local and spottily distributed as the Theano Alpine, the Colorado Alpine covers much of the central Rockies summitland, particularly in Colorado. Yet its overall range is peculiar—populations occur in North America, Asia, and the Middle East, with none in between. All the alpines have been pushed about over once-arctic landscapes by glaciers and shifting climates. Some species invaded the far North in the wake of the melting glaciers. Others, like the Colorado Alpine, were left behind on alpine islands. Why this species occurs nowhere between Mongolia and Montana is a mystery.

755 Red-bordered Brown
(*Gyrocheilus patrobas*)

Description: 2¼–2⅞″ (57–73 mm). Wings rounded, with HW markedly scalloped and FW less so. *Above, dark brown with small white dots near FW tip; prominent russet-red band around HW margin* may cover a third of wing. *Below,* similar except lighter brown; *HW band maroon with white and brown-violet mottling,* and vague eyespots on inner edge; margin has soft blue markings.

Similar Species: Red-eyed Wood Nymph has russet on FW rather than HW. Crimson-banded Black has similar pattern but is black and crimson, not brown and russet.

Life Cycle: Unreported.

Flight: 1 brood in Arizona; August–October.

Habitat: Pine forests and canyons.

Range: Central Arizona south to Mexico.

The Red-bordered Brown, unique in North America, represents a group of butterflies which have a stronghold in the Andes Mountains of South

America. Known as pronophiline satyrs, many have been discovered only recently.

742, 745 Riding's Satyr
(*Neominois ridingsii*)

Description: 1½–1⅞" (38–48 mm). *Above, graduated shades of gray* (lead, putty, or sandy brown-gray dominate in different populations); *bands of oblong orbs of cream-white or milk-white cross all wings.* 1, 2, or 3 black eyespots with white pupils lie in FW bands, more rarely 1 small eyespot appears on HW. Below, grayish-tan, heavily speckled and striated, with upperside pattern repeated less distinctly.

Life Cycle: Egg chalk-white, keg-shaped. Caterpillar light reddish, banded with green on sides and back, covered with minute bumps and hair. Host plants probably grasses (Poaceae); after feeding in clumps of grass, caterpillar pupates in soil. Chrysalis color similar to that of caterpillar.

Flight: 1 brood; June–August, emerging and remaining later at higher altitudes. Occasionally 2nd brood.

Habitat: Dry, sunny prairies and other grasslands; sage flats and subalpine sagebrush summits.

Range: S. Alberta and Saskatchewan south through Rockies, along western edge of Great Plains to Nebraska, Great Basin, central Arizona, and New Mexico, and E. California and S. central Oregon.

No other American butterfly looks at all like the Riding's Satyr, with its pattern of spotted gray. This species is the only New World representative of a Eurasian group of satyrs that includes many Himalayan butterflies. North American populations vary in appearance from dark gray with clean white spots in Colorado and California, to pale and sandy-colored with cream-colored spots in the Great Basin, to very pale in Arizona. Wherever it lives, this satyr's colors and patterns elegantly blend with sandy, stony backgrounds. When disturbed, it flies short distances close to the ground, resembling a moth.

734, 738, 741 California Arctic
(Oeneis ivallda)

Description: 2–2⅛" (51–54 mm). *Large. Above, male pale ashen gray-brown with yellowish highlights and overtones; female whitish-gray with pale yellowish cast.* 1 or more small eyespots along FW margins, sometimes also on HW. *Beneath, line shaped like a bird's beak separates darker inner part of FW from lighter marginal area;* HW minutely and complexly mottled with brown, crossed by wavy brown median band.

Similar Species: Chryxus Arctic much darker and tawnier; male has prominent FW sex patches. Great Arctic larger, tawnier, lacks bird's beak design on FW.

Life Cycle: Undescribed. Caterpillar green; feeds on grasses (Poaceae).

Flight: 1 brood; July–August.

Habitat: Arctic-alpine ridges, summits, and outcrops, drier meadows, and fell-fields; always above 10,500′ (3202 m).

Range: Sierra Nevada of California from Donner Pass south to Mineral King.

California is the only state with its own endemic species of arctic. Although the California and Chryxus arctics fly together in some parts of the High Sierra, they have not been observed to interbreed. Hikers along the John Muir Trail or through the high country of Sequoia and Yosemite national parks should watch for this butterfly. Apparently, the pale California Arctic has adapted to the light backgrounds of its white granite habitats.

727 Great Arctic
"Nevada Arctic"
(*Oeneis nevadensis*)

Description: 2–2½" (51–63 mm). *Large. Above, pale to bright tawny with darker, somewhat scalloped margins.* Male has dark patches at base and along FW costa, *on FW 1 or 2 small black eyespots with pupils; female* lacks dark patch and *has more and larger eyespots;* both sexes usually have 1 eyespot on HW margin. *Below, HW uniform barklike brown and gray striated,* with darker, irregular median band in some populations; FW tawny.

Similar Species: Chryxus Arctic smaller, lighter, with a more prominent median band and less uniform striation on HW below.

Life Cycle: Egg dirty-whitish, oblong, flat-topped. Caterpillar tan with black stripes along back, brown and whitish stripes along sides; feeds on grasses (Poaceae); takes 2 years to mature.

Flight: 1 brood; late May–August. Normally only even years in North.

Habitat: Forest clearings and roads, dry meadows and mountain slopes near forests, and canyons.

Range: Vancouver and San Juan islands; British Columbia to California in Cascades and Coast Ranges.

Despite its Latin name and its alternate common name, this species has not recently appeared in Nevada. Great Arctic is a far more appropriate name, particularly since this butterfly is the largest western arctic. Vancouver Island populations are particularly large. Groups of adults, especially males, tend to fly to prominences, and in small forest glades males patrol their territory. The Great Arctic flies in mountain forests, on wooded bluffs, and on bare mountain summits near sea level, but not at all in arctic conditions. Near the Pacific its distribution is very spotty but is more continuous in the mountains.

744, 747 Canada Arctic
"Macoun's Arctic"
(*Oeneis macounii*)

Description: 2–2¼" (51–57 mm). *Large. Above, rich yellowish-brown, normally with 2 black eyespots with white pupils on FW, 1 on HW; veins dark* and margins dark and somewhat scalloped. *Below, FW yellowish-brown; HW very dark brown with heavy striation, broad, wavy darker brown median band* has some whitish fogging just within and beyond.

Similar Species: Chryxus Arctic smaller, lighter; male has dark FW patch, more mottled on darker HW below. Jutta Arctic much darker. Canada Arctic male lacks prominent FW sex patches of 2 other species.

Life Cycle: Egg gray-white. Caterpillar striped with alternating light and dark bands of brownish-yellow, greenish-brown, and sometimes reddish; tapered to forked rear, head rounded. Host plants are grasses (Poaceae). Caterpillar takes 2 years to mature; overwinters twice before pupating. Chrysalis yellow-brown.

Flight: 1 brood; late June–early July, in odd years west of Lake Winnipeg, even years to east.

Habitat: Jack pine forests and nearby glades and fields.

Range: S. Northwest Territories, Alberta, Saskatchewan, Manitoba, and Ontario east to Algonquin, N. Minnesota, and Isle Royale in Michigan.

The fact that this arctic flies in alternate years may be an adaptation to the short northern growing season, although the reason for its appearance in odd years in the western part of its range and in even years in the eastern portion is unknown. Haunting small, sunlit glades and clearings among pinewoods, male Canada Arctics assume territorial posts on prominent boughs or twigs. Sorties lead them into aerial tussles

with other males, in search of potential
mates, and for occasional nectar visits.

743, 746 **Chryxus Arctic**
(*Oeneis chryxus*)

Description: 1¾–2" (44–51 mm). *Above, tan or
tawny-brown to dark brown; male has dark
sex patch across cell and surrounding dark
region* sometimes covering much of FW,
leaving tawny marginal band. *Below,
FW tawny; male's FW crossed by dark line
shaped like a bird's beak* (female has line
but much vaguer), mottled and striated
brown and white; *HW crossed by broad
and prominent brown median band.*
Usually 2 or 3 eyespots above and
below on FW, 1 on HW outer angle
(tornus); eyespots may be obscure or
well-developed.

Similar Species: Great Arctic and Canada Arctic larger,
have dark margins, bigger eyespots,
and plainer undersides. Alberta Arctic
smaller, duller. Uhler's Arctic smaller,
lacks bird's beak marking and male's
sex patch. California Arctic much paler
gray-brown or whitish.

Life Cycle: Egg white. Mature caterpillar, to 1¼"
(32 mm), striped lengthwise with
straw-color, olive, green, brown, and
brownish-yellow; overwinters when
young, some taking 2 years to mature.
Host plants are grasses (Poaceae),
possibly Idaho fescue (*Festuca idahoensis*)
in Washington. Chrysalis pale
yellowish-brown with dark brown head
and wing cases.

Flight: 1 brood; May–August, exact period
depending upon local conditions.

Habitat: Arctic and alpine tundra, evergreen
forest clearings, mountain meadows and
sage flats, northern prairies and
parklands, and shaly steep slopes.

Range: Alaska and Yukon south to central
California, E. Nevada, and New
Mexico (but absent from Oregon), and
east to eastern edge of Rockies and

Ontario, N. Wisconsin, N. Michigan, Quebec, and Gaspé Peninsula.

The Chryxus has the broadest tolerances of any arctic, dwelling in a wide array of habitats. Some cut-off populations, such as the ones in the Olympic Mountains of Washington and the northern Sierra Nevada, have become physically distinctive. Its absence from Oregon may be due to recent volcanic activity in the southern Cascade Mountains. The related Sentinel Arctic (*O. excubitor*) flies on lower slopes in Alaska, the Northwest Territories, and the Yukon. It has 2 eyespots on its hindwing outer angle above.

728 Uhler's Arctic
(*Oeneis uhleri*)

Description: 1½–1⅞" (38–48 mm). *Above,* some shade of orange-brown from *pale tawny to deep brandy-colored to olive-ocher* with margins narrowly dark, and *usually 2 FW eyespots and ring of 2–5 HW eyespots. Below,* similar to upperside; *HW conspicuously striated in dark arc concentric with margin.* Many populations lack true median band; *all lack dark bird's beak line on FW below.*

Similar Species: Chryxus Arctic larger, with prominent FW sex patches on male. Alberta Arctic smaller. Chryxus and Alberta arctics both have bird's beak design on FW below.

Life Cycle: Caterpillar greenish, striped; overwinters when nearly mature; feeds on and pupates in grass (Poaceae).

Flight: 1 brood; May–July or August, early in milder climates. Rarely 2nd brood.

Habitat: Pine forest openings in mountains, dry, grassy glades, shrubby foothills, sage flats near mountain streams, prairie steppes, and drier taiga.

Range: Northwest Territories, Alberta, Saskatchewan and Manitoba, Montana,

Dakotas, Wyoming, Nebraska, Colorado, and New Mexico.

Although Uhler's Arctic and the Chryxus Arctic share many Rocky Mountain locales, Uhler's tends to flutter in a hovering fashion over some fixed point on a grassy slope, while the Chryxus flits from spot to spot, seldom flying for long at any time. The unusual, in-place flights of male Uhler's Arctics permit the butterflies to reconnoiter their surroundings for potential mates. Grassy sagebrush flats near middle elevation mountain streams offer good opportunities for finding Uhler's and Chryxus arctics, Riding's Satyrs, Dark Wood Nymphs, and Ocher Ringlets in late spring. Unlike other arctics, this species extends for some distance east of the mountains onto the high plains.

737, 740 Alberta Arctic
(*Oeneis alberta*)

Description: ⅜–1⅞" (35–48 mm). *Small. Above, gray-brown, pale tawny-brown or reddish-tan; FW has dark bird's beak design above, more pronounced below. Below, paler; HW striated brown on white, always crossed by more or less prominent, dark brown and wavy median band.* Small eyespots, with or without white pupils, on FW above and below; often 1 on HW.

Similar Species: Chryxus Arctic larger and tawnier; male has prominent FW sex patches. Uhler's Arctic lacks bird's beak design and defined median band on HW below.

Life Cycle: Caterpillar striped lengthwise with light and dark earth tones; feeds on grasses (Poaceae), including fescues (*Festuca*).

Flight: 1 brood; April–June, from shortly after snowmelt until early summer.

Habitat: High mountain dry plateaus, alpine

meadows in White Mountains of
Arizona, and northern prairies.

Range: Prairie Provinces, North Dakota,
Montana, Colorado, S. Utah, Arizona
in White Mountains, and N. New
Mexico.

The Alberta Arctic has a spotty
distribution, occurring in Alberta and
adjacent provinces and states, then in
the central Rockies (but mysteriously
missing from Wyoming), and finally in
the southwestern White Mountains.
Between these different locales its
habitat, as well as appearance, changes.
The Alberta Arctic flies earlier than
most other high altitude and northern
butterflies. On the dry, sandstone
plateaus and ridges around Colorado's
South Park, this arctic can be found
bracing itself against the wind in
earliest spring.

731 White-veined Arctic
"Labrador Arctic"
(*Oeneis taygete*)

Description: 1¾–1⅞" (44–48 mm). Above, soft,
dull brownish-gray, somewhat
translucent; browner or slightly tawny
in some populations; sometimes bearing
small, whitish eyespots on any wing
surface. *Below, HW 2-toned with heavy,
broad, dark and outwardly zigzagged
median band running across whitish HW
disk;* darker again toward base and
margin. *HW veins below usually sharply
delineated with white scaling.*

Similar Species: Chryxus Arctic tawnier, has more
eyespots. Arctic Grayling lacks white
vein-scaling. Melissa and Polixenes
arctics smaller, more translucent, less
sharply banded below.

Life Cycle: Unknown in North America. Host
plants probably grasses (Poaceae) or
sedges (Cyperaceae).

Flight: 1 brood; mid-June to August according

to location, usually peaking in July.

Habitat: Arctic and arctic-alpine tundra, high stony and shrubby ridges, and mountain summits.

Range: Arctic Alaska and Canada to Greenland, southward discontinuously to Alberta, Montana, Wyoming, Colorado, Labrador, and Gaspé Peninsula; also Siberia.

Extremely hardy, the White-veined Arctic is one of very few butterflies native to Greenland. It occurs southward only where high mountains duplicate its northern environs. It was scattered by the Pleistocene glaciers onto the Beartooth, Uinta, Sawatch, and San Juan ranges of the Rocky Mountains. In Colorado at Cumberland and Cottonwood passes the White-veined Arctic may be seen together with the Melissa Arctic.

732 Arctic Grayling
(*Oeneis bore*)

Description: 1¾–1⅞″ (44–48 mm). Above, gray-brown quite unmarked except for darker scaling on FW sex patch of male and lighter, tanner areas toward margin of female. *Below, FW tan; HW darker at base and margin, with bold, dark median band running across frosty-whitish area;* veins are not conspicuously white.

Similar Species: White-veined Arctic has white veins. Melissa and Polixenes arctics paler gray, more translucent. Males of all 3 species lack prominent sex patches.

Life Cycle: Undescribed for North America. Caterpillar striped; feeds on grasses (Poaceae), including fescues (*Festuca*).

Flight: 1 brood; June–July.

Habitat: Rocky alpine slopes, arctic fell-fields, and gravelly and shaly tundra.

Range: Alaska and N. Canada to Hudson Bay; also Scandinavia and Siberia.

Known for 2 centuries in Europe, the Arctic Grayling needs much more study in North America. The construction of the Alaska Pipeline has apparently increased the number of habitats favorable to Arctic Graylings by increasing the amount of bare, stony land on which they prefer to bask.

733 Jutta Arctic
(*Oeneis jutta*)

Description: 1⅞–2⅛″ (48–54 mm). *Variable between populations and sexes. Above, basically dense grayish or olive-brown with more or less yellow to ocher between veins* near margins, forming patches or even bands; *within the yellow are prominent black eyespots with or without white pupils;* male has dark sex patch in FW cell. *Below,* FW similar to upperside; *HW striated with barklike tones of gray and brown,* with darker gray, highly irregular median band (not obvious in all populations); eyespots on HW, if present, are small.

Similar Species: Canada Arctic much lighter; male lacks FW sex patches.

Life Cycle: Egg yellowish-white. Caterpillar buff- and olive-striped; may feed on cotton grass (*Eriophorum spissum*), a sedge. Chrysalis yellow-green with greener wing cases and brown dots on abdomen.

Flight: 1 brood; June–July.

Habitat: Black spruce and tamarack sphagnum bogs and northern taiga in East; lodgepole pine forest glades in Rocky Mountains; cotton grass tundra.

Range: Circumpolar: Alaska and most of sub-Arctic Canada east to Maritimes; south down Rockies to Utah and Colorado; N. Minnesota, Wisconsin, Michigan, and Maine; also N. Europe and Asia.

The range of the Jutta Arctic takes in much of the boggy boreal forest belt of

the Northern Hemisphere. It finds
younger, damper sphagnum bogs with
cotton grass more favorable than older,
drier muskeg with denser trees. This
species is more accessible in the cold,
high lodgepole pine forests of the
Rocky Mountains, where it flies
frequently with the Meadow Fritillary.
In Alaska, the Jutta Arctic tends to be
smaller and duller, while it reaches its
largest size and brightest markings in
Nova Scotia.

735 Melissa Arctic
"White Mountain Butterfly"
(*Oeneis melissa*)

Description: 1⅝–1⅞" (41–48 mm). Above and
below, lightly scaled; *smoke-gray to dull
dark brown,* sometimes slightly tan,
especially females. *Below, FW similar to
upperside; HW heavily striated with gray,*
either uniformly or with a vague to
pronounced darker median band.
Translucent; generally lacking eyespots.

Similar Species: Polixenes Arctic smaller, more
translucent and browner; often with
small eyespots or whitish dots and more
pronounced median band below.
White-veined Arctic larger, yellower,
with noticeably whitened veins. Arctic
Grayling darker gray-brown, less
translucent.

Life Cycle: Egg pale greenish, oval and flattened on
back. Mature caterpillar moss-green
with blue-green side stripes and back
stripes and broader, spotted yellow
stripes down side. Host plants are
grasses (Poaceae) and sedges
(Cyperaceae). Chrysalis blackish-olive,
small and rounded; hangs from curved,
pointed tail end.

Flight: 1 brood; June–August, most commonly
in July.

Habitat: Arctic tundra, alpine summits, ridges,
and swales; arctic-alpine meadows and
rockslides.

Range: Alaska, N. Canada east to Hudson Bay and Gulf of St. Lawrence; south in Cascades and Rockies to N. Washington, NE. Utah and N. New Mexico; entering eastern U.S. only in White Mountains of New Hampshire; also Old World Arctic.

This small, smoke-colored arctic flies on many of the same Colorado boulderfalls frequented by the Magdalena Alpine. When the Melissa settles against lichen-encrusted granite or black quartzite, it is completely camouflaged. The New Hampshire population, known as the White Mountain Butterfly, has been long recognized as a glacial relict.

729, 730 Polixenes Arctic
"Katahdin Arctic"
(*Oeneis polixenes*)

Description: 1½–1¾" (38–44 mm). Above and below, gray-brown and quite *translucent,* sometimes virtually transparent; *small light dots may occur just inside margins. Below,* FW similar to upperside; *HW has dark median band,* surrounded by lighter, striated areas. Wings may be infused with yellowish scaling especially submarginally above.

Similar Species: White-veined Arctic larger and has white veins. Melissa Arctic and darker Arctic Grayling lack light spots and yellow scaling above, and have less pronounced band beneath.

Life Cycle: Caterpillar feeds on arctic and alpine grasses (Poaceae); overwinters in early stages, resumes feeding the following spring; may take 2 years to mature.

Flight: 1 brood; July–August.

Habitat: Moist true arctic and arctic-alpine tundra, windswept summits, ridges, well above and north of timberline.

Range: Alaska and Yukon east across Northwest Territories and Canadian

Arctic Archipelago to both shores of Hudson Bay. Isolated colonies in Wyoming, Colorado, New Mexico, Quebec, and on Mt. Katahdin in Maine.

The Polixenes Arctic often has highly translucent wings; the most sparsely scaled individuals resemble lightly smoked glass. When they alight, these and other arctics often tip their clasped wings at an angle close against the ground, possibly to cast as little shadow as possible, improving their already superb camouflage. Another reason for tilting their darkened undersides toward the sun may be to gather solar warmth necessary for activity under arctic conditions, or perhaps this posture is an adjustment to the perpetual strafing of the wind. On Mt. Katahdin, Maine, an isolated and distinctive race thrives above timberline. Long known as the "Katahdin Arctic," its stony, windy habitat is protected as wilderness within Baxter State Park.

MILKWEED BUTTERFLIES
(Danaidae)

300 species worldwide; only 4 in North America. Although this family contains the familiar Monarch, most of its members are found in tropical Asia. North American species, with wingspans of 3–4″ (76–102 mm), are orange or brown with black markings, but many tropical milkweed butterflies are larger, with brown, black, and blue coloring. The milkweed butterflies are often considered a subfamily of the Nymphalidae because, like the nymphalids, they have reduced fore legs. But unlike the nymphalids, milkweeds have scaleless antennae. Males possess sex pouches on the hind wings and brushlike hair pencils within the abdomen. In courtship, the males of some species extend these brushes, releasing scents that subdue the female during mating.

Milkweed butterfly eggs are raised and blimplike. The brightly striped caterpillars have pairs of long filaments at the head and rear. Chrysalises are compact and studded with metallic spots. Because the caterpillars of most North American species feed on toxic milkweeds, the adult butterflies are distasteful to birds, thereby gaining some protection from predators. For this reason, the Viceroy and other species mimic the milkweeds' coloring. Milkweed butterflies are known for their strong, soaring flight, which finds its greatest expression in the Monarch's remarkable migrations.

2, 35, 596 **Monarch**
(*Danaus plexippus*)

Description: 3½–4″ (89–102 mm). *Very large, with FW long and drawn out. Above, bright, burnt-orange with black veins and black margins sprinkled with white dots;* FW tip broadly black interrupted by larger white and orange spots. Below, paler, duskier orange. 1 black spot appears between HW cell and margin on male above and below. Female darker with black veins smudged.

Similar Species: Viceroy smaller, has shorter wings and black line across HW. Queen and Tropic Queen are browner and smaller. Female Mimic has large white patch across black FW tips.

Life Cycle: Egg, ³⁄₆₄″ h × ⁹⁄₂₅₆″ w (1.2 × 0.9 mm), pale green, ribbed, and pitted, is shaped like lemon with flat base. Caterpillar, to 2″ (51 mm), is off-white with black and yellow stripes; 1 pair of fine black filaments extends from front and rear. Chrysalis, to ⅛″ (28 mm), pale jade-green, studded with glistening gold; plump, rounded, appears lidded, with lid opening along abdominal suture. Host plants are milkweeds (*Asclepias*) and dogbane (*Apocynum*).

Flight: Successive broods; April–June migrating northward, July–August resident in North, September–October migrating southward, rest of year in overwintering locales. Year-round resident in S. California and Hawaii.

Habitat: On migration, anywhere from alpine summits to cities; when breeding, habitats with milkweeds, especially meadows, weedy fields and watercourses. Overwinters in coastal Monterey pine, Monterey cypress, eucalyptus groves in California, and fir forests in Mexican mountains.

Range: Nearly all of North America from south of Hudson Bay through South America; absent from Alaska and Pacific Northwest Coast. Established in the Hawaiian Islands and Australia.

One of the best known butterflies, the Monarch is the only butterfly that annually migrates both north and south as birds do, on a regular basis. But no single individual makes the entire round-trip journey. In the fall, Monarchs in the North begin to congregate and to move southward. Midwestern and eastern Monarchs continue south all the way to the Sierra Madre of middle Mexico, where they spend the winter among fir forests at high altitudes. Far western and Sierra Nevada Monarchs fly to the central and southern coast of California, where they cluster in groves of pine, cypress, and eucalyptus in Pacific Grove and elsewhere. Winter butterflies are sluggish and do not reproduce; they venture out to take nectar on warm days. In spring they head north, breed along the way, and their offspring return to the starting point. Both Mexican and international efforts are underway to protect the millions of Monarchs that come to Mexico. In California, nearly all of the roosting sites face threatening development.

1, 33, 594 Queen
(*Danaus gilippus*)

Description: 3–3⅜" (76–86 mm). *Large. Deep fox-brown above and below with black margins and finely lined black veins. Fine white dots speckle margins,* larger ones occur on FW tip and along margin. West of Mississippi, populations have white scaling along HW veins above and are somewhat lighter.

Similar Species: Monarch larger, orange with blacker veins. Tropic Queen orange-brown, pale toward outer edge. Dark Florida Viceroy has strong black line arcing across HW.

Life Cycle: Egg, ¾₄" h × ⁹⁄₂₅₆" w (1.2 × 0.9 mm), light, oval; laid on milkweeds and related plants (Asclepiadaceae). Caterpillar, to 2" (51 mm), brownish-white, with dark brown-black and light yellow crossbands, and yellow or yellow-green side stripes; has 3 pairs of dark filaments, 2 longer pairs toward front and 1 shorter pair near rear. Blunt chrysalis, to 1⅛" (28 mm), green with gold spots and prominent abdominal suture. Host plants blunt-leaved milkweed (*Asclepias amplexicaulis*) and rambling milkweed (*Sarcostemma hirtellum*).

Flight: Successive broods; April–November, briefer in North, perhaps all year in Texas.

Habitat: Deserts, coasts, prairies, watercourses, and other open places with milkweeds.

Range: Nevada and S. California east to Kansas and Texas, around Gulf to Florida and S. Georgia, south to South America.

The Queen cannot withstand cold winters. Records for northwestern Utah, Nebraska, and Kansas represent temporary immigrations. The Viceroy, an accomplished mimic resembling the Monarch in its northern, summer range, has a dark russet race in the South, which mimics the Queen. Male Queens possess brushes, or hair pencils, within the tips of their abdomens. As courtship begins these brushes are extended, releasing a compound that subdues the female during mating.

595 Tropic Queen
(*Danaus eresimus*)

Description: 3⅛–3¼" (79–83 mm). *Large. Above, base of wings deep russet, outer portion much paler orange-brown with white-dotted,* brown-black *margins and dark veins.* Often *white patch in and about HW* cell. Dark veins and more markings below.

Similar Species: Monarch larger, longer wings, deeper orange with black, light-spotted FW tips. Queen darker, almost brown, veins not heavily blackened above, no white patch on HW cell. Female Mimic has broadly black FW tips crossed by white patch.

Life Cycle: Feeds on milkweed (*Asclepias*).
Flight: Fall and winter in S. Florida; successive broods in Antilles.
Habitat: Subtropical forest edges, disturbed areas, and gardens.
Range: Extreme S. Texas to South America, rarely to S. Florida; also Antilles.

One race of the Tropic Queen becomes common in Texas late in the year. The Caribbean variety appears in Florida.

Tropical Milkweed Butterfly
(*Lycorea cleobaea*)

Description: 3¼–3¾" (83–95 mm). *Large, with long, rounded wings. Above and below, transversely striped* with orange and black, yellow and black at FW tip; HW margin black with white dots.
Life Cycle: Caterpillar greenish with yellow and black stripes; bears 1 pair of black filaments near head. Red-flowered milkweed (*Asclepias curassavica*), papaya (*Carica*), and figs (*Ficus*) are host plants in Tropics
Flight: Successive broods; any month possible, mostly late fall and winter.
Habitat: Subtropical forests, scrub, hammocks.
Range: S. Texas and S. Florida south to Bolivia; also Antilles.

The Tropical Milkweed Butterfly is unique in North America. A Caribbean race wanders to the Florida Keys, and the Central American population extends regularly into Texas. This species is a member of a tropical mimicry complex involving longwings and other nymphalids. The Tropical Milkweed Butterfly bears a much greater resemblance to these relatives than to other milkweed butterflies. Birds come to recognize and avoid the really unpalatable members; thus all the look-alikes gain mutual protection. Formerly known as *Lycorella ceres*.

TRUE SKIPPERS
(Hesperiidae)

Approximately 3000 species worldwide;
nearly 250 in North America. Named
for their rapid, skipping flight, true
skippers differ from true butterflies in
their proportionately larger bodies,
smaller wings, and in numerous other
structural details. True skippers are
small to medium-sized, with wing
spans of ½–2½" (13–64 mm). They
have broad heads and hooked antennae,
and somewhat resemble moths with
their robust and often hairy bodies and
triangular wings.

All adult true skippers have 6 well-
developed legs. The eggs are less than
$\frac{1}{256}$" (0.1 mm) wide. Most caterpillars
are green and tapered, and appear
to have a constricted neck. They weave
silk and leaf shelters for protection
during the day. Most later pupate in
loose cocoons. The streamlined
chrysalises are often covered with a
waxy powder or bloom; they or the
caterpillars may overwinter.

True skippers are represented by 3
subfamilies. The tawny-orange or
brown skippers belong to the subfamily
Hesperiinae. When basking, these
skippers hold their fore wings and hind
wings at different angles, rather like
folded paper airplanes. Males of many
species have stigmas, or sex patches, on
the fore wings. Hesperiine caterpillars
feed mostly on grasses and sedges. The
dark, checkered, and long-tailed
skippers comprise the subfamily
Pyrginae. These skippers tend to bask
with wings spread open or partly open
in the same plane. Males lack clearly
defined sex patches but may possess less
conspicuous sex scales, and many
species have a costal fold on the fore
wing. Pyrgine caterpillars feed on a
wide variety of plants. The third
subfamily, Pyrrhopyginae, is a
subtropical group that includes

generally robust, red-spotted skippers. The group has a single species in North America, the Araxes Skipper.

296 Araxes Skipper
(*Pyrrhopyge araxes*)

Description: 1¾–2½" (44–64 mm). *Large.* Brown above with several large, glassy white spots in median band across FW, smaller spots toward FW tip. *Below,* FW similar, orange near base; *HW orange or tawny-orange with vague brown marks* and margin. Fringes white and brown checkered. Abdomen orange on sides. *Antennal clubs entirely bent back.*

Similar Species: Other large brown skippers with glassy spots lack orange underside and totally bent back antennal club.

Life Cycle: Caterpillar reddish-brown with narrow yellow stripes and black head. Chrysalis orange and rust-colored in patches, covered with white, flaky bloom or powder. Host plants are oaks (*Quercus*), including Arizona oak (*Q. arizonica*).

Flight: Probably 2 broods; June or late July–September.

Habitat: Canyons and oak woodlands.

Range: SE. Arizona, SW. New Mexico, S. Texas, and Mexico.

The Araxes Skipper is drabber than its tropical relatives, which have scarlet-spotted bodies and wings with a blue or green sheen. Araxes lacks these but has the broad, robust thorax typical of its subfamily.

300 Mangrove Skipper
(*Phocides pigmalion*)

Description: 2–2½" (51–64 mm). Robust and long-winged. Deep, shiny black with *submarginal band of metallic blue bars on HW above* and other blue reflections

around them; *FW above has* more or less
obvious blue rays along veins from base.
Black below with faint blue lines across
HW; white checkered fringe.

Life Cycle: Caterpillar stout, brownish; exudes a
white powder which masks its
markings. Chrysalis robust, green,
with off-white patch; bears 2 horns
between prominent eye cases. Host
plant is red mangrove (*Rhizophora
mangle*).

Flight: 2 or more broods; November–May,
most of year farther south.

Habitat: Coastal hammocks and swamps.

Range: S. Florida south to Argentina: also
Bahamas and Greater Antilles.

With no close look-alikes, the
Mangrove Skipper visits the flowers of
mangrove as an adult and devours the
leaves as a caterpillar. A close relative,
the Urania Skipper (*P. urania*), has
more blue bars on the hind wing, and
blue-green streaks and white bands on
the fore wing. It wanders to Texas
and Arizona from Mexico.

299 Guava Skipper
(*Phocides polybius*)

Description: 2–2½" (51–64 mm). Wings long,
drawn out at tip. Black above with
metallic green streaks and reflections,
and *2 crimson spots at tip of FW cell;*
white fringes on HW. Similar below.
Scarlet collar just below head; palpi red.

Similar Species: Other greenish *Phocides* and *Astraptes*
lack crimson patches on FW above.

Life Cycle: Young caterpillar crimson or maroon,
with yellow crossbands or rings,
maturing to white, with brown head
and yellow spots on face. Chrysalis
smooth, whitish to green. Host plants
are guavas (*Psidium*), including
common guava (*P. guajava*).

Flight: Successive broods; most months.

Habitat: Tropical forest openings, gardens.

Range: Southern Texas around Brownsville;
Mexico south to Argentina.

This brightly colored skipper has only
recently been recognized as a breeding
resident within our range. Tropical
populations have orange fringes.

292, 301 Mercurial Skipper
(*Proteides mercurius*)

Description: 2¼–2⅝″ (57–67 mm). *Large; FW very
long and drawn-out;* lobed, slightly
tailed HW. *Above, dark brown with gold
at base and series of squarish white spots
crossing middle of FW.* Below, deep
maroon to rust-colored; FW has white
spots repeated and HW has broad, rust-
colored and whitish median band; both
have frosty blue-gray marginal spots.
Fringe checkered buff and dark. Head
and thorax broad, golden-orange.
Similar Species: Zestos Skipper smaller, has yellow FW
spots, lacks gold at base.
Life Cycle: Caterpillar amber to greenish with
brown bands and reddish stripes; feeds
on sennas (*Cassia*) and other shrubs.
Chrysalis brown with bluish dusting.
Flight: Year-round in Tropics; mostly in spring
in North America.
Habitat: Subtropical and tropical scrub,
streamsides, and along canals.
Range: Latin America to Argentina; rarely
wandering to Arizona, Texas, New
Mexico, and Florida; also West Indies.

This skipper is rare north of Mexico. It
drinks from damp ground, taking in
moisture, salts, and amino acids.

308 Silver-spotted Skipper
(*Epargyreus clarus*)

Description: 1¾–2⅜″ (44–60 mm). *Large;* long-
winged, with stubby lobe on HW.

Above, dark brown with a broad band of *squarish, glassy yellow-orange spots* across middle of FW, 1 smaller spot beyond FW cell, and minute dots near tip. *Below,* FW paler with similar spots, frosty purplish margin; HW has frosted, purplish scaling outwardly; *HW disk mostly filled by large, irregular patch of silver.* Fringes checkered buff.

Similar Species: Golden-banded and Gold-spotted skippers both lack extensive silver patch below. Gold-spot Aguna lighter in color, orange band on FW lighter, silvery mark on HW below narrower, linear, and irregular. Hoary Edge has lighter yellow band, silver marginal frosting below, not cell patch.

Life Cycle: Egg green, globular. Mature caterpillar light yellow-green with darker green lines, patches, and speckles, and rust-red head; builds shelter of leaves on host plant and forms dark brown chrysalis with darker and lighter markings in a loose cocoon among ground litter. Host plants include wisteria (*Wisteria*), locusts (*Gleditsia, Robinia*), beggar's tick (*Desmodium*), beans (*Phaseolus*), and licorice (*Glycyrrhiza*).

Flight: 1 brood in North; May–September. 2 broods or more in South; most of year.

Habitat: Riparian forests, open woods, canyons, grassy hillsides, parks, and gardens.

Range: British Columbia east to Quebec and south to Baja California, N. Mexico, and Florida.

The Silver-spotted Skipper occupies one of the most extensive ranges of any North American butterfly. It adapts readily to suburbs and parks and often puts on spectacular aerial displays. The Zestos Skipper (*E. zestos*) flies throughout the West Indies and southern Florida. It has large, glassy, golden spots on the fore wing above, but lacks the large glassy spot on the hind wing. Its caterpillars feed on woody legumes.

298 Hammock Skipper
(*Polygonus leo*)

Description: 1¾–2″ (44–51 mm). Long-winged;
HW curved into a stumpy lobe. Warm
brown above, with *several square, glassy
white spots on FW disk,* smaller pair near
tip. *Below,* FW tip and HW frosty
*violet with brown bands; HW has 1 small
brown dot near base.* Purple in sun.

Similar Species: Manuel's Skipper has larger white spots
and blue gloss above, more contrasting
HW band below. Purple-washed
Skipper smaller, has bluish-purple
sheen beneath.

Life Cycle: Caterpillar yellowish-green with
yellow stripes and patches along its
sides. Host plants are Jamaica dogwood
(*Ichthyomethia*) and pongam (*Pongamia*).

Flight: Several broods in Florida; year-round.
Fewer broods from Texas westward;
mainly late summer and fall.

Habitat: Shady glades in hardwood hammocks.

Range: S. California, Arizona, W. Texas, S.
New Mexico, Florida south to
Argentina; also West Indies.

The Hammock Skipper's name derives
from its strict habitat preference in
Florida, where it is common. This
species is far less common in southern
California, but on occasion the fall
Santa Ana Wind blows it from the
deserts to the coast.

Manuel's Skipper
(*Polygonus manueli*)

Description: 1⅝–1¾″ (41–44 mm). Above, dark
brown, blackish toward FW tip, with
strong blue gloss at base; 3 large, square,
bright white glassy spots in FW disk, 3
minute dots at FW tip; HW has dark
margin and submarginal spot band.
Below, FW bluish-charcoal with white
spots repeated from above. *HW grayish-
brown, with a rust-colored anal fold near*

abdomen and 3 dark brown spot bands.

Similar Species: Hammock Skipper larger, purplish, with dull white spots above, lacks rust-colored hue toward abdomen below.

Life Cycle: Caterpillar mottled green and yellow with reddish head; feeds on muelleria (*Muelleria moniliformis*) in South America.

Flight: February–November in Florida.

Habitat: Hardwood hammocks, tidal mud flats.

Range: S. Florida and Antilles to Argentina.

Manuel's Skipper often perches among foliage with its wings closed, making it difficult to spot. While rare in North America, it flies in the Everglades.

314 White-striped Longtail
(*Chioides catillus*)

Description: 1¾–2″ (44–51 mm). *HW has long tail* directed away from body at 90° angle to FW costa when wings are closed. *Warm brown above,* with cluster of pale, amber, glassy spots on FW disk, and a few dots near the tip. *Below,* FW paler brown and frosty, gray-brown at tip, with dots repeated and a large black spot on costa just below tip; HW rich chestnut-colored, with black costal spot and *long white band from costa to tail* (silvery or smoky brown-gray).

Similar Species: Zilpa Longtail has white patch near tail, not long band.

Life Cycle: Egg white, becoming yellowish. Young caterpillar dull green, maturing to rose-colored with orange stripes on sides. Chrysalis dark brown with waxy white bloom. Host plants include a number of legumes (Fabaceae), including sensitive plants (*Mimosa*) and beans (*Phaseolus*).

Flight: Several broods; year-round in Texas.

Habitat: Tropical forest clearings, woodland edges, canals, sparsely wooded hills.

Range: S. Arizona and Texas south to Argentina; also West Indies.

White-striped Longtails, like many of the larger subtropical skippers, sometimes wander and leave eggs north of their permanent breeding range. As a resident butterfly, the range of this species is restricted by the average winter temperature. Only in southern Texas can the White-striped Longtail survive most winters in our area.

313 Zilpa Longtail
(*Chioides zilpa*)

Description: 1½–1⅞" (38–48 mm). *Large, long-winged and long-tailed.* Above, warm brown FW has median row of bright, glassy, gold spots, 1 spot beyond the row, and minute dots near tip. *Below,* FW similar, paler toward tip; *HW* mahogany, frosted with whitish scales, and *large, crisp whitish patch in lower disk with white bar below patch near tail base.*

Similar Species: White-striped Longtail has paler yellow spots above, white stripe on HW below.

Life Cycle: Unknown.

Flight: Spring in Arizona, autumn in Texas.

Habitat: Tropical forest and subtropical scrub.

Range: SE. Arizona, S. Texas to Ecuador.

The Zilpa Longtail is not a year-round resident north of Mexico, but wanders infrequently to Arizona and Texas.

302 Gold-spot Aguna
(*Aguna asander*)

Description: 2–2¼" (51–57 mm). *Large, long-winged;* HW drawn out into tail stump. Warm *golden-brown above,* darker toward FW tip; FW bears *median band of yellow spots* and smaller spots toward tip. *Below,* FW similar, violet-brown at tip; *HW frosted violet-brown or whitish with long, straight bright white band* from costa to abdomen.

Similar Species: Silver-spotted Skipper has irregular silvery patch on HW beneath and FW band above more orange. Hoary Edge darker brown, with white along margin below.
Life Cycle: Unreported. Host plants belong to legume family (Fabaceae).
Flight: August–November in U.S.
Habitat: Shady forests.
Range: S. Texas south to Argentina; also Greater Antilles.

The Gold-spot Agunas tend to remain in or near forests. When not taking nectar or flying through sun-dappled clearings, they position themselves on the bottoms of large leaves. The white band beneath varies greatly.

---

297 Short-tailed Arizona Skipper
(*Zestusa dorus*)

Description: 1½–1⅝″ (38–41 mm). *Triangular wings; short, lobelike tail* on HW. Brown above with slightly reddish and leaden-gray cast toward base; *several pale yellowish, glassy spots on FW,* short row of spots across HW disk. *Below,* FW light orange on lower third, spots same as upperside; *HW frosted bluish-gray, with 2 brown bands* meeting at costa.
Similar Species: Chestnut-spotted Skipper has shorter HW lobes, striking reddish marking below. Arrowhead Skipper smaller, has white cell spot and HW band above.
Life Cycle: Mature caterpillar pale yellow-green with light longitudinal stripes and orange-brown head. Chrysalis overwinters. Host plants are oaks (*Quercus*), including Emory oak (*Q. emoryi*) and Arizona oak (*Q. arizonica*).
Flight: Several broods in S. Arizona; March–May and July–September. 1 brood in Colorado; April–May.
Habitat: Oak-covered flats and hills.
Range: Arizona, SW. Colorado, New Mexico, and W. Texas, south into Mexico.

Breeding populations of this uncommon skipper are local, yet the butterfly is known to wander, perhaps after breeding. Males gather at mud puddles and perch frequently on oak twigs, especially those over streams.

305 Arizona Skipper
(*Codatractus arizonensis*)

Description: 1⅝–2¼″ (41–57 mm). *Large; long-winged, mostly blackish-brown.* Dark brown above with tawny-brown hair over FW base, much of HW, and on body; *large, square, milk-white spots (sometimes bluish) cross FW in median band* and continue sparsely out toward tip. *Below* FW similar to upperside, with some purplish scaling; *HW mostly purplish-brown, crossed by dark brown bands.* Fringe buff and dark.

Similar Species: Araxes Skipper has broader wings, orange-washed below.

Life Cycle: Unknown.

Flight: Several broods in U.S.; April–October.

Habitat: Canyons, desert edges, and dry woods.

Range: SE. Arizona, SW. New Mexico, W. Texas south to S. Mexico.

In the Southwest, this species is often confused with the chiefly Mexican Melon Skipper (*C. melon*), which can only be distinguished by experts.

8, 47, 310 Long-tailed Skipper
(*Urbanus proteus*)

Description: 1½–2″ (38–51 mm). *Large;* robust body, long wings with *HW tail projecting* ½″ (13 mm). *Above,* dark brown with band of square white spots across FW, smaller spots toward tip; *intense bluish-green or green iridescent sheen on FW base, on HW, head, and thorax.* Below, pale brown; FW spotted, HW has 2 dark bands and dark near tail.

Similar Species: Other long-tailed skippers in North America do not have green iridescence.

Life Cycle: Egg, about ⅟₂₅₆″ (0.1 mm), pale yellow. Caterpillar, to 1¼″ (32 mm), yellow-olive or greenish with brownish lines and yellow and black speckling; reddish toward front and rear. Chrysalis, to ⅞″ (22 mm), cocoa-brown with bluish and yellowish shading and a white, powdery surface. Host plants include many families of plants with 2 seed leaves (dicots) such as legumes (Fabaceae) and crucifers (Brassicaceae), and more rarely, plants with a single seed leaf (monocots), such as canna lilies (*Canna*).

Flight: 3 or more broods in Florida; year-round. Summer and fall in North.

Habitat: Watercourses, shores, and gardens.

Range: Resident from S. California, Arizona, Texas, and Florida south to Argentina; strays to Mississippi Valley and East Coast to Connecticut and Midwest.

The very beautiful Long-tailed Skipper is considered a pest by farmers and gardeners, who call it the "Bean-leaf Roller." In southern California, the loss of bean fields has made this species less common there, but in Florida it occurs prolifically. Because the caterpillars feed on many host plants, the species is often abundant, and huge emigrations result. Most years some adults reach the North, but they cannot survive hard winters.

315 Lilac-banded Longtail
(*Urbanus dorantes*)

Description: 1½–2″ (38–51 mm). *Long-winged and long-tailed.* Grayish-brown above with a loose series of glassy white spots on FW disk and near FW tip. *Below,* FW similar but paler, purplish near tip; *HW purplish frosty-gray, with dark brown spots at base, 2 dark bands across disk, and 1 band on margin.* Checkered buff fringe.

Similar Species: Long-tailed Skipper has green head, thorax, and wing bases.

Life Cycle: Egg shiny green. Caterpillar rose-orange to chartreuse, reddish toward ends, with light spots, short hair, and black head. Chrysalis light brown. Host plants include beans (*Phaseolus*), butterfly-pea (*Clitoria*), and beggar-weed (*Desmodium tortuosum*).

Flight: 3 or more broods; June–October.

Habitat: Hardwood hammocks, edges of woodlands, and saw grass marshes.

Range: S. California (rarely), S. Arizona, S. New Mexico, S. Texas, and S. Florida south to Argentina and Antilles.

Well-established in Florida, the Lilac-banded Longtail prefers shade to open sunshine, and darts back and forth across paths and clearings until landing beneath a leaf. It takes nectar from lantana, Spanish needles, and ironweed.

311 Teleus Longtail
(*Urbanus teleus*)

Description: 1⅝–1⅞" (41–48 mm). *Long, narrow wings with fairly long HW tail.* Dark sienna-brown *above with thin, whitish line from costa most of way across FW,* and a series of 4 small whitish dots near FW tip. Below, tan, white marks repeated on FW; HW has 2 dark brown bands and dark brown basal spot.

Similar Species: Brown Longtail lacks white FW line.

Life Cycle: Little known. Brown caterpillar feeds on schrankia (*Schrankia*), a legume, in Brazil.

Flight: 2 or more broods; May–December.

Habitat: Tropical forest clearings and edges.

Range: S. Arizona, S. Texas to Argentina.

The Teleus Longtail appears only occasionally in Arizona and Texas and may not breed north of Mexico.

312 Brown Longtail
(*Urbanus procne*)

Description: 1⅝–1⅞″ (41–48 mm). *Long wings, long tails. Normally unmarked brown above,* sometimes with a poorly developed white line across FW. Tan *below,* white line present or absent but seldom prominent; 2 brown spot bands across HW, the *basal band not reaching the outermost of 2 brown spots on costa.* Fringes buff, uncheckered.

Similar Species: Teleus Longtail has narrow white band across FW. Other longtails have yellow or amber spot bands across FW above and some white on HW below.

Life Cycle: Early stages undescribed. Caterpillar feeds on grasses (Poaceae), including Bermuda grass (*Cynodon dactylon*). Before pupating, caterpillar weaves grass litter and silk to make shelter.

Flight: 3 or more broods in Texas; year-round, most common March–October.

Habitat: Banks of old watercourses and irrigation canals in Texas.

Range: S. Texas, perhaps Arizona, south to Argentina.

Nearly all of the Pyrginae use plants with 2 seed leaves (dicots) as their host plants. But the caterpillar of the Brown Longtail feeds on grasses and is an interesting exception.

306 White-tailed Skipper
(*Urbanus doryssus*)

Description: 1⅛–1¾″ (28–44 mm). *Stocky, with short, thick, white tail* projecting outward from HW. Dark brown above with a narrow, glassy, white line from costa more than halfway across FW; *HW has white tail and margin* halfway to costa. Fringe light.

Life Cycle: Unknown.

Flight: Several broods in Mexico; Texas records from March and November.

Habitat: Scrub patches, plantations and edges among tropical deciduous forests.
Range: S. Texas south to Argentina.

With its snow-white tails, white stripe, and deep brown wings, this skipper is unmistakable. It flies rapidly, giving the impression of a flying streak.

309 Flashing Astraptes
(*Astraptes fulgerator*)

Description: 1⅞–2⅜" (48–60 mm). *Large, robust body,* long triangular wings. Basically black above with *brilliant metallic turquoise-blue scaling on head, body, and bases of wings.* Beyond blue, *band of squarish, glassy, white spots crosses FW* and series of minute white dots lies near FW tip. Below, FW blue along basal costa, otherwise shades of brown; HW has bright white patch along costa.
Life Cycle: Caterpillar velvety black with yellow bands on each segment and large rust-black head. Chrysalis blackish, covered with loose, whitish powder. Chaste-tree (*Vitex*) is a Texas host plant.
Flight: 2 broods; March–May and August–December.
Habitat: Shady, shrubby places, roadsides, and canal banks in subtropical woodlands.
Range: S. Texas south to Argentina.

One of the most brilliant butterflies found north of Mexico, the Flashing Astraptes loses its iridescent scales soon after emerging, and older individuals tend to look rather drab.

307 Dull Astraptes
(*Astraptes anaphus*)

Description: 2–2¼" (51–57 mm). *Large;* long wings. *Coffee-brown above,* lighter below; crossed above and below by a number of

darker brown bands parallel to body.
Usually *bright yellow patch at HW outer*
angle above and on margin below.

Life Cycle: Caterpillar yellow, brown head has
eyespots. A wild bean (*Phaseolus*) serves
as host plant in South America.

Flight: Successive broods in Mexico; most of
year. Early autumn in U.S.

Habitat: Unreported.

Range: Latin America and Greater Antilles,
rarely north to S. Texas.

One population found in Texas has a
prominent yellow patch on the hind
wing, but in other races it is missing.

253, 676   Golden-banded Skipper
(*Autochton cellus*)

Description: 1⅝–2″ (41–54 mm). Blackish-brown
above with *broad, golden-yellow bars*
across FW, and a small white spot bar
near FW tip. HW above yellowish near
outer angle. *Below,* FW similar to
upperside with paler grayish or brown
near margins; *HW gray-frosted toward*
margin with row of submarginal black
spots, reddish-brown over disk crossed
by *2 bands of dark brown spots without*
noticeable black outlines. HW fringe
checkered on upper ⅔, brown below that
to HW outer angle. *Antennae all black.*

Similar Species: Hoary Edge has gold above in broken
patch, white-frosted below. Gold-
spotted Skipper has longer wings, lobed
HW, orange spots in band above.

Life Cycle: Eggs yellow, deposited in rows of 2–7,
like those of most skippers. Caterpillar
chartreuse with yellow stripes and
speckles, reddish head; feeds on hog
peanut (*Amphicarpa*) in Maryland and
lives in shelter of silk-bound leaves,
venturing out to feed at night.
Chrysalis dark greenish-brown with
waxy bloom; overwinters.

Flight: 2 broods in North; May–August. More
broods in South; February–September.

Habitat: Watersides, woodland ravines, grassy
spots.

Range: Ohio to New York, south to N.
Florida, Alabama, SE. Missouri and
Gulf States; also W. Texas, SW. New
Mexico, and SE. Arizona south into
Mexico.

Many of the big, attractive skippers are
limited to the southern states, but this
one ranges well into the northeast in
small, well-separated colonies. Adults
clamber over hollyhocks, bramble
blossoms, ironweed, and buttonbush in
search of nectar. The smaller False
Golden-banded Skipper (*A. pseudocellus*)
lacks the yellow on the hind wing costa
and grayish scaling below. It is rare in
southeastern Arizona.

294 Hoary Edge
(*Achalarus lyciades*)

Description: 1½–1¾" (38–44 mm). *Wings
triangular, outer edge of HW quite
straight.* Blackish-brown above with a
*cluster of 4 or 5 glassy, yellow-orange,
squarish spots on FW disk.* Beneath, FW
has dark brown toward base with
yellow to tan spot band, area beyond
spots toward tip is paler brown,
clouded with light scales. *HW below*
banded black and brown at base; outer
half of wing *heavily frosted with white,*
also near abdomen. HW fringes white,
lightly checkered; HW fringes heavily
checkered.

Similar Species: Golden-banded Skipper lacks hoary
edge below. Silver-spotted Skipper and
Gold-spot Aguna have white on disk
below, longer orange FW bands above.

Life Cycle: Egg off-white. Caterpillar grass-green,
has bluish back stripe and yellowish
speckling. Chrysalis light brown with
dark and yellowish patches. Host plants
include tick trefoil (*Desmodium*), wild
indigo (*Baptisia*), bush clover (*Lespedeza*).

Flight: 1 brood in North; May–July. Many
broods in South; April–December.

Habitat: Scrub oaklands, woodland edges, roads,
gardens, disturbed areas, and meadows.

Range: Minnesota east to New Hampshire and
south to Florida and N. Texas.

The Hoary Edge is named for its frosty-
white edge below. It is generally
common east of the Mississippi Valley.
Males select small recesses or openings
in the woods, which they defend from
other males or where they find mates.

Mexican Hoary Edge
(*Achalarus casica*)

Description: 1½–2″ (38–50 mm). Dark brown
above with a *cluster of small, glassy white
spots* on outer FW. Below, FW frosted
with grayish toward tip; HW finely
striated and broadly banded with dark
brown on inner ⅔, *margin heavily frosted
with white.* Fringes white, with
checkering light on HW, heavier on
FW. Male has costal fold.

Similar Species: Caicus Skipper smaller, less white along
HW edge below; male lacks costal fold.
Drusius Cloudywing blacker, less
whitish on HW margin below.

Life Cycle: Unknown.

Flight: 2 or more broods; May–October.

Habitat: Scrub, wood edges, and weedy areas.

Range: SE. Arizona, SW. New Mexico, Texas,
and Mexico.

It has been suggested that the whitened
hind wings break up the wing pattern,
making them less easy prey for birds.

290 Coyote Skipper
(*Achalarus toxeus*)

Description: 1⅝–2″ (41–51 mm). *Large,* with
narrow and very pointed wings. Chocolate-

brown on both surfaces, slightly lighter on FW below. *FW* above and below *crossed by very vague brown bands* of a different shade. *HW below crossed by 2 more dark brown bands. HW fringe white.*

Similar Species: Hermit has rounded wings. Other large tailless skippers have more prominent markings. Drusius Cloudywing blacker with glassy FW spots.

Life Cycle: Unreported. Host plant is Texas ebony (*Pithecellobium flexicaule*) and probably other species.

Flight: Probably several broods; most of year.

Habitat: Chaparral, desert and mountain scrub forests, and savannahs.

Range: Arizona and Texas south to Panama.

In the Santa Ana National Wildlife Refuge of southern Texas, the Coyote Skipper flies about Texas ebony trees. Males visit the flowers while females deposit their eggs deep in the foliage. The very similar Jalapus Skipper (*A. jalapus*) has a more clearly lobed hind wing. It sometimes strays to Texas.

285 Southern Cloudywing
(*Thorybes bathyllus*)

Description: 1¼–1⅝″ (32–41 mm). Brown *above and below, with FW white spots pronounced and joined together in bars, including 1 hourglass-shaped glassy spot across cell.* HW forms rudimentary tail; fringes are light buff and checkered. *Dark brown bands cross HW beneath,* separating paler brown outer region and darker center. Male lacks costal fold.

Similar Species: Northern Cloudywing lacks glassy bar across FW cell; males have costal fold.

Life Cycle: Egg pale green. Caterpillar dull mahogany, lightly lined; uses various legumes (Fabaceae) as host plants before pupating in stout, moss-brown chrysalis.

Flight: 1 brood in North; June. At least 2 broods in South; March–December.

Habitat: Roadsides, dry meadows, clearings, burns, and upland barrens.

Range: Minnesota and Nebraska, east to New England, south to Texas and Florida.

Even in its expanded northern range, this species is often overlooked. Its flight is faster and more erratic than that of the Northern Cloudywing.

283, 286 Northern Cloudywing
(*Thorybes pylades*)

Description: 1¼–1¾" (32–44 mm). Rounded wings. Above, brown with checkered fringes; *white glassy spots on outer FW usually small, triangular, and not connected in band nor present in cell. Beneath, frosted gray on outer edges.* Angular spots in 2 HW bands darker than background color, outlined in darker scales. Male has costal fold.

Similar Species: Southern Cloudywing has hourglass glassy bar across FW cell. Eastern, Southern, and Western cloudywing males all lack costal fold.

Life Cycle: Egg light green, becoming white before hatching. Caterpillar purplish-green, striped with pink and purple; constructs silken nest among leaves of legume hosts (Fabaceae). Chrysalis dark brown, paler on abdomen and wing cases.

Flight: 1 brood in North and high altitudes; May–July. 2 or more broods in South; March–December.

Habitat: Open woods, edges, fields, brush, roadsides, meadows, and clearings.

Range: British Columbia to Quebec, south throughout U.S. into Baja California and S. central Mexico.

The Northern Cloudywing is the most common and broadly distributed *Thorybes* skipper in North America. The similar Western Cloudywing (*T. diversus*) has a larger, well-developed glassy spot below the cell. It flies in mountains from southern Oregon to central California during the summer.

288 Mexican Cloudywing
(*Thorybes mexicana*)

Description: 1¼–1½" (32–38 mm). Dark to golden-brown *above with numerous glassy white spots on outer FW, and dirty-colored fringes,* sometimes checkered. *Underside finely striated,* sometimes has darker bands across HW beneath; *outer half of both wings below distinctly frosted* and paler than brown inner half.

Similar Species: Drusius Cloudywing darker, less 2-toned below.

Life Cycle: Unrecorded. Host plants include clovers (*Trifolium*), vetch (*Vicia*), and peas (*Lathyrus*).

Flight: 1 brood; May–August, for about 1 month in a single location.

Habitat: Alpine or subalpine areas in Oregon and California, mountains elsewhere.

Range: Oregon Cascades and N. California through W. Nevada, SW. Colorado, Arizona, and New Mexico into Mexico.

The Donner Pass in California is a good place to seek the Mexican Skipper. A related species also found in the American Southwest and Mexico is the Valeriana Cloudywing (*T. valeriana*). This large tawny-brown skipper has checkered fringes and dark spot bands on the frosty-violet hind wing below.

284 Eastern Cloudywing
(*Thorybes confusis*)

Description: 1¼–1⅝" (32–41 mm). Rounded wings. *Above,* dark brown with *linear glassy spots. Below, brownish HW* crossed by darker bands; outer FW gray-brown. Fringes checkered.

Similar Species: Northern Cloudywing has more triangular glassy spots; male has costal fold. Southern Cloudywing has more pointed FW and larger glassy spots.
Life Cycle: Undescribed.
Flight: 1 brood in North; July and August. Several broods in South; February–October.
Habitat: Abandoned pastures, hay meadows, and other open spaces, often near woods.
Range: Kansas and Pennsylvania, south to Florida and Texas.

Many of the cloudywings (*Thorybes*) are similar, with individuals within a species quite variable, but each species has distinctive behavior and habitat.

Drusius Cloudywing
(*Thorybes drusius*)

Description: 1½–1⅞" (38–48 mm). *Blackish-brown above with several loosely arrayed glassy, white small dots in outer half of FW and bright white, uncheckered fringes on HW.* Lighter brown below with vague, black-outlined brown bands across HW; outer part of HW grayish.
Similar Species: Coyote Skipper larger, more pointed, lacks FW glassy spots. Mexican Hoary Edge more whitish along HW edge below, has striations and obvious bands on HW below. Caicus Skipper smaller, has more white on HW edge below.
Life Cycle: Unknown.
Flight: Probably one brood; April–June in Texas, June–August in Arizona.
Habitat: Moist areas in arid mountains.
Range: SE. Arizona, SW. New Mexico, W. Texas south into Mexico.

Because of its snow-white hind wing fringes and contrasting dark wings, the Drusius Skipper differs dramatically from other cloudywings. It displays these fringes as it perches on flowers and damp earth with wings spread.

287 Potrillo Skipper
(*Cabares potrillo*)

Description: 1⅛–1½" (28–38 mm). *Above*, cocoa-brown with several *white chevrons pointing toward FW tip;* darker brown bands vaguely cross wings, and *HW margin has distinctive bend. Below, frosty-white and cloudy-violet,* crossed by dark brown clouds and white chevrons.

Life Cycle: Egg pea-green. Caterpillar brown with orange speckles and lines. Chrysalis tan, with violet tint and ashen powder. Host plant unknown.

Flight: 2 or 3 broods in Texas; March–May, June, and November.

Habitat: Tropical and semitropical open areas.

Range: Extreme S. Texas to Costa Rica and Greater Antilles.

Visitors to the Santa Ana National Wildlife Refuge in Texas may see this skipper taking nectar on verbena.

303 Falcate Skipper
(*Spathilepia clonius*)

Description: 1⅝–1¾" (41–44 mm). Extremely distinctive with *FW curved outward and nipped off squarely.* Above and below, brown-black with very dark FW crossed by prominent white band of spots; slightly scalloped margins; dark checkered fringes.

Life Cycle: Caterpillar yellow; bears brown network on back. Host plants are a member of the mimosa family (*Inga edulis*) and wild beans (*Phaseolus*). Chrysalis stubby, tan, covered with sheath of waxy, white, powdery spikes.

Flight: May and November in Texas.

Habitat: Open spaces near towns.

Range: S. Texas south to Argentina.

The Falcate Skipper is not uncommon throughout much of the American Tropics but it rarely wanders to Texas.

304 Mimosa Skipper
(*Cogia calchas*)

Description: 1¼–1¾″ (32–44 mm). Long, pointed wings rounded on margins. Mat-brown *above with cocoa-brown fringes and minor light spotting* about FW tips. *HW beneath pale bluish-gray with violet-brown band;* inner margins below silvery.

Similar Species: Others in genus similar; Mimosa Skipper has contrasting bands below.

Life Cycle: Caterpillar compact, yellow, speckled with white; feeds on mimosa (*Mimosa pigra*) in Texas.

Flight: 3 broods; March–November.

Habitat: Thorn savannah and forest edges.

Range: S. Texas to Argentina.

When seen near mimosa bushes, there should be no doubt about this skipper's identity.

291 Acacia Skipper
(*Cogia hippalus*)

Description: 1½–1⅞″ (38–48 mm). Long, triangular wings. Brown with *conspicuous white spots on FW above* and below; HW gray-brown *beneath crossed by zigzag bands of brown*. FW fringes checkered, *HW fringes conspicuously white. Violet cast beneath.*

Similar Species: Others in genus very similar.

Life Cycle: Unrecorded. Caterpillar feeds on acacias (*Acacia angustissima*) in Texas.

Flight: 2 or more broods; April–September.

Habitat: Acacia thorn scrub and hilly drylands.

Range: Arizona, New Mexico, and W. Texas through Baja California and Mexico to N. South America.

The Acacia Skipper may be numerous around its host plant as well as some distance away at other favorable nectar sources. A good place to look for it is Olmos Park Bird Sanctuary in San Antonio, Texas.

289 Caicus Skipper
(*Cogia caicus*)

Description: 1⅜–1⅝″ (35–41 mm). *Dark sienna-brown* above with a number of small, glassy, white dots and bars on outer FW. *Below*, FW lighter toward trailing edge; HW cocoa-brown *distinctly banded with dark brown, the bands outlined with black; HW margin frosted violet-whitish with brown striations, white HW fringe.*

Similar Species: Mexican Hoary Edge larger with a broader whitish margin on HW below and no obvious black outlines to HW bands below. Other members of genus very similar. Drusius Cloudywing larger, darker, with fainter bands beneath and without striations.

Life Cycle: Unknown.

Flight: 2 or more broods; March–August.

Habitat: Tropical and subtropical thorn forests.

Range: Arizona south to Honduras.

It is likely that this species feeds on leguminous thorn trees as do its close relatives, but it is poorly known here.

281 Purplish Black Skipper
(*Nisoniades rubescens*)

Description: 1¼–1½″ (32–38 mm). *FW pointed, HW quite rounded* although long. Above and below, *black with obscure purplish banding* and mottling (or deep plum with black banding) and a short series of white dots from FW costa near tip.

Life Cycle: Unrecorded.

Flight: November in U.S.

Habitat: Watercourses among tropical forests.

Range: Mexico to Brazil, rarely S. Texas.

The Purplish Black Skipper is not a regular member of the Texas butterfly fauna, but it occasionally wanders to Pharr, Texas, in the lower Rio Grande Valley, where many tropical species enter the United States.

264   Golden-headed Sootywing
(*Staphylus ceos*)

Description: 1–1⅛" (25–28 mm). *Small.* Wings
very rounded with 1 slight scallop on
HW. Above and below, dark brown to
black with vague dark and light marks.
Bright golden to orange head; long palpi.

Similar Species: Red-headed Roadside Skipper has
narrow, pointed wings, orange fringes.

Life Cycle: Unknown.

Flight: Overlapping broods; spring–autumn.

Habitat: Gulches, canyons, and bottoms of
narrow valleys.

Range: S. New Mexico, S. Arizona, W. and S.
Texas, and N. Mexico.

Golden-headed Sootywings resemble
true sootywings (*Pholisora*). But they
behave differently—the 3 *Staphylus*
spend much of their time resting with
their wings spread on or under leaves.

263   Southern Scalloped Sootywing
(*Staphylus mazans*)

Description: 1–1¼" (25–32 mm). *Purplish-brown or
blackish* above and below, with black
bands, more apparent in female. Tiny
white dots on FW, 1 below cell, 2 near
tip. *Margins irregularly scalloped.*

Similar Species: Scalloped Sootywing lighter.

Life Cycle: Recorded host plants in Texas are
goosefoot (*Chenopodium album, C.
ambrosioides*), and wild beet (*Amaranthus
retroflexus*). Immature stages unknown.

Flight: 3 broods; March–November.

Habitat: Tropical and subtropical forests.

Range: S. Texas through E. Mexico perhaps to
Brazil.

The Southern and Scalloped sootywings
are best distinguished by distribution.
The Southern only ranges north to the
latitude of Austin, and south of San
Antonio has no look-alikes.

262 Scalloped Sootywing
(*Staphylus hayhurstii*)

Description: 1–1¼" (25–32 mm). *Above, charcoal to dark brown* with conspicuous black bands and minute white dots on FW above, and 1 dot below cell, 2 near tips. *Deeply scooped-out margins of HW* look nibbled. Dark and buff checkered fringes. Female more contrasting.

Similar Species: Southern Scalloped Sootywing darker with less distinct dark bands.

Life Cycle: Egg orange. Mature caterpillar green with pinkish sheen and maroon head. Chrysalis olive-brown except on orange abdomen; bears powdery, white bloom. Goosefoot (*Chenopodium*) and amaranth (*Alternanthera*) are host plants.

Flight: 3 broods; March–November.

Habitat: Moist places with shade: lanes, trails, well vegetated areas, urban vacant lots, and willow and cottonwood stands.

Range: Mississippi Drainage, ranging from S. Canada west of Great Lakes and east of Rockies, to Pennsylvania, central Texas, and S. Florida.

This species and the Southern Scalloped Sootywing overlap between Austin and San Antonio, Texas, where they are difficult to distinguish. North of Austin, only this species occurs.

261 Variegated Skipper
(*Gorgythion begga*)

Description: ⅞–1⅛" (22–32 mm). *Small; FW indented* slightly just below tip, otherwise wings rounded. Above and below, complexly *variegated with bands and bars of cocoa-brown and chocolate-brown*. Prominent dark brown bar across light brown FW cell. Minute white dots near tip on FW costa. HW below has more distinct dark bands and striations, sometimes also white scaling around HW outer angle (tornus).

Upperside may appear violet-tinged.

Similar Species: Duskywings larger with buff spots on brown HW; FW margin regular. Alphaeus Sootywing much less banded-looking, has checkered fringe.

Life Cycle: Unknown.

Flight: Early spring in U.S.

Habitat: Roadsides, watercourses, and valley bottomlands in wooded areas.

Range: Mexico to Argentina, rarely S. Texas.

The Variegated Skipper is one of only 4 members of its tropical American genus, *Gorgythion*. It has been recently seen in the lower Rio Grande Valley.

256 Hoary Skipper
(*Carrhenes canescens*)

Description: 1⅛–1⅜" (28–35 mm). *Above, dust-brown to smoke-gray,* somewhat browner on HW, crossed by several rows of darker lines and partial pale bands and *submarginal row of brown, blurry chevrons; 8 or 9 small glassy white spots,* more or less pronounced, *form 2 broken, crooked rows* from costa to middle of FW. Below, lighter buff with 5 dark concentric rows of darker scales.

Similar Species: Duskywings larger, darker with fewer glassy spots.

Life Cycle: Unknown.

Flight: Early spring in Texas.

Habitat: Subtropical woodland borders.

Range: S. Texas south to Argentina.

Hoary Skippers rarely enter Texas, but may be more common than has previously been thought.

257 Window-winged Skipper
(*Xenophanes trixis*)

Description: 1⅛–1⅜" (28–35 mm). *Slightly sickle-shaped, pointed FW.* Above, buff-brown with *numerous white glassy spots in centers of both FW and HW, which give bright pearly reflections.* Below, FW spot in band and that of HW expand to fill entire wing out to brown margin.

Life Cycle: Caterpillar dotted, brown-headed; feeds on mallows (*Malachra fasciata*) in Brazil. Chrysalis tan and speckled, white-powdered, and hairy.

Flight: July in U.S., longer in Tropics.

Habitat: Floodplains and clearings.

Range: Extreme S. Texas to Argentina.

This singular butterfly merely grazes our region; it may be a stray, a periodic colonist, or a species whose range is slowly advancing northward.

---

260 Texas Powdered Skipper
(*Systasea pulverulenta*)

Description: 1–1¼″ (25–32 mm). *Above,* tan, powdered with black scales and *variegated* patches and bands of brown, white, rust-colored, and olive. *FW above has series of white glassy spots aligned with inner margin. HW margins scalloped;* fringes dirty-white, checkered. HW disk especially scaled; HW submarginal band above is usually bright copper-colored. Similar but duller below.

Similar Species: Arizona Powdered Skipper larger, more olive-gray, has wavier HW margin; inner margin of FW spots not aligned.

Life Cycle: Unrecorded. Host plants are mallows (*Abutilon, Pseudabutilon, Wissadula, Sphaeralcea*).

Flight: February–November in Texas.

Habitat: Mountain streamsides and well-drained slopes and ridges in hilly country.

Range: Texas to Guatemala.

This is the principal species in Texas and one of only 2 powdered skippers. Caterpillars can often be found feeding on mallows next to caterpillars of the Laviana Skipper.

259 Arizona Powdered Skipper
(*Systasea zampa*)

Description: 1–1½" (25–38 mm). *Above,* olive-tan powdered with sparse, black scales; *variegated* patches in tan, gray, and olive; *FW glassy spots form zigzag stripe;* sometimes copper-colored patches beyond cell of FW; HW submarginal band above olive-brown. Similar but paler below. HW edge wavy.

Similar Species: Texas Powdered Skipper smaller, more copper-colored; glassy spots in line.

Life Cycle: Unknown.

Flight: Successive, overlapping broods; April–October in Arizona.

Habitat: Desert canyons, arroyos, and oases.

Range: W. Texas, S. New Mexico, Arizona, S. California, Baja California, and NW. Mexico.

Rather common in parts of Arizona and Baja California, this powdered skipper is considered a rarity in southern California. Colonies occur along the western edge of the Colorado Desert.

258 Sickle-winged Skipper
(*Achylodes thraso*)

Description: 1⅝–1⅞" (41–48 mm). Wings full, rounded; FW indented below slightly hooked tip. *Above,* blackish-brown with a *violet sheen; FW has sickle-shaped black band just inside margin.* Male has clouds and patches of purplish or bluish-brown; female paler brown overall, sometimes blue-gray or olive-gray. Below, similar but lighter.

Life Cycle: Caterpillar yellow-green with a darker head and middle stripe. Chrysalis green with white powder; hangs from silken girdle. Host plants are citruses (Rutaceae); in Texas also lime prickly-ash (*Zanthoxylum fagara*).

Flight: Successive broods; year-round in Texas.

Habitat: Rolling, hilly country and open spaces.

Range: Texas south to Argentina; also Greater
Antilles.

The hooked tips, together with the
violet coloring, separate this species
from any other skipper.

255 Hermit
(*Grais stigmaticus*)

Description: 1¾–2¼" (44–57 mm). *Large* with long
wings. Brown, with understated darker
spots; female has small cluster of glassy
spots near FW tip. Below, lighter,
with more diffuse markings. *Antennae
bent back like shepherd's crook; palpi bright
orange below;* underside of abdomen buff
with dark midline.

Life Cycle: Unrecorded.
Flight: August–September.
Habitat: Woods, thickets, and savannah.
Range: S. Texas to Argentina; also Jamaica.
Wandering occasionally to N. Texas.

The Hermit and the Brown-banded
Skipper populate southern Texas in
autumn, when the parched countryside
finds many other butterflies between
broods. The Hermit is named for its
behavior: seeking protection or perhaps
shade, it crawls beneath large leaves.

Brown-banded Skipper
(*Timochares ruptifasciatus*)

Description: 1⅝–1¾" (41–44 mm). *Mothlike.*
Above, mostly gray-brown on FW
crossed by rows of generally separated
brown spots; tawnier on *HW with 3
bands of joined, square brown spots.* Quite
golden-orange beneath with shadow
bands of markings above.

Life Cycle: Egg pale, darkens to golden. Green
caterpillar bears yellow speckles and
lines dotted with orange. Caterpillar

feeds on vines (Malpighiaceae).
Chrysalis glossy green.

Flight: August—October.

Habitat: City gardens in Texas; where host vines grow elsewhere.

Range: S. Texas, perhaps to Arizona, and Mexico.

This butterfly resembles certain migrant, day-flying, noctuid moths.

244 White Patch
(*Chiomara asychis*)

Description: 1¼—1½" (32—38 mm). *Very variable;* basically chalk-white with gray, brown, and blackish patches, chevrons, dashes, and marks above and below. FW sharply pointed. *HW above usually has outstanding clear white patch.* HW below often largely white except for margin. Long gray hair partially covers HW.

Similar Species: Checkered and White skippers are more neatly patterned.

Life Cycle: Caterpillar and chrysalis green. Host plants are tropical shrubs of the family Malpighiaceae.

Flight: 2 broods, March—June and August—December.

Habitat: Urban gardens and open spaces.

Range: S. Nevada, Arizona and Texas through Argentina and Lesser Antilles.

The White Patch may be found on the Santa Ana National Wildlife Refuge in the lower Rio Grande Valley.

279 Blue-banded Skipper
(*Gesta gesta*)

Description: 1⅛—1⅜" (28—35 mm). Relatively short, blunt wings. Dark brown above and below, crossed by a *series of bands; the innermost FW band blackish-brown, the next frosty-blue, and the next violet* with darker purple on outer margin; FW dark at base and in cell. HW fringes white and vaguely checkered; those of FW much darker. Below similar.

Similar Species: Duskywings lack bluishness, dark basal area, blunt FW, and checkered HW fringe. Florida Duskywing has glassy spots on FW, more pointed wings.

Life Cycle: Egg orange, turns whitish. Young caterpillar yellow-green, turns darker, with whitish ends, green median stripe and yellow spots along sides. Chrysalis shiny green. Host plants are legumes, senna (*Cassia*) and indigo (*Indigofera*).

Flight: March–November in Texas.

Habitat: Hills, coastal flats, and shrub areas.

Range: Arizona, Texas, Mexico to Argentina, and West Indies.

The Blue-banded Skipper can be seen in Aransas National Wildlife Refuge on the coast of Texas.

265, 280 Florida Duskywing
(*Ephyriades brunnea*)

Description: 1¼–1⅝" (32–41 mm). *Fairly large, with long wings. Above, male mostly brown* with inner half dark chocolate-brown; *female violet-brown with 3 dark brown bands across wings alternating with lavender bands.* Both sexes have glassy, white or greenish-white spots on FW, more numerous and larger in female. Below, similar but more subdued.

Life Cycle: Caterpillar yellow-green with 2 yellowish stripes and 1 gray stripe along each side; feeds on Barbados cherry (*Malpighia glabra*) in Florida. Chrysalis stout, pale green.

Flight: Overlapping broods; year-round.

Habitat: Gardens and open spaces.

Range: S. Florida, Keys, and West Indies.

Not really a duskywing, this species nevertheless bears a striking similarity and close relationship to that group.

274 Dreamy Duskywing
(*Erynnis icelus*)

Description: 1–1⅜" (25–35 mm). *Small. Rounded, stumpy wings. FW patterned above;* lacks glassy spots; *chains of black scales across FW* enclose gray-brown spots within links. HW above fairly plain, gray-brown or blackish-brown, sometimes violet at base; HW has pale spots in 2 rows. Below, similar but less strongly marked. Palpi long and conspicuous. Male has hair tufts on hind leg (tibia). Antennal club long and pointed.

Similar Species: Sleepy Duskywing is larger, has clear chain link marks on FW, shorter palpi, and stout, blunt antennal club.

Life Cycle: Egg ribbed; changes from green to pinkish before hatching. Mature caterpillar light green, speckled with white beneath short, profuse hair; black head may have red, yellow, or orange patterning. Caterpillar overwinters. Chrysalis dark green or brownish. Host plants are willows (*Salix*), poplars (*Populus*), locusts (*Robinia*), and birches (*Betula*).

Flight: 1 extended brood; April–August.

Habitat: Roadsides, trails and clearings among northern forests, along streams, and in foothill mesas and canyons.

Range: British Columbia to Nova Scotia, south to Georgia, and west to Indiana, New Mexico, and California; not in most of southern Great Plains and Gulf states.

The Dreamy Duskywing probably uses a variety of trees as host plants. The eggs are deposited low on new shoots and saplings. Adults favor hawkweeds and clovers as nectar sources.

268, 271 Sleepy Duskywing
(*Erynnis brizo*)

Description: 1⅛–1⅝" (28–41 mm). *Above, FW patterned,* HW fairly plain; dark brown with bands of gray to frosty spots outlined in black running across upper FW, sometimes giving impression of chainlike pattern; lacks obvious glassy spots near tips of FW. Small buff spots on HW; fringes brown. Below, similar, paler with vaguer marks. Palpi short.

Similar Species: Persius Duskywing smaller, with shorter wings, has tibia hair. Other duskywings have glassy FW spots above.

Life Cycle: Egg green at first, turning pink. Caterpillar pale green with white specks and covered with short hair; head has red, yellow, or orange pattern; overwinters. Chrysalis brown or dark green. Host plants are variety of oaks (*Quercus*) and occasionally American chestnut (*Castanea dentata*), but rarely since blight.

Flight: 1 brood in North; usually May. 2 broods in South; early spring–autumn.

Habitat: Scrub oak flats, foothills and chaparral, and acid sand and serpentine barrens.

Range: 4 subspecies distributed separately: Manitoba to Massachusetts, south to N. Florida and west to E. Texas; peninsular Florida; Rockies and Great Basin; and California mountains down into Baja California.

The Sleepy Duskywing flies in several kinds of places, and often in company with Edwards' Hairstreak.

269 Juvenal's Duskywing
(*Erynnis juvenalis*)

Description: 1¼–1¾" (32–44 mm). *FW patterned*, HW fairly plain. *Above, male* blackish-brown on FW with bands of black spots and chevrons and many glossy white spots; *HW* lighter brown *with buff spots* somewhat obscured by *long, gray hair*. Below, similar but lighter. Female much paler overall with more distinct markings. Fringes white or brown.

Brown-fringed populations have 2 pale spots near HW tip below.

Similar Species: Mournful Duskywing has white spots on HW inside white fringes. Scudder's and Pacuvius duskywings best separated by location.

Life Cycle: Egg green, turning pink. Mature caterpillar light green with white specks, thick hair, and red, yellow, or orange head; overwinters. Chrysalis dark green or brown. Host plants are black oak (*Quercus velutina*), northern red oak (*Q. rubra*), and white oak (*Q. alba*).

Flight: Usually 1 brood; spring. 2 broods in SE. Arizona; April–September.

Habitat: Oak woodlands, scrub oak chaparral, roadsides, and fields near oaks.

Range: Manitoba to Nova Scotia, E. United States west to Wyoming, Texas, and SE. Arizona into Mexico.

Like other duskywings, these sleep with wings folded rooflike over the back, in the manner of a moth. During the day they bask with wings spread.

266 Rocky Mountain Duskywing
(*Erynnis telemachus*)

Description: 1¼–1⅝″ (32–41 mm). *FW patterned; HW fairly plain. Gray-brown FW above crossed by 2 rows of outwardly pointing blackish arrowheads,* several nearest costa having tiny white glassy spots in centers. HW above and below warmer brown with 1 or 2 rows of blurred buff spots near margin; HW below usually with 2 prominent spots at costa near base. FW below paler than above, less distinctly marked. Female lighter, more contrasting. *Fringes buff-gray. Male has long FW hair.*

Similar Species: Juvenal's in Southwest, Funereal, and Mournful duskywings have white fringes. Persius Duskywing smaller with less definite pattern.

Life Cycle: Early stages unreported. Host plant is Gambel oak (*Quercus gambelii*) and probably other species.

Flight: 1 brood; April–July.

Habitat: Oak woodland in mountains.

Range: S. Wyoming, Utah, Colorado, SE. Nevada, Arizona, New Mexico, and W. Texas.

A closely related species, the Southern Duskywing (*E. meridianus*), is extremely similar, but dingier and usually less contrasting. It flies in California, Nevada, and New Mexico.

267 Propertius Duskywing
(*Erynnis propertius*)

Description: 1¼–1¾" (32–44 mm). *FW patterned; HW fairly plain.* Above, medium brown with glassy spots on FW, often lying within prominent dark brown patches; HW lighter brown, fringes cocoa-brown. *Below,* lighter; *HW has 2 pale spots.* Abundant, long, whitish hair looks furry. Female paler than male.

Similar Species: Dingy Duskywing darker, lacks 2 pale spots on HW below near tip. Rocky Mountain Duskywing paler.

Life Cycle: Egg pinkish before hatching. Mature caterpillar hairy, pale green, speckled with white; has red, yellow, or orange head; overwinters. Chrysalis green or brown. Host plants are coast live oak (*Quercus agrifolia*) and northwest Oregon white oak (*Q. garryana*).

Flight: 1 brood; early spring.

Habitat: Oak openings, glades, sunny slopes and ridges; mountains and hills but usually not lowlands.

Range: Vancouver Island to Baja California, discontinuously. Also western edges of Okanogan, Great, and Columbia basins.

The large, light duskywing of Pacific arid lands, Propertius occurs spottily, depending upon the presence of oak.

But colonies in the rain shadow of the Olympic Mountains on the San Juan Islands and in southern British Columbia may be many miles from one another and from the main oak belts.

254, 270   Horace's Duskywing
(*Erynnis horatius*)

Description: 1¼–1¾" (32–44 mm). *FW patterned, HW fairly plain.* Above, brown with *little contrast in male, more in female,* and variably-sized glassy FW spots and brown fringes. Below, paler; HW mottled light and dark brown, lacks white markings or pale spots near tip.

Similar Species: Juvenal's Duskywing has 2 pale spots near HW tip below in brown-fringed populations. Dingy Duskywing very similar. Rocky Mountain Duskywing paler; mostly found farther west.

Life Cycle: Egg green at first, becoming pinkish. Mature caterpillar hairy, light green, speckled with white; has red, yellow, or orange head; overwinters before forming dark green or brown chrysalis. Host plants are several species of oaks, including Muhlenberg's oak (*Quercus muehlenbergii*) in East, Gambel oak (*Q. gambelii*) in Rockies, and Texas oak (*Q. texana*) in Texas.

Flight: 2 broods in Rockies and North, 3 broods in South; January–October.

Habitat: Warm, sunny spots in clearings or along edges of woodlands; oak scrub and adjacent open spaces; roadsides.

Range: E. Rockies, across northern U.S. to Massachusetts, south to Florida, and west to E. Texas; absent from almost all of Great Plains.

Horace's Duskywing has become well adapted to many secondary growth habitats such as sunny, well-drained powerline rights-of-way that are maintained by cutting rather than by herbicides.

278 Mournful Duskywing
(*Erynnis tristis*)

Description: 1¼–1⅝″ (32–41 mm). *FW patterned;*
HW fairly plain. Male dark brown,
especially on HW above; female
warmer brown, mottled with lighter
brown on HW above. Both sexes have
rows of darker wedge-shaped spots
crossing FW, with small glassy dots
along median row and a few small
white spots in the submarginal row.
Below, generally lighter; *HW has white
patches near margin* in populations
outside California. *White HW fringes.*

Similar Species: White-fringed populations of Juvenal's
and Pacuvius duskywings have partial
brown fringe checkering. Funereal
Duskywing has narrower, longer FW
and squared HW.

Life Cycle: Several species of oaks (*Quercus*) are host
plants.

Flight: Up to 3 broods; March–November in
Arizona-New Mexico; spring, fall, and
sometimes January in California.

Habitat: Valleys, mountains, and foothills
among oak scrub and savannah;
redwood canyons and hills.

Range: California, Arizona, New Mexico,
Texas, and Mexico south to Colombia.

Mournful Duskywings seek hilltops and
perch on low vegetation. Scudder's
Duskywing (*E. scudderi*) of southeastern
Arizona and Mexico is rather similar
and also white-fringed. It has faint
brown checkering in the fringes,
however, and the hind wing below
lacks whitish patches near the margin.

273 Mottled Duskywing
(*Erynnis martialis*)

Description: 1–1⅜″ (25–35 mm). Above and
below, light brown with lavender cast;
*very strongly contrasting dark patches on all
wings make it appear almost banded;* tiny

glassy spots on FW; brown fringes.

Life Cycle: Egg pale green, turning pink before hatching. Mature caterpillar light green with white specks and covered with short hair; head has red, yellow, or orange pattern. Caterpillar overwinters. Chrysalis dark green or brown. Host plant is red root (*Ceanothus americanus*) in East and buckbrush (*C. fendleri*) in Colorado.

Flight: 2 broods in most of range; May and July. 1 brood in Rockies; May–June.

Habitat: Wooded uplands often on acid soils in East; open woods and thickets; clumps of vegetation on plains.

Range: South Dakota east to New England, south through Georgia, and west to central Texas; also into eastern foothills of Rockies along High Plains.

Although it occupies a diverse set of habitats, the Mottled Duskywing is seldom abundant. It frequently flies with both the Northern and Southern cloudywings.

275 Pacuvius Duskywing
(*Erynnis pacuvius*)

Description: 1⅛–1½" (28–38 mm). *FW usually distinctly patterned,* HW fairly plain. *Variable:* warm tan, dark brown, frosty lilac-gray, or velvet black; glassy spots on outer portion of FW vary in size. *Fringes usually light brown;* in S. Rockies population, fringes may be whitish with brown checks at vein tips. Nearly black in central California Coast Ranges.

Similar Species: Brown-fringed duskywings are often inseparable in the field. Persius and Afranius duskywings very similar. Mournful Duskywing white-fringed.

Life Cycle: Egg green at first, becoming pinkish. Mature caterpillar light green, speckled with white at base of short hair; black head with red, yellow, or orange patterning. Caterpillar overwinters.

Chrysalis dark green or brownish.
Host plants may be buckthorns
(*Ceanothus*), including buckbrush
(*C. Fendleri*) in Colorado.

Flight: 1 brood in North; spring. 2 broods in
South; spring and fall.

Habitat: Mountains; fairly moist slopes
between subalpine areas and hot
foothills.

Range: Connecticut south to Florida,
Mississippi, and Louisiana

Like several other duskywings, this
species' distribution has large gaps
and many questionable areas. Some
older records are probably unreliable,
adding to the uncertainty of the
range.

276 Zarucco Duskywing
(*Erynnis zarucco*)

Description: 1⅛–1¾″ (28–44 mm). Basically
brown above and below with glassy
dots on long, pointed FW; usually
has *conspicuous light brown patch*
(sometimes quite russet especially
on female) near base and *pale patch
on outer costal edge. Male has costal fold
on FW. HW large and squarish* with
brown fringes.

Similar Species: Wild Indigo Duskywing has pale
patch on FW but much stubbier
wings.

Life Cycle: Egg changes from green to pinkish
before hatching. Mature caterpillar
light green with white specks, hairy;
head has red, yellow, or orange pattern.
Chrysalis dark green or brownish. Host
plants are legumes such as locusts
(*Robina*).

Flight: Several broods; all year in Florida.

Habitat: Open areas, roadsides, fields, and hills.

Range: Connecticut south to Florida,
Mississippi, and Louisiana.

Largely a coastal butterfly, the Zarucco
Duskywing is common along Florida

roadsides, and a predictable visitor of
flowers, especially Spanish needles.

277   Funereal Duskywing
(*Erynnis funeralis*)

Description: 1⅛–1¾" (28–44 mm). *FW very long
and narrow; HW rather square. FW
patterned,* HW plain. *Above, dark brown*
with blacker, mottled markings across
FW; HW almost uniform blackish-
brown. Female lighter than male with
some buff spots on HW above. Small
glassy white spots from costa to middle
of FW. Below, lighter overall.

Similar Species: Other white-fringed duskywings have
shorter, broader FW, are lighter with
more prominent dark markings.
Fringes of Juvenal's and Pacuvius
duskywings have brown checkering.

Life Cycle: Caterpillar greenish with distinctive
yellow-spotted yellow lines along sides.
Host plants are woody and herbaceous
legumes (Fabaceae). Chrysalis green.

Flight: Successive broods in S. Texas; year-
round. 3 broods in S. California;
February–October.

Habitat: Mountains with pine, juniper, oak;
moist valleys; prairie and desert edges.

Range: California, S. Nevada, S. Utah,
Colorado and W. Kansas south to
Argentina and Chile.

Named for its sober coloring, the
Funereal Duskywing wanders far afield
to colonize such pioneer legumes as
vetch, lotus, and alfalfa. It is
therefore common over a large range
and during all seasons.

Columbine Duskywing
(*Erynnis lucilius*)

Description: ⅞–1¼" (22–32 mm). *Small,* with
short, rounded wings. *FW patterned*

above and below, *with bands of darker marks* on FW (male hoarier between bands, female warmer brown), and several glassy white dots near FW tip and margin. HW plain dark brown, paler near margins, with buff spots.

Similar Species: Persius and Wild Indigo duskywings usually larger with longer wings.

Life Cycle: Egg green, becoming pink before hatching. Young caterpillar pale greenish-white; mature caterpillar hairy, light green with white specks and red, yellow, or orange patterned head; overwinters. Chrysalis dark green or brown. Host plant is wild columbine (*Aquilegia canadensis*) and garden columbine (*A. vulgaris*).

Flight: 2 or 3 broods; May–August.

Habitat: Edges of woodlands, glades, gorges, ravines, and shaly slopes.

Range: Ontario and Quebec to Minnesota, Pennsylvania, and New Jersey.

Rather uncommon overall, the Columbine Duskywing may become locally numerous under favorable conditions.

272 Wild Indigo Duskywing
(*Erynnis baptisiae*)

Description: 1⅛–1⅝″ (21–41 mm). *FW patterned above,* dark chocolate brown on inner part, outer part purplish to light brown with darker markings and lighter bands; glassy spots near FW tips; *pale brown patch between spots and dark base of wing.* HW brown with lighter patches. Fringes brown. Below, similar to upperside with fainter markings.

Similar Species: Zarucco Duskywing has pale brown patch and large HW. Columbine Duskywing smaller. Persius Duskywing best separated by host plant.

Life Cycle: Egg green at first, becoming pinkish. Young caterpillar orange-white; overwinters. Fully grown caterpillar

hairy, light green with white specks,
and red, yellow, or orange patterned
head. Chrysalis dark green or brownish.
Host plants are wild indigos (*Baptisia
tinctoria, B. laevicollis*); possibly also
crown vetch (*Coronilla varia*).

Flight: Several broods; spring–early fall.

Habitat: Sandy, acid soils, along brushy edges,
lanes, and woodlands.

Range: E. Nebraska, Kansas, and Texas, east
to Massachusetts and E. New York,
south to Florida, with stronghold north
and east of Ohio River.

The habitats favored by the Wild
Indigo Duskywing, such as barrens, are
commonly considered wastelands, and
too often have been used in ways which
have destroyed them as habitat. Since
this species feeds on crown vetch, a
legume planted along roadsides, it may
develop a more continuous distribution.

Persius Duskywing
(*Erynnis persius*)

Description: 1–1⅜″ (25–35 mm). *Above, FW
patterned* with small but conspicuous
glassy dots, pattern obscured by long
gray hair on male FW. *HW above*
plain; light or dark brown with *rows of
understated black markings* and rows
of pale brown or buff spots. Below,
similar but less strongly marked.
Female lighter, more marked.

Similar Species: Wild Indigo and Columbine
duskywings best separated by hosts. In
Southwest, Afranius Duskywing has
whitish fringes but farther north is very
similar, and cannot be separated in
field. Pacuvius may have more distinct
marks.

Life Cycle: Egg changes from green to pinkish
before hatching. Mature caterpillar
hairy, light green with white specks,
and red, yellow, or orange head;
overwinters. Chrysalis dark green or

brownish. Host plant is probably lupine (*Lupinus*) in East, definitely golden banner (*Thermopsis*) in West; old records for willows (*Salix*) and poplars (*Populus*) are doubtful.

Flight: 1 brood; April—June in East, later in Alaska, July in Rockies. 2 broods in California; March—September.

Habitat: In East, generally mountainous sites such as willow swamps and sandy aspen flats; an array of habitats farther west.

Range: Alaska to Maritimes, south in mountains to central California, Arizona, New Mexico, and Tennessee.

An uncommon, sparsely distributed butterfly in the eastern portion of its range, the Persius Duskywing is more common in the West. It is enormously versatile, yet usually quite uniform in appearance. The similar Afranius Duskywing (*E. afranius*) is somewhat lighter but lacks the hairy fore wings of the Persius. Some populations, especially in the Southwest, have white-tipped fringes. The Afranius Duskywing's caterpillars feed on lupine (*Lupinus*), lotus (*Lotus*), and other legumes. It flies from Alberta to Mexico.

250 Alpine Checkered Skipper
(*Pyrgus centaureae*)

Description: ⅞–1¼" (22–32 mm). Dark brown-black *above, with 2 bands of bright white spots across FW. Pair of tiny white spots a third out from base of FW above are adjacent; uppermost spot, if present, does not extend slightly along, nor even touch, cell.* Spot bands become suffused on HW. Olive below; HW crossed by bands of white, the outermost arrayed in zigzag chevrons. Fringes boldly checkered.

Similar Species: All other *Pyrgus* species lack adjacent spots on FW above and/or have upper spots extending to cell.

Two-banded Checkered Skipper is rust-tinged on HW below.

Life Cycle: Blackberry (*Rubus chamaemorus*) is host plant in Europe.

Flight: 1 brood; May in East, June and July in West and North.

Habitat: Hills, heath acid barrens in East; scrub oak openings in Lakes States; also subarctic taiga clearings in far North; above timberline in western mountains.

Range: Alaska, Labrador and Gaspé Peninsula, in mountains to New York, North Carolina, Michigan, and Washington. Farther south only in boreal habitats.

The Alpine Checkered Skipper varies from small and dark in the East to large and pale in the West, and occupies drastically different habitats, from Long Island, New York to Long's Peak, Colorado. In the West, it is strictly an alpine skipper.

247 Two-banded Checkered Skipper
(*Pyrgus ruralis*)

Description: ¾–1⅛" (19–28 mm). *Small.* Blackish-brown with *white spots in 2 separate bands; pair of white spots closest to FW base not adjacent, uppermost spot extends slightly along cell.* Spot bands pronounced on HW above; submarginal spots on HW above are crescent-shaped; *white spot at HW base* (absent in other checkers). *Below,* variable, *usually with some reddish* on HW, *always variegated* and with some bright white spots on both wings, forming band on HW. HW margin has zigzag of chevrons, white above, rusty below. Fringes black and white checkered. Male has costal fold.

Similar Species: Northern Checkered Skipper has 2 FW basal spots that are adjacent. Small Checkered Skipper lacks basal HW spot above. Both lack reddish below.

Life Cycle: Egg green to yellow. Caterpillar unrecorded. Host plant is horkelia

(*Horkelia*) and cinquefoil (*Potentilla*).

Flight: 1 brood; April—August, depending on altitude, for 4—6 weeks.

Habitat: Normally mountains; also down to sea level in Pacific Northwest. Open meadows among ponderosa pines, fire clearings, and glades.

Range: British Columbia and Alberta south to S. California in Cascades, Sierra Nevada, and Coast Ranges; east to N. Utah and Colorado in Rockies.

While not uncommon in western mountains, this species is seldom found in large numbers. It is one of the earlier butterflies to appear in the Northwest.

Southern Checkered Skipper
(*Pyrgus xanthus*)

Description: ⅞—1″ (22—25 mm). *Small.* Black *above* with 2 bands of white squarish spots, the submarginal *HW spots are reduced,* not crescent-shaped; *usually has basal white spot on HW above.* Often some *bluish scaling.* Below pale olive with white bands. White and black fringes. Male lacks FW costal fold.

Similar Species: Small Checkered Skipper lacks HW basal white spot above and blue tint. Two-banded Checkered Skipper has buff checkered fringes as well as chestnut patches beneath, crescent-shaped submarginal spots on HW above, and male has FW costal fold.

Life Cycle: Unknown.

Flight: May—July.

Habitat: Restricted to cooler montane zones; glades, clearings, and streamcourses.

Range: S. Colorado, New Mexico, Arizona.

Of all the American checkered skippers (*Pyrgus*), the Southern Checkered Skipper may be the most habitat-specialized. It replaces the Two-banded Checkered Skipper nearly completely in the southern Rockies.

248 Small Checkered Skipper
(*Pyrgus scriptura*)

Description: ⅝–1″ (16–25 mm). *Tiny.* Shiny dark gray to dark brown above, with small and separate white checks (no checks on margins), and *no basal spot on HW above. Long white fringes with little dark checkering.* Below, FW gray-olive; HW crossed by alternating rows of clean white and olive-tan, without black outlines. Male lacks costal fold.

Similar Species: Southern and Two-banded Checkered skippers have basal spot on HW above; Two-banded reddish below. Southern usually has heavier checkering on fringes. Early spring Small Checkered Skipper cannot be distinguished in the field from Southern and Two-banded. Common Checkered Skipper usually larger, has broader white patch bands and black edges to olive spots below. Desert Checkered Skipper gray below.

Life Cycle: Host plants are alkali mallow (*Sida hederacea*), globe mallow (*Sphaeralcea coccinea*), and probably other mallows.

Flight: Successive broods; most of year in California and Southwest, April–August further north.

Habitat: Open grasslands, prairies, high plains, abandoned fields, and canal sides; marshes in California.

Range: SE. Alberta and North Dakota south to Mexico and west to Arizona, S. Nevada, California, and Baja California.

This species is one of the few butterflies to emerge just after the snow melts in foothill canyons. It also flies in parched weedy patches in late summer.

251 Common Checkered Skipper
(*Pyrgus communis*)

Description: ¾–1¼″ (19–32 mm). *Extremely variable.* Above can be quite blackish with little white checkering (especially in female) or very pale with broad bands of white spots (particularly males); usually some black at base and *often considerable bluish hairy scaling. Below,* FW similar to upperside but paler; *HW* pale eggshell-white to yellowish, crossed by 2 major and 2 minor rows of *olive-tan to olive-green spots* normally linked into solid bands, *outlined finely with black or brown scales.* Fringes checkered with gray and white.

Similar Species: White Checkered Skipper similar. Small Checkered Skipper lacks black outlines to olive spots below, usually has less white above. Tropical Checkered Skipper has little black above, no olive below. Desert Checkered Skipper gray or silvery below. Large White Skipper larger, white, lacks black at bases above.

Life Cycle: Egg changes from green to cream-color before hatching. Caterpillar tan with darker median line, brown and white side lines, and black head. Chrysalis greener toward head and browner toward tip; has dark speckles and dashes in bands. Host plants are available mallows (Malvaceae).

Flight: Successive broods; year-round in far South, April–October in Midwest.

Habitat: Foothills, weedy plains, fields, roadsides, riverbanks, valley bottoms, gardens, vacant lots, and parks.

Range: S. Canada to Argentina, but absent from northwest states and north of Massachusetts.

The Common Checkered Skipper is considered by many to be the most common skipper in North America. Highly aggressive, males of this species patrol tightly circumscribed territories, darting out at passing objects. Because this species can exploit hollyhock, hibiscus, or cheeseweed in towns as well as wild mallows in the countryside, it is found in many lowlands across the continent.

245   White Checkered Skipper
(*Pyrgus albescens*)

Description: ⅞–1⅛″ (22–28 mm). Grayish-black to dark gray-brown above, with crowded bands and spots of *white dominating most of wing, whitish scaling over much of black,* especially near base. Below, FW grayish near base followed by white and black spot bands and an olive tip; HW has alternating irregular bands of white and olive with *fine black borders in between.* White fringes finely checkered.

Similar Species: Desert Checkered Skipper paler and vaguely marked on HW below. Common Checkered Skipper tends to be larger, darker, but often similar.

Life Cycle: Egg greenish-white. Caterpillar cream-colored with greenish lines down back and sides; makes tent from rolled leaf. Chrysalis brown, with freckles and streaks. Host plants are various weedy mallows (Malvaceae).

Flight: Successive broods; year-round, except in freezing periods.

Habitat: Disturbed lowlands, agricultural areas, hot, arid lands.

Range: S. California east to Texas Coast.

The White Checkered Skipper has different ecological preferences from the Common Checkered Skipper, and is usually paler. But where the 2 species occur together, variation among both makes microscopic examination of genitalia necessary for identification. In general, however, the White Checkered Skipper flies in drier habitats.

243   Tropical Checkered Skipper
(*Pyrgus oileus*)

Description: 1–1⅜″ (25–35 mm). Checkered above; dark, especially toward FW tip and margin; bluish-gray hair over body and much of wings gives *gray appearance above. Below,* FW similar to upperside but paler; HW cream-colored to warm tan with *rows of dark brown and dark gray spots outlined in black;* spotting lacks olive or green.

Similar Species: Common Checkered Skipper usually more black above, olive below.

Life Cycle: Egg pale. Caterpillar greenish with black head and light stripes above. Chrysalis also green but yellower on abdomen, covered with minute white hair. Host plants are mallows (*Sida, Malva, Abutilon*) and hollyhock (*Althaea*).

Flight: Overlapping broods; year-round.

Habitat: Wherever mallows grow.

Range: E. Texas and Arkansas through Deep South to S. Florida; also throughout Caribbean islands and Latin America.

In North America, the common Tropical Checkered Skipper hugs the Gulf. Also called *P. syrichtus.*

246 Desert Checkered Skipper
(*Pyrgus philetas*)

Description: ⅞–1⅛" (22–28 mm). Wings above blackish or grayish-brown, checkered with uniform pattern of many white spots with *some bluish-gray basal hair. Marginal spots on HW conspicuous above,* with clear white or mildly checkered fringes. *Below, dull gray or quite silvery,* with vague bands or slightly dark HW.

Similar Species: Other pale arid-land species larger and more strongly marked beneath.

Life Cycle: Unrecorded.

Flight: Several successive broods; February–December in Texas.

Habitat: Arroyos, valley bottomlands, and desert waterholes.

Range: S. New Mexico and Arizona, central Texas to Baja California and Mexico.

This silvery skipper whirrs up and down small gullies in the desert Southwest.

White-banded Skipper
(*Heliopetes domicella*)

Description: 1–1⅜" (25–35 mm). Broad black borders and blue-gray bases above, giving impression of *broad, clear white bands crossing all wings*. Below mirrors above, except for beige-pink bands on FW base and olive-green HW.

Similar Species: Checkered skippers lack pinkish bands below. Other white skippers (*Heliopetes*) lack black bands above.

Life Cycle: Unrecorded.

Flight: 2 broods; April and August–October.

Habitat: Forest clearings, edges, watercourses, and hilly scrub.

Range: S. Arizona, Texas and perhaps New Mexico, south to Argentina.

Unlike other members of its genus, the White-banded Skipper is not common.

59, 82, 242 **Large White Skipper**
(*Heliopetes ericetorum*)

Description: 1⅛–1⅝" (28–41 mm). *Milk-white above with black and white chevrons in rows along borders;* blue-gray at base. (Female may have considerable bluish-gray above.) *Obscure pinkish-brown areas cloud FW tip and HW below,* otherwise whitish. Fringes checkered.

Similar Species: Common Checkered Skipper has black markings near base, olive bands below. Other *Heliopetes* have more distinct patterns beneath.

Life Cycle: Egg yellow, becoming white before hatching. Caterpillar dull chartreuse-green with lengthwise rows of yellow bands. Chrysalis tan with blue flush. Globe mallow (*Sphaeralcea*) and other mallows (Malvaceae), as well as hollyhock (*Althaea*), are host plants.

Flight: 5 broods in S. California; most of year. Fewer in north; May–September.

Habitat: Vegetated spots in arid lands; rocky watersides in Columbia Basin.

Range: Washington south to Baja California, across Great Basin to Colorado, New Mexico, Arizona, and N. Mexico.

The Large White Skippers colonize bush mallow in recently burned or disturbed portions of the southern California chaparral. Although they fly in alfalfa fields, uniform wheat production excludes them from farms.

72, 241 Arsalte Skipper
(*Heliopetes arsalte*)

Description: 1⅜–1⅝" (35–41 mm). Clear *white above* with a few dark markings near border, especially around FW tip. *Below,* white *unmarked except for black veins of HW and outer FW; FW has orange area on costa near base.*

Similar Species: Other *Heliopetes* mottled below.

Life Cycle: Unknown.

Flight: October in U.S.

Habitat: Well-flowered arroyos and chaparral.

Range: Mexico to South America; rarely S. Texas.

A relatively common skipper in Mexico, the Arsalte Skipper was not recorded in the United States until 1973, at Brownsville, Texas.

293 Laviana Skipper
(*Heliopetes laviana*)

Description: 1¼–1⅝" (32–41 mm). *Wings rather long, triangular and straight-edged. Above,* cool white with dark, chainlike *border pattern on FW,* broadest around tip, *rather faint or absent on HW.* Male has lustrous sheen above. *Below,* FW white with some olive near tip; HW deep olive with white where wing joins abdomen (anal fold), beige patches near base and beige suffusion of scales on margin; *large beige band crosses olive HW disk, irregular inwardly, straight-edged outwardly,* not curved with margin.

Similar Species: Large White Skipper has very vague pattern beneath. Macaira Skipper smaller, pale beige below with more HW marginal markings.

Life Cycle: Immature stages undescribed. Host plants are mallows (Malvaceae), including sida (*Sida filipes*) and mallows of the genera *Abutilon* and *Malvastrum*.

Flight: All months in S. Texas.

Habitat: Chaparral, brush areas with mallows, canal sides, and lakeshores.

Range: Arizona east to Texas, south to Argentina.

Although a regular resident of Texas, the Laviana Skipper is much more common in Mexico.

249 Common Streaky Skipper
(*Celotes nessus*)

Description: ¾–1″ (19–25 mm). *Overall pleated appearance* given by alternating cocoa-tan and dark brown streaks radiating out on wings and *true scalloping of HW margins*. Streaks form dark chevrons on HW beneath. Small glassy spots occur above and below in short rows on costa, near tip, and beyond FW cell. Fringes checkered.

Life Cycle: Caterpillars have been found on mallows (*Abutilon, Althaea, Sida, Sphaeralcea, Wissadula*).

Flight: Several broods; March–November.

Habitat: Dry washes in foothills, open hill country, gulches, and roadsides.

Range: New Mexico and Arizona through Texas into Mexico.

A weak flier and frequent percher, the Common Streaky Skipper is unlike any other but the Scarce Streaky Skipper (*C. limpia*) of west Texas. The 2 species are nearly indistinguishable.

282 Common Sootywing
(Pholisora catullus)

Description: ⅞–1¼″ (22–32 mm). Fairly long, rounded wings. Shiny black or dark brown crossed by crescent-shaped rows of white spots on FW above and, less often, on HW above. Amount of spotting varies widely. *Fringes and HW below may be more brown than black.*

Similar Species: Other sootywings usually lighter, with HW spots vague or absent above.

Life Cycle: Egg exceptionally small, yellow-white. Caterpillar pale green with straw-colored flecks, stripes on front, and dark head; makes shelter by bending leaves toward midveins and fastening them with silk, eventually overwinters in stronger, larger tent. Chrysalis purplish-brown. Host plants include pigweed (*Amaranthus*), cheeseweed (*Malva rotundifolia*), and lamb's quarters (*Chenopodium album*).

Flight: 2 separate broods in North, at least 3 in South; March–November.

Habitat: Various weedy, disturbed sites, low mountains, and cultivated landscapes.

Range: Throughout North America between mid-Canada and N. Mexico; absent from Florida.

The Common Sootywing thrives in many habitats, even waste areas. The Mexican Sootywing (*P. mejicana*) is nearly identical above but its hind wings below have a distinctive bluish-gray luster against which the black veins stand out crisply. It occasionally wanders into Texas, New Mexico, and Colorado, where its caterpillars feed on pigweeds (*Amaranthus*).

295 Great Basin Sootywing
(Pholisora libya)

Description: 1–1⅜″ (25–35 mm). Brassy-black to dark brown above, with crescents of white spots on FW, smaller on male, larger on female. Below, brown to brownish-gold, with spots only at tip; *HW* sooty gray-brown to black *with large, clean white orbs aligned in rows*—1 median row and basal spot in Great Basin or 2 rows of median and submarginal orbs in Southwest.

Similar Species: Sooty Hairstreak lacks white dots on FW, normally has much less distinct spots on HW below.

Life Cycle: Eggs pale orange to cream-colored. Caterpillar robin's-egg blue with black dots along sides and white bristles; head black with reddish fur. Chrysalis stubby, curved, tan, with black wing pads. Caterpillar feeds on common saltbush (*Atriplex canescens*) and perhaps other species of saltbush.

Flight: 1 brood in Great Basin; June–July. 2 broods in S. California; March–May and September–October.

Habitat: Alkaline sage flats and deserts.

Range: SE. Oregon, E. Montana, and W. North Dakota south to W. Colorado, Arizona, S. California, Baja California, and Sonora, Mexico.

Chiefly a denizen of the Great Basin, this skipper also inhabits saltbush stands in the Mojave and Sonoran deserts. Reaching across the Red Desert of Wyoming, one population lives in the plains east of the northern Rockies.

Saltbush Sootywing
(*Pholisora alpheus*)

Description: ¾–1" (19–32 mm). *Small;* rounded wings. Brown with grayish scales on *FW above* setting off *submarginal row of dark arrowhead-shaped marks,* dark or light-centered, more dark marks near FW base and a short row of white dots from costa near FW tip. Often with pale, well defined row of submarginal spots on HW above and below, or HW may be vaguely grizzled. Buff and brown checkered fringes.

Similar Species: Small duskywings have less contrast below and uncheckered fringes.

Life Cycle: Unrecorded. Host plant in Mojave Desert is saltbush (*Atriplex canescens*).

Flight: 1 brood in Mojave Desert and Great Basin; April–June. 2 broods elsewhere; March–July and September.

Habitat: Semiarid open country, shaly deserts.

Range: Basin-and-Range country of S. Oregon across Great Basin to Colorado, south to Texas, Arizona, and Mojave Desert.

In an inhospitable landscape, these skippers are limited to areas where their host plants find enough moisture to survive. This sootywing is paler in the Great Basin and Mojave Desert. The MacNeill's Sootywing (*P. graceliae*) is smaller, with more prominent cell spots and more obscure submarginal markings that are not shaped like arrowheads above. Named for C. Don MacNeill, who first described it, this species uses lenscale (*Atriplex lentiformis*) as its host plant. It flies in 2 or 3 broods from April to October in California, Arizona, and probably Mexico.

252 Arctic Skipper
(*Carterocephalus palaemon*)

Description: ¾–1¼" (19–32 mm). *Above,* FW and HW dark brown with crisply outlined, *clear orange orbs in submarginal and cell bands parallel to margin. Below,* FW yellow-orange with open black checks; HW brandy-tan or russet with several *silver or pale yellow, black-rimmed orbs.*

Life Cycle: Egg greenish-white. Caterpillar dusky-green or ivory with dark stripe above and yellow stripe on sides above rows of dark spots; overwinters. Host plant is purple reedgrass (*Calamagrostis purpurascens*) in California. Spring chrysalis resembles fragment of faded grass.

Flight: 1 brood; late May–early August.
Habitat: Streambanks, bogs, mixed upland forests, sedge lowlands, woodland trails, and sub-Arctic meadows.
Range: Alaska, Canada, and N. United States south to Pennsylvania, Minnesota, Wyoming, and central California.

Despite its name, this species is more sub-Arctic than Arctic and it is well known in many lands. In England, where it is known as the "Chequered Skipper," it is an Endangered Species. But this species is still fairly common in northern North America and is one of the few true boreal species of the Adirondacks and Catskills. The Arctic Skipper takes nectar at wild iris and Jacob's ladder.

208   Russet Skipperling
"Pirus Skipperling"
(*Piruna pirus*)

Description: ⅞–1″ (22–25 mm). *Small.* Plain velvet brown *above with 3 minute white spots in tight row near FW tip,* 1 or 2 more below cell. Below, inner ⅔ of FW blackish, outer part russet; HW disk below violet when fresh.
Similar Species: Spotted Skipperling has spots on both FW and HW above.
Life Cycle: Unknown.
Flight: 1 brood; late May–July.
Habitat: Valley floors and gulches, foothills with oaks and adjoining plains, and prairie watercourses.
Range: Wyoming, extreme S. Idaho, Colorado, Utah, New Mexico, and Arizona.

The Russet Skipperling lives along cottonwood-lined canals, flying among long grasses overhanging the banks. It takes nectar at snowberry or dogbane.

200 Spotted Skipperling
(*Piruna polingii*)

Description: ¾–1″ (19–25 mm). *Small.* Brown *above* with *2 submarginal rows of 5 white spots on FW* and 2 or 3 white spots on HW. *Below,* rust-colored with *row of 3 silvery spots on HW,* more near base.

Similar Species: Russet Skipperling has unspotted HW.

Life Cycle: Unknown.

Flight: 1 brood; late June–August.

Habitat: Lower and middle elevations of mountains, mainly near streams.

Range: Arizona, New Mexico, and Mexico.

The Spotted Skipperling, while not overly common, ranges well into the southwestern Rockies.

230 Malicious Skipper
(*Synapte malitiosa*)

Description: 1–1¼″ (25–32 mm). FW oblong, drawn out at tip. *Above,* FW warm, dark brown with *pale salmon cloud in cell, crossed by black streak* reaching nearly to margin; HW dark with tawny orange tinge. Below, FW paler, similar to above; HW tan with dark striations.

Life Cycle: Unknown.

Flight: 2 or more broods; May–November.

Habitat: Riverbanks and flood plains.

Range: Lower Rio Grande Valley of Texas and south to Argentina and Cuba.

Unless disturbed, this skipper perches quietly for prolonged periods, and then flies up unexpectedly. A related species, the Salenus Skipper (*S. salenus*), arrived in Texas with Hurricane Beulah in 1967; since then it has been reported to occupy shaded, grassy lowlands in the Santa Ana National Wildlife Refuge. Hurricanes are a substantive factor in the movements of many southern butterflies.

238 Violet-patch Skipper
(*Monca tyrtaeus*)

Description: ¾–1" (19–25 mm). *Small* with narrow, rounded wings. Above, FW dark blackish-brown with 2 rows of yellowish-white spots; HW unmarked. *Below,* FW blackish, costal margin violet near base and tip, rust-colored in between; HW rust-brown, strikingly marked with *large, triangular, violet to bluish patch pointing toward base,* and a violet band outside of that.

Similar Species: Clouded Skipper larger, broader, with violet below deeper and less discrete. Fawn-spotted Skipper lighter brown, with purple less pronounced.

Life Cycle: Unknown.

Flight: 2 broods; January–May and October–December.

Habitat: Openings among deep shade: glades, trails, and streams through woods.

Range: Lower Rio Grande Valley of Texas, south to Colombia.

Flying slowly and close to the earth, the Violet-patch Skipper is easy to follow until it stops and instantly disappears in the dappled shadows.

220, 222 Swarthy Skipper
(*Nastra lherminier*)

Description: ⅞–1" (22–25 mm). *Small;* FW triangular, HW distinctly rounded. *Dull gray-brown above,* sometimes with 2 vague pale areas below end of FW cell. Below, olive-brown with no spots but *distinct yellow veins on HW*.

Similar Species: Julia Skipper has distinct spots on FW.

Life Cycle: Egg pearly-white. Young caterpillar opaque white with brown head. Prairie beardgrass (*Andropogon scoparius*) is known host plant; others suspected.

Flight: 2 broods; May and September in North, March–June and August–October in South.

Habitat: Fields, barrens, beaches, and meadows.
Range: New York south to Florida, west to Missouri and Texas.

Small knots of Swarthy Skippers fly low over meadows, attracting little notice.

221 Julia Skipper
(Nastra julia)

Description: 1–1⅛″ (25–28 mm). Triangular wings. Gray-brown to warmer brown *above with small whitish spots near end of FW cell* and on FW costa. *Below, pale yellow-orange,* especially on HW, often with pale ray extending from base to margin near abdomen (anal angle); brown on lower half of FW.

Similar Species: Swarthy Skipper lacks distinct FW spots but may be similar.

Life Cycle: Caterpillar feeds on St. Augustine grass *(Stenotaphrum secundatum)* in Texas, a common lawn grass.

Flight: Overlapping broods; all months in Texas.

Habitat: Grassy borders, lawns, and chaparral.

Range: SE. California east to Alabama, south into Mexico.

The Julia Skipper is common and widespread in Texas. A very similar species, the Neamathla Skipper (*N. neamathla*), flies in much of the same range; it is smaller and duller yellowish-brown below.

215 Three-spotted Skipper
(Cymaenes tripunctus)

Description: ⅞–1⅛″ (22–28 mm). Slate-brown *above with 6 tiny spots near FW tip,* 3 clustered together and 3 more dispersed; HW unmarked brown. *Below,* FW blackish at base, with spots above repeated and suffused with tan

near tip; smooth tan *HW has row of
faint yellow spots across disk. Long
antennae nearly half FW length.*

Similar Species: Eufala Skipper paler, grayish. Clouded
Skipper darker, violet. Both have
shorter antennae.

Life Cycle: Egg white or pale green. Caterpillar
aqua with darker green longitudinal
stripes. Chrysalis thin, tubelike, green
with pink tongue case; bears head horn.
Panic grass (*Panicum maximum*) and
sugar cane (*Sacchinarum officinarum*)
noted as host plants.

Flight: Several broods; February–October.

Habitat: Tropical and subtropical grassy sites.

Range: S. Florida, West Indies, and Mexico to
Argentina.

The genus *Cymaenes* is well distributed
in the Antilles and Latin America but
only reaches the subtropical portions of
North America.

180 Fawn-spotted Skipper
(*Cymaenes odilia*)

Description: 1–1¼″ (25–32 mm). Unmarked, light
brown *above, usually with 4 minute white
spots near FW tip* (3 in cluster and 1
separate); HW unmarked light brown.
Below, FW plain light brown; *HW
lighter brown with fawn patches.* Long
antennae.

Similar Species: Violet-patch Skipper smaller, with
more rounded wings, and sharper,
brighter violet patches. Clouded
Skipper darker above and deeper violet
below; male has dark stigma.

Life Cycle: Undescribed.

Flight: Several broods; most of year, especially
April–December.

Habitat: Grassy glades and woodland edges.

Range: S. Texas to Argentina.

The Fawn-spotted Skipper is very local
in southern Texas, but is common and
well-established where it occurs.

213, 239 Clouded Skipper
(*Lerema accius*)

Description: 1–1⅜" (25–35 mm). Pointed wings. Blackish-brown *above with row of small white spots near FW tip,* 1 in cell, *larger spot below end of cell;* HW unmarked blackish-brown; male has black stigma. *Below,* FW blackish near base, outer half dusted with violet; *HW banded or clouded by fields of dark and light purple or bluish-purple and brown.* Long antennae.

Similar Species: Other dark southeastern skippers lack purplish or violet bands.

Life Cycle: Caterpillar whitish, flecked and lined with dark brown. Chrysalis greenish with head drawn out into long, narrow beak and tongue case free to end of abdomen. Host plants include woolly beard grass (*Andropogon*), St. Augustine grass (*Stenotaphrum secundatum*), Indian corn (*Zea mays*), water grass (*Echinochloa*), and paspalum (*Paspalum*).

Flight: Several broods; February–November in South.

Habitat: Grassy watersides, damp open lowland fields, hilly glades, and wastelands.

Range: Illinois east to Massachusetts, south to Florida, west to Texas, and south to Central America.

The Clouded Skipper, while a common breeding resident in the South, visits the North only as an immigrant.

236 Green-backed Skipper
(*Perichares philetes*)

Description: 1¾–2⅛" (44–54 mm). *Large* with long, rounded wings. *Above,* dark blackish-brown; *FW has angular yellow spots, HW iridescent green at base,* brown toward margin. *Male has violet-gray stigma. Below,* FW similar to above but less distinct; *HW marbled with brown and lilac-purple.* Fringes checkered. *Abdomen above iridescent green.*

Similar Species: Long-tailed Skipper has white spots and tails. Hoary Edge white-frosted below.

Life Cycle: Caterpillar light yellowish-green with glossy back and downy hair on sides. Chrysalis darker green. Millet (*Panicum*), panic grass (*P. maximum*), and sugar cane (*Sacchinarum officinarum*) are host plants in West Indies; the palm *Desmonicus* in Brazil.

Flight: November–December in Texas.

Habitat: Sugar cane fields and similar places.

Range: S. Texas and West Indies south to Argentina.

This skipper and other butterflies which feed on human crops seldom occur in sufficient numbers to cause serious economic hardship.

142 Least Skipperling
(*Ancyloxypha numitor*)

Description: ¾–1″ (19–25 mm). *Small;* wings rounded. Above, orange and black; FW variable, entirely black or with heavy orange-gold at base, along costa, and near cell; *HW above gold with broad black border. Below,* FW black with gold tip, costa, and border; *HW clear orange-gold.*

Similar Species: Tropical Least Skipperling smaller, paler.

Life Cycle: Egg glossy yellow. Caterpillar light green with brown head. Chrysalis brown and white with blunt head; often found in marsh millet (*Panicum*) over water. Host plants also include bluegrass (*Poa*) and rice (*Oryza sativa*).

Flight: 2 broods in Colorado; late June to mid-July and August–September. 3 broods in New England; May–October. 4 broods in Texas; April–November.

Habitat: Reservoirs, ditches, ponds, and streams; moist pastures, meadows, and marshes.

Range: Saskatchewan east to Nova Scotia, south to Florida, and west to Texas, barely reaching western edge of Great Plains in Colorado.

Very common in much of the East, the Least Skipperling flies feebly among tall grasses, and is easy to spot because of its distinctive, 2-toned wings.

Tropical Least Skipperling
(*Ancyloxypha arene*)

Description: ½–¾″ (13–19 mm). *Small. Bright orange above with vague dark border* running in along FW veins a short distance; *HW dark-bordered on margin and costa. Below,* FW black at base with orange disk and yellow tip; *HW bright yellow-orange or golden.*

Similar Species: Southern Skipperling has dark veins and brighter ray on HW below; male has stigma. Least Skipperling larger, much darker. Orange Skipperling larger; male has stigma.

Life Cycle: Unknown.

Flight: Perhaps 2 broods; May–August in U.S.

Habitat: Marshy margins and spring edges.

Range: S. Arizona, S. New Mexico, and S. Texas south through Costa Rica.

The Tropical Least Skipperling is far smaller than its relatively robust relative, the Least Skipperling.

226 Poweshiek Skipperling
(*Oarisma poweshiek*)

Description: 1–1¼″ (25–32 mm). *Wings sharply pointed at tip.* Above, dark brown with bright orange FW costa above cell; HW dark brown, with some gold near base. *Below,* FW dark at base with some orange on costa and at tip; *HW dark brown on inner marginal third without orange ray,* rest of *HW below has silvery veins;* light brown near abdomen.

Similar Species: Garita Skipperling orange below on FW cell, and on inner third of HW.

Life Cycle: Unrecorded.

Flight: 1 brood; June–July.
Habitat: Marshy lakeshores and wetlands.
Range: Dakotas eastward probably to Lake
           States, Iowa, Illinois, and Nebraska.

Local and known from rather few sites,
this species can be numerous in the
right places. Drainage of the prairie
wetlands may be one important factor
in this butterfly's overall scarcity.

227 Garita Skipperling
(*Oarisma garita*)

Description: ¾–1" (19–25 mm). *Small, variable.*
                 Brownish above with golden dusting.
                 Below, FW sometimes quite reddish;
                 HW usually olive-gold *with silvery veins,*
                 inner third of HW usually uniformly
                 orange or brownish, *edged with faint
                 orange ray from base to margin.*
Similar Species: Poweshiek Skipperling has black FW
                 cell below, lacks orange streak on HW.
Life Cycle: Egg white. Caterpillar green with 1
                 white line on back and 3 lines on each
                 side; feeds on various grasses (Poaceae).
Flight: 1 extended brood; June–July,
                 occasionally until August at high
                 elevations.
Habitat: Mountain meadows, fields, roadsides,
                 and hillsides; especially moist sites.
Range: British Columbia to Saskatchewan,
                 through Dakotas, E. Washington,
                 Montana, and Idaho, to Arizona, New
                 Mexico, and Mexico.

Variable in appearance yet always rather
golden in the sun, this is the common
skipperling of the Rockies. It flies
among the grasses of the montane
rangelands, mountainsides, and
meadows, where it is a frequent
associate of Butler's Alpine. The
common name may come from the La
Garita Mountains of southwestern
Colorado, since the species was described
from the Colorado Territory in 1866.

Edwards' Skipperling
(*Oarisma edwardsii*)

Description: ⅞–1⅛″ (22–28 mm). *Above, unmarked, varying from bright orange to dusky gold.* Below, FW has black lower margin; HW grayish-olive outwardly, orange nearer base. Veins not lightened.

Similar Species: European Skipper lacks black border above.

Life Cycle: Unknown.

Flight: 1 brood; April–July, earlier southward.

Habitat: Open areas, including grassy flats and roadsides, shrubby valley bottoms.

Range: Colorado to Arizona, Texas, Mexico.

This attractive little skipper is named for William Henry Edwards, a 19th-century collector and author of the monumental *Butterflies of North America.*

137 Orange Skipperling
(*Copaeodes aurantica*)

Description: ¾–⅞″ (19–22 mm). *Small;* wings angular. *Above,* tawny-orange; *male almost totally unmarked* except for pencil-line dark border and long, slender, obscure stigma; *female often has dark line below FW cell,* dark lower margin, and broad border on outer margin of FW and HW. Below, both sexes have black FW base; HW gold.

Similar Species: Southern Skipperling smaller, has yellow streak on HW below.

Life Cycle: Egg cream-colored and flattened. Caterpillar green, purple-striped, with 2 horns on head. Chrysalis light tan marked with pink and brown; hangs by silken girdle. Bermuda grass (*Cynodon dactylon*) is host plant in Texas.

Flight: Successive broods; most of year in Southwest.

Habitat: Hill country, canyons, and washes in chaparral near deserts.

Range: S. California, Nevada, Utah, Colorado, to Kansas, Southwest, and Panama.

Quite a rapid flier, the Orange
Skipperling is considered a common
butterfly in Texas, sometimes occurring
in large numbers. It is nearly extinct in
Orange County, California, but can be
found in the canyons of the Colorado
and Mojave deserts.

136 Southern Skipperling
(*Copaeodes minima*)

Description: ½–¾″ (13–19 mm). *Tiny.* Above,
clear bright orange with *minute black
stigma on male,* darkened veins, especially
on female. *Below,* orange; FW plain,
HW has yellow streak from base to margin.

Similar Species: Orange Skipperling larger, lacks yellow
ray below. Tropical Skipperling has
longer wings; male lacks stigma.

Life Cycle: Unreported; has been known to use
Bermuda grass (*Cynodon dactylon*).

Flight: Successive broods; March–October,
year-round some places.

Habitat: Open fields and grassy margins.

Range: Arkansas and Georgia through Florida
and Texas to Panama.

Our smallest skipper, the Southern
Skipperling often flies along with the
larger Least Skipperling. Its flight
among the grasses is weak and erratic.

145 Sunrise Skipper
(*Adopaeoides prittwitzi*)

Description: ⅞–1⅛″ (22–28 mm). *Above,* bright
lustrous orange, with thin dark border;
veins darkened near border in male and
all across wings in female. *Below,* light
yellowish to reddish-gold; *HW has
yellow ray running from base to outer
margin. Fringes conspicuously orange.*

Similar Species: Tropical and Southern skipperlings
smaller, brighter, lack dark veins above.

Life Cycle: Unrecorded.

Flight: 2 broods; May–June, September.
Habitat: Near springs in dry settings.
Range: W. Texas and SE. Arizona to Mexico.

Known in the United States for just over 30 years, the Sunrise Skipper as yet remains limited in range and population. Its name refers to the yellow ray on the hind wing beneath.

143, 228 European Skipper
(*Thymelicus lineola*)

Description: ¾–1″ (19–25 mm). *Small.* Above, brassy-orange with narrow dark border, male has tiny stigma; *veins darkened above,* especially on female. Below, FW clear orange, HW olive-ocher, greenish-gray, or copper-colored.

Similar Species: Delaware Skipper larger, male lacks slender black stigma, female has broader dark border. Orange Skipperling has light streak on HW below. Edwards' Skipperling lacks black border.

Life Cycle: Egg whitish; laid in strips of 30 or 40; overwinters. Caterpillar green with brown head and dark midstripe. Chrysalis yellow-green with down-curved horn. Timothy (*Phleum arvense*) is a host plant.

Flight: 1 brood; June–August.

Habitat: Meadows, pastures, and grassy waste ground.

Range: Minnesota, New Brunswick, and W. central British Columbia south to Maryland and Illinois.

The European Skipper was introduced to North America in London, Ontario, in 1910, and has since spread dramatically. The population of the butterfly fluctuates markedly. In a northwestern New Brunswick city, European Skippers are known to roost together by the hundreds in tall grasses.

154 Fiery Skipper
(*Hylephila phyleus*)

Description: 1–1¼" (25–32 mm). *Above, male has*
pattern of pale, clear yellow-orange,
large black stigma and *zigzagged border;*
female above has long wings, large
tawny spots. *Below, male* yellow-orange
with scattered brown dots on tan HW;
paler bands bounded by dark flecks
along veins; *female has orange along FW
costa. Extremely short antennae.*

Similar Species: Other small orange skippers lack dots
below and have longer antennae.

Life Cycle: Egg glossy, pale turquoise. Caterpillar
tan with 3 lengthwise dark stripes.
Chrysalis lighter tan with black dorsal
line tip to tip. Host plants are grasses
(Poaceae), including bent grass
(*Agrostis*), sugar cane (*Sacchinarum
officinarum*), Bermuda grass (*Cynodon
dactylon*), and St. Augustine grass
(*Stenotaphrum secundatum*).

Flight: 2 or more broods, April–December in
S. California; shorter period northward.

Habitat: Lawns, hedges, grassy lowlands,
clearings, edges and glades in forests.

Range: California east through New Mexico
and north through Illinois to Michigan
and Connecticut; throughout Southeast
and Southwest but not north into
Rockies or Great Basin. South to Chile.

The Fiery Skipper is established in the
American South but is only emigrant
northward, and then only in the East.
The caterpillars make their shelters
among the roots of grasses, where they
are safe from lawn mowers.

Plains Gray Skipper
(*Yvretta rhesus*)

Description: 1–1¼" (26–32 mm). *Above, warm gray
with sparse white spotting* (conspicuous in
female); male stigma inconspicuous.
Below, FW similar to above; *HW*

yellowish-gray with white veins and black patches next to white spot band. Fringes white.

Similar Species: Desert Gray Skipper has yellowish spots and lacks large black patches on HW below. Uncas Branded Skipper larger, has yellow-orange flush above.

Life Cycle: Undescribed. Caterpillars feed on blue grama grass (*Bouteloua gracilis*) in Colorado.

Flight: 1 brood; May–June.

Habitat: Prairies, mesa tops; rarely flatlands.

Range: Dakotas south through Colorado and Kansas to New Mexico and Arizona.

The mothlike Plains Gray Skipper varies greatly in abundance from year to year, probably as a result of moisture and spring temperatures. Pawnee National Grassland in northeastern Colorado is a good place to see it.

188 Desert Gray Skipper
(*Yvretta carus*)

Description: ⅞–1¼″ (22–32 mm). *Soft grayish-yellow above with 2 rows of buff-yellow spots* on FW, 1 row on HW; male has strong, slender stigma. *Below, yellow-gray with 2 bands of white spots, more suffused on FW;* female has white scales along HW veins, male black between FW and HW veins. Fringes gray.

Similar Species: Uncas Gray Skipper has orange cast above. Plains Gray Skipper has white spot band, white veins on HW below.

Life Cycle: Unknown.

Flight: April–September, year-round in Texas.

Habitat: Moist areas in desert.

Range: S. California, Arizona, New Mexico, and Mexico to Panama.

The drab coloring of the mothlike gray skippers may have evolved to help conceal them from predators in their desert environments. This skipper is found near waterholes and sand bars.

Salt-grass Skipper
(*Pseudocopaeodes eunus*)

Description: ⅞–1⅛" (22–28 mm). *Small.* Bright orange above with thin black border and black veins near border; *male has very minute stigma. Below,* FW orange, dark at base and at ends of veins; *HW pale orange with 2 bright yellow rays.*

Similar Species: Sunrise Skipper larger, duller, with darkened veins above, and 1 light streak below. Orange and Edwards' skipperlings lack light streaks below.

Life Cycle: Egg, cream-colored. Host plant is desert saltgrass (*Distichlis stricta*).

Flight: 2 or more broods; April–September.

Habitat: Inland stands of desert saltgrass, in or near desert seeps.

Range: Central California and Nevada south to Baja California and Mexico; probably also Utah, Arizona, and perhaps SE. Oregon.

Even the forbidding alkaline flats of the arid Great Basin have their special indigenous butterfly fauna, which includes the Salt-grass Skipper.

Morrison's Silver Spike
(*Stinga morrisoni*)

Description: 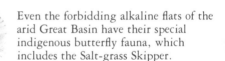 1–1¼" (25–32 mm). Above, tawny-orange; FW has dark, toothed borders, weak stigma on male, dark stigmalike spot on female; HW light brown at base on male, dark on female, making dog's head pattern. *Below,* FW similar to upperside; HW rust-brown with tawny inner edge, crossed by *silver-white chevron of spots enclosing silver bar running from base along cell out toward, but not meeting, white chevron band.*

Similar Species: Branded skippers (*Hesperia*) lack silver bar below. Snow's Skipper darker.

Life Cycle: Unknown.

Flight: 1 brood; May to mid-June in Rockies, earlier farther south.

Habitat: In Colorado, mountain valleys with willows, streams, and moist meadows.

Range: Mountainous parts of Colorado, New Mexico, Arizona, and W. Texas.

While others were seeking gold and silver around Pike's Peak in the 1880's, the lepidopterist H. K. Morrison prospected gold-spotted and silver-spotted butterflies. The silver spike of this skipper's name is the shiny white bar along the hind wing cell beneath.

189 Uncas Skipper
(*Hesperia uncas*)

Description: 1–1⅝" (25–41 mm). Tawny-orange above with brown at margin and base and whitish spots near FW tip (larger on female); male has black stigma, female dark patch on cell. *Below,* FW duller with more white spots; HW olive-tinged with *curved submarginal silver-white spot band, basal silver-colored crescent, and white scaling running out along HW veins,* very noticeably on darker part of wing. (E. California populations lack white scaling.)

Similar Species: Plains Gray Skipper smaller, darker, lacks tawny-orange above. Other western *Hesperia* lack whitened veins; E. California populations of Uncas cannot be distinguished in the field. Sandhill Skipper has yellow veins.

Life Cycle: Egg, to ¹⁄₁₆" (1.6 mm), greenish-white. Caterpillar undescribed; feeds on needle grass (*Stipa nevadensis*) and blue grama grass (*Bouteloua gracilis*).

Flight: 2 or more broods; May–September.

Habitat: Dry grasslands and sagebrush country, prairie mesas; cooler mountains.

Range: Alberta and Saskatchewan through all western states except Washington.

In an extremely difficult genus to identify, this species is usually quite distinctive because of its white veins.

Juba Skipper
(*Hesperia juba*)

Description: 1–1½″ (25–38 mm). *Wings large, pointed. Male bright orange above* with some brown toward HW base; *female tawny*-orange with more brown on HW; *both sexes have sharply contrasting saw-toothed, broad, brown border scalloped by orange on veins running out to margin.* Male has slender black FW stigma, *female has 2 brown bars* in same area. Below, FW orange, black near base, and olive at tip with white spots; HW olive with silver-white submarginal chevron of spots angled outwardly, with hindmost spot offset inwardly and small silver chevron near base often in 2 parts. *Antennae half FW length.*

Similar Species: Nevada and Sierra skippers smaller, with less distinct FW border above. Other *Hesperia* females may lack FW bars.

Life Cycle: Egg pinkish, becomes grayish before hatching. Mature caterpillar dull green with black head. Chrysalis brown with waxy bloom and dark abdominal dashes. Grasses (Poaceae) are host plants.

Flight: 2 broods; April–June, August–October.

Habitat: Gullies among sagelands, canyons, grassy flats, and drier mountains.

Range: British Columbia south along coast to Great Basin and Rockies.

The Juba Skipper is common and ecologically versatile. Both males and females tend to return again and again to concentrated sources of moisture or nectar, darting away when disturbed.

156, 183, 185 **Common Branded Skipper**
(*Hesperia comma*)

Description: ⅞–1″ (22–25 mm). *Small, variable.* Tawny-orange above with brownish border and spots; male brighter than female, has *black stigma often with lighter streak down middle. Below,* FW similar to above; HW ocher to olive or brass-colored, with curve of *spots enclosing inner comma, semicolon, or "U" of spots;* HW spots generally bright white, silver, or light yellow. *Antennae have long clubs, a quarter to a third shaft length.*

Similar Species: All silver- or yellow-spotted *Hesperia* skippers can be extremely similar.

Life Cycle: Egg whitish or pinkish, hemispherical. Young caterpillar yellowish or cream-colored with dark head, matures to dull green; makes silken shelter. Chrysalis green or brown. Host plants are pine bluegrass (*Poa*) and red fescue (*Festuca*) in California, and grasses (*Stipa, Lolium, Bromus*) elsewhere. Loose cocoon spun of silk and debris. Either chrysalis or caterpillar may overwinter.

Flight: 1 brood in early, mid-, or late summer.

Habitat: Boreal openings, tundra, subalpine meadows, sagelands, and foothills.

Range: Alaska east to Labrador, south to Maine, Dakotas, and Southwest.

Many subspecies of the Common Branded Skipper have been named, some formerly considered separate species; currently all are treated as members of a large, evolving species complex. A similar species is Lindsey's Branded Skipper (*H. lindseyi*). It flies in May or June in California and Nevada, July in southwest Oregon. It is largely light orange above, yellowish-orange below, with a whitish spot band. Its host plants are fescue (*Festuca*) and oat grass (*Danthonia*).

201 Apache Skipper
(*Hesperia woodgatei*)

Description: 1⅛–1⅝" (28–41 mm). *Male reddish-orange above* with broad brown border, black stigma. *Female FW above* orange *with very broad brown border and brown patch* on disk as well as *several white spots* toward tip; HW above mostly brown, with submarginal orange spot band. *Both sexes below,* FW orange at costa, yellow toward outer angle, olive at tip; *HW brown to olive with a warm yellowish tinge;* submarginal white spot band bent, *broken into distinct, somewhat rounded small spots. Antennae extremely long with tiny clubs.*

Similar Species: Other branded skippers in range are greener below, with shorter antennae.

Life Cycle: Egg cream-colored to chalk-colored. Caterpillar olive; feeds on grasses (Poaceae). Chrysalis brown.

Flight: 1 brood; September–October.

Habitat: Grassy slopes in high, cool mountains.

Range: Arizona, New Mexico, S. central Texas and N. Mexico.

The Apache Skipper occupies the mountains where the Apache Indians once roamed. It visits waterholes in the moist, high hills.

133, 149 Ottoe Skipper
(*Hesperia ottoe*)

Description: 1¼–1⅝″ (32–41 mm). Tawny-golden above with *narrow, toothed dark borders on male and black- or gray-centered stigma; female borders broader,* but vague and with rounded white spot toward hind margin from end quarter of FW cell. *Usually no spot band on HW below;* female may have hint of band.

Similar Species: Pawnee Skipper male has yellow-centered stigma, female has well-defined borders. Dakota Skipper has stubbier wings and is smaller.

Life Cycle: Eggs laid on purple coneflower (*Echinacea*) in Minnesota. Caterpillar feeds on different grasses (Poaceae), including fall witchgrass (*Leptoloma cognatum*); nearly grown caterpillar passes winter in shelter.

Flight: June–early August.

Habitat: Undisturbed prairies, rangeland oak scrub and swamps, and nearby roads.
Range: Montana and Michigan south to Colorado and Texas.

Much more widespread and adaptable to altered grasslands than the Dakota Skipper, the Ottoe avoids weedy conditions, and is rather rare.

153 Leonardus Skipper
(*Hesperia leonardus*)

Description: ⅞–1⅜″ (22–35 mm). Above, dark brown; male has broad dark border, tawny-orange toward base, and large stigma; female all dark brown except for tawny-orange bands on each wing. *Below,* FW orange at base, brown along margin, yellow across disk; *HW rust-colored on male, chestnut on female, both with offset whitish spots in limited band.*
Similar Species: Bunchgrass Skipper larger, with tawny area more restricted above on male, and with spots less distinct on HW below.
Life Cycle: Young caterpillar overwinters; mature caterpillar maroon with green highlights. Host plants are bent grass (*Agrostis*), panic grass (*Panicum*), and tumble grass (*Eragrostis*).
Flight: 1 brood; August–September.
Habitat: Fields, meadows, and nearby roads, pine-oak barrens, and oak openings.
Range: Ontario and Nova Scotia to Carolinas, Missouri, Alabama, and Louisiana.

The Leonardus Skipper can be quite abundant in the Adirondacks after most others are gone. A western relative, the Pawnee Skipper (*H. pawnee*), may well be a subspecies of the Leonardus Skipper. Large and tawny with few markings, the Pawnee Skipper male has a yellow-centered stigma and the female has well-defined dark borders. It flies in August and September from Minnesota to Montana and south to Kansas and

Colorado, where its caterpillars feed on bunchgrasses (*Tridens*). A dark race lives in the Rockies in the Platte River Canyon; the main colony may be threatened by a pending dam that would inundate its hillside habitat. The plains population may be sought in Pawnee National Grasslands in northeastern Colorado, named for the same plains Indians as the butterfly.

186 Pahaska Skipper
(*Hesperia pahaska*)

Description: 1⅛–1½" (28–38 mm). *Rich tawny-orange above, with narrow brown border* except broad toward FW tip. *Male has black stigma with yellow center,* female may have some blurred brown markings on FW disk. Below, FW paler than above; HW brownish to greenish-olive, with well-developed submarginal and basal crescents of silver-white spots, the outer band broken toward costa, inner one usually fragmented.

Similar Species: Common Branded, Nevada, and others in range cannot be separated except by center color of male stigma.

Life Cycle: Caterpillar olive-green; feeds on grasses (Poaceae), including blue grama grass (*Bouteloua gracilis*) and desert bunchgrass (*Tridens pulchella*). Caterpillar forms partially underground shelters of silk and ground litter.

Flight: 1 brood in northern range; June–July. 2 broods southward; May–September.

Habitat: Ridges in plains and basins; valleys in mountains; desert seeps and mesas.

Range: S. Saskatchewan, W. Dakotas, and Montana south to New Mexico and west to S. Nevada and S. California.

Specialists distinguish the very similar and variable western silver-spotted skippers by examining the structure of the genitalia and the color of a feltlike material at the center of the stigma.

152 Columbian Skipper
(*Hesperia columbia*)

Description: ⅞–1⅜" (22–35 mm). *Bright tawny-orange above, with* brown suffused inward from *deeply toothed border. Male* has broad *black stigma with yellow center,* female has brown patch on FW cell above. *Below,* FW black at base, orange along costa, pale peach toward outer angle, and greenish at tip; HW yellowish or greenish-golden with *silver-white spot bands reduced to 1 short basal and 1 longer submarginal arm, both narrow and straight.*

Similar Species: Other skippers in range and genus have broader, curved spot bands below.

Life Cycle: Egg white. Caterpillar cream-colored with brown head. Chrysalis light brown with darker mottling and long tongue case. Host plant is chaparral bunchgrass (*Koeleria cristata*) in N. California.

Flight: 2 broods; early spring–summer, fall.

Habitat: Chaparral, clearings, brush, barrens.

Range: S. Oregon south to Baja California in Coast Ranges and Sierra Nevada.

The Columbian Skipper is a chaparral insect of the Pacific Slope. It is often found with the Sleepy Duskywing.

192 Cobweb Skipper
(*Hesperia metea*)

Description: 1–1⅜" (25–35 mm). Wings *stubby. Above,* medium brown to olive-ocher with *tawny-orange restricted to small patches* near base and on either side of male stigma on FW, and with sparse orange overscaling. Female and, to lesser extent, male have small buff spots near FW tip and on bent HW band. *Below,* FW olive-tinged with light spots near tip and pale outer angle, orange on costa of male; HW olive-gray, with white basal spot in male, less often in female, and *white connected*

spot band consisting of "V" pointed
outward at midmargin. *White scales
below often along veins.* Females of Texas
and Arkansas populations very dark,
lack spot band below.

Similar Species: Sandhill Skipper yellow-veined, with
more orange above.

Life Cycle: Egg white. Caterpillar brown with
greenish back stripe. Chrysalis drab
green; probably overwinters. Host plant
is bluestem beardgrass (*Andropogon
scoparius*).

Flight: 1 brood; April–June, but only a few
weeks in any locality.

Habitat: Fields and clearings, hillsides, rocky
sites often among oak scrub, limestone
outcrops, burns and barrens.

Range: Minnesota east to Maine and south to
Texas and Florida.

The name "Cobweb Skipper" comes
from the weblike pattern of veins
below. This skipper follows new grass
growth into recent burns and denuded
or cleared sites. Another apparent sand
barrens dweller is the Dotted Skipper
(*H. attalus*), which is orange or
brownish above and mustard-colored on
the hind wing below, with very little
trace of a white spot band, if any. The
Dotted Skipper has been found in few
localities in its range, which stretches
from New Jersey south to Florida along
the coastal plain and west along the
Gulf to central Texas.

184 Green Skipper
(*Hesperia viridis*)

Description: 1–1⅜" (25–35 mm). *Dusky orange
above,* with broad, diffuse brown
borders, less apparent in brighter
female; male has stigma on FW disk.
Below, FW paler, blackish at base,
greenish toward tip; *HW below bright
ocher-green when fresh,* fading to
brownish-olive. *White-silver spot band*

prominent below; basal crescent broken
into 2 or 3 spots, *submarginal band bent,*
with lower arm complete and curved,
turning toward margin at ends, upper arm
broken; *spots often black-rimmed.*

Similar Species: Common, Nevada, and other branded
skippers can be very similar.

Life Cycle: Host plant thought to be blue grama
grass (*Bouteloua gracilis*).

Flight: 2 broods; April—June and August—
October. Spring brood only in North.

Habitat: Low mountains, bases of mesas, prairie
ridges, canyons, and gullies.

Range: Wyoming and Nebraska south to W.
Texas, E. Arizona, and Mexico.

Green Skippers can weather chilly
nights and last into a mild fall.
They take nectar on rabbit brush.

178 Dixie Skipper
(*Hesperia meskei*)

Description: 1⅛–1⅜" (28–35 mm). Dark above
with *tawny-orange quite restricted;* male
has FW stigma and broadly bordered
outwardly by dusky patch of raised
scales. *Below, deep, rich golden-orange;
with sooty-colored overscaling* everywhere
on HW except costa and inner margin;
black base on FW, with pale, vague, or
absent spot band on HW.

Similar Species: Crossline and Tawny-edged skippers
smaller, more olive, and darker. Indian
Skipper tawnier above, with larger
spots below. Palmetto Skipper larger,
has narrower stigma; female browner.

Life Cycle: Unknown.

Flight: 2 broods likely; May—June and
September—October.

Habitat: Clearings and roadsides among open
woods at low elevations.

Range: Arkansas, North Carolina, Texas,
Georgia, and Florida.

The Dixie Skipper's obscurity matches
that of the Dotted Skipper. That 2

members of a much studied genus are comparatively poorly understood may indicate that they find the present state of ecological affairs inhospitable, or perhaps they are extremely local. This species has the most narrowly restricted southern range of any *Hesperia*. It may replace the Leonardus in some areas.

177 Dakota Skipper
(*Hesperia dacotae*)

Description: 1–1⅜" (25–35 mm). *Compact, with short, rounded wings.* Male above bright tawny-orange with short but prominent stigma, little border; *clear yellow-orange below,* usually with only vague suggestion of pale spot band. Female above olive-brown, with tawny-orange mostly restricted to upper marginal area of FW, and square glassy spot below end of cell; often traces of light yellow spot band on HW above and whitish band on golden-gray HW below.

Similar Species: Ottoe Skipper larger, with longer wings. Pawnee Skipper larger, more sharply bordered; flies in fall.

Life Cycle: Female lays single egg quite near ground on grass blade. Host plants include grama grass (*Bouteloua*), beard grass (*Andropogon*), needle grass (*Aristida*), and other grasses (Poaceae). Caterpillar constructs shelter partly or wholly beneath soil; overwinters in 5th instar.

Flight: 1 brood; June–early July.

Habitat: Well-drained, unplowed, ungrazed or lightly grazed remnant prairie.

Range: S. Manitoba and Dakotas to Minnesota, Iowa, and Illinois.

Like the other 2 prairie *Hesperias,* the Dakota Skipper is severely limited by its need for intact grasslands, but it is even more fragile or site-specific than either the Ottoe or Pawnee skippers. The only hope for the survival of this species lies in the preservation of its native prairie habitat.

166 Indian Skipper
(*Hesperia sassacus*)

Description: 1–1⅜″ (25–35 mm). *Long, triangular wings.* Above, male has slender stigma and broad, jagged dark border; female dark brown or black, with orange confined to sharply-bordered patches. *Both sexes below,* FW orange and tawny, black at base; *HW light tan with pale band of large, connected spots.*

Similar Species: *Polites* skippers often have blunter wings, broader stigmas, more olive-brown above, duller on HW below.

Life Cycle: Mature caterpillar reddish-brown, mottled with green highlights and light speckling. Host plants are prairie grass (*Panicum*), fescue (*Festuca*), and crabgrass (*Digitaria*).

Flight: 1 brood; May–July. Perhaps 2 broods in South.

Habitat: Dry, old fields and pastures, acid soil scrub, and damp meadows and roadsides.

Range: S. Ontario east to Maine, south to Virginia and Tennessee, and west to Wisconsin and Iowa.

The Indian Skipper is more often seen in May, but flies until mid-July. It is one of the few springtime skippers, and the time of its appearance will separate it from most others with which it might be confused.

193 Sierra Skipper
(*Hesperia miriamae*)

Description: 1⅛–1⅜″ (28–35 mm). *Male cool, tawny-orange above,* with diffuse brown border and black FW stigma. Female paler above, with more defined border and light dots near tip. Both sexes below similar to upperside but lighter and greenish at tip; *HW below grayish or golden-blue with a frosty sheen,* crossed by a crescent-shaped band of *clear white, connected spots.*

Similar Species: Nevada Skipper brighter above, more olive below, with broken spot band. Other branded skippers lack blue below.

Life Cycle: Undescribed.

Flight: 1 brood; June–early July.

Habitat: Fell-fields, stony summits and high ridges and benches.

Range: Central Sierra Nevada of California and White Mountains of E. California and W. Nevada.

The Sierra Skipper is one of the most strictly arctic-alpine of the branded skippers.

182 Nevada Skipper
(*Hesperia nevada*)

Description: 1–1⅜" (25–35 mm). Wings fairly short and rounded. *Male bright tawny-orange above,* with a brownish border either broad or narrow and jagged, not sharply defined; has black FW stigma. Female above may be quite brownish, with tawny areas limited to patches. Both sexes *below* have FW black near base, yellowish-tawny at middle, and greenish near tip; HW lead-green or gray-amber, *silver-white spot band prominent, consisting of crescent-shaped, irregular, and broken-up band, especially on upper arm, end spot on lower spot-band arm very offset toward base, almost separate.* Normally also white chevron with opposing spot at HW base below.

Similar Species: Juba Skipper larger, has more pointed FW and more sharply scalloped border. Common Branded Skipper darker ocher above, with longer stigma and antennae. Sierra Skipper pale gold

above, bluish on HW below. Juba,
Common Branded, Green, Pahaska,
and Nevada skippers are very similar.

Life Cycle: Egg dull white. Caterpillar olive-tinged
with cream-marked black head.
Western needlegrass (*Stipa occidentalis*)
is a host plant in California; fescue
(*Festuca ovina*) in Colorado.

Flight: 1 brood; late May–July.

Habitat: High sagelands and forest edges, alpine
slopes, and highland meadows.

Range: British Columbia east to Saskatchewan,
south through eastern parts of coastal
and mountain states to S. Nevada,
Arizona, and New Mexico.

Although similar looking, branded
skippers exhibit definite ecological
specializations. The Nevada Skipper is a
high-country butterfly that only flies in
the cooler parts of the basinlands.

139, 168 Sachem
(*Atalopedes campestris*)

Description: 1–1⅜″ (25–35 mm). Male and female
tawny-orange and brown above; *male*
mostly tawny-orange with *massive,
square black stigma; female* mostly brown
*with pair of bright spots below FW cell, 1
glassy spot prominent.* Below, FW mostly
tawny, with short black bar outward
from base, and yellow-orange disk;
HW dusky yellow, with band of paler
spots, more contrasting in female.

Similar Species: Bunchgrass Skipper has tawny-yellow
spots on FW above. Whirlabout has
larger patch above, dark spots below.

Life Cycle: Egg greenish-white. Caterpillar olive-
green, darkened by black hair borne on
dark tubercles. Chrysalis blackish-
brown with white patches on sides of
thorax. Host plants are various grasses
(Poaceae), including Bermuda grass
(*Cynodon dactylon*).

Flight: 3 broods in South; most of year.

Habitat: Rural areas in pastures and fields; also

disturbed sites, lawns, and gardens.

Range: Resident throughout in southern
third of U.S.; emigrants to Oregon,
Colorado, Iowa, and New York.

The Sachem is one of relatively few
emigrant skippers; it can be numerous
in summertime in areas far from where
it breeds. Its caterpillar spins a
protecting tent at the grass roots; from
there it ventures up the blade, clips off
tender tips of grass leaves, and returns
to eat them in the safety of its tent.

161 Yellowpatch Skipper
(*Polites coras*)

Description: ¾–1″ (19–25 mm). Male above dark
brown with tawny-orange patch along
costal margin, sometimes extending
through much of wing base; broad
gray-brown patch next to outside of
male stigma. Female above dark brown
with discrete tawny spots stretching
into broad discal patch on HW. *Both
sexes have yellow-gold spots on several wing
surfaces:* on male HW above and FW
below, on female FW and HW above
and FW below; spots are well-defined,
square bars within dark veins. HW
below dark rust-colored with gold
toward margin, and *square, sharply edged
yellow spots expanding into striking massive
patches.*

Similar Species: Other small, tawny skippers lack
prominent yellow patches below except
male Zabulon, which lacks stigma.

Life Cycle: Egg green, almost spherical. Caterpillar
maroon with brown mottling. Chrysalis
maroon with white appendage cases.
Overwinters as caterpillar or chrysalis.
Host plants are grasses (Poaceae).

Flight: 1 or more overlapping broods. May–
September or July–August in Colorado
and Arizona mountains.

Habitat: Weedy, grassy open spaces, meadows,
marshes, pastures, and roadsides.

Range: British Columbia east to Maritimes,
south to Oregon, E. Arizona, Colorado,
and Georgia.

The Yellowpatch is one of the most
common skippers in the East. Formerly
known as Peck's Skipper (*P. peckius*),
this species is a generalist in its habitat
requirements, and therefore its range is
still expanding.

140, 181 Sandhill Skipper
(*Polites sabuleti*)

Description: ¾–1⅛″ (19–28 mm). *Highly variable;*
generally pale orange above, with
deeply-toothed outer HW margins.
Female only slightly darker than male,
with yellow to whitish spots inside of
FW margin above. *Below,* FW orange
at costa, black at base, olive-yellow
toward margin and tip. HW blackish-
olive with *coalesced yellow spot band on
lower HW that runs out along veins to
margin.* Desert populations may be
almost unmarked below.

Similar Species: No other *Polites* has yellow veins below
except Draco Skipper, which has much
broader stigma on male and darker
female. Uncas Skipper has silver spots
and scaling along veins.

Life Cycle: Egg light blue-green. Caterpillar
variable gray, green, or brown.
Chrysalis green with mottled rust-
colored abdomen. Host plants are
grasses (Poaceae).

Flight: 2 broods in North; May–June and
August–September; longer in South,
June–September in mountains.

Habitat: Coastal dunes, desert alkaline seeps and
flats, urban lawns, and watercourses.

Range: S. British Columbia and Washington
south to Baja California and southeast
through Idaho and Colorado.

Although this skipper's name reflects
its affinity for dunes, it is hardly

restricted to sandhills. While autumn
nectar can be a limiting factor for
butterflies in the North, the Sandhill's
adaptation to European knotweed as a
late-summer nectar source has allowed
this insect to colonize in British
Columbia.

Mardon Skipper
(*Polites mardon*)

Description: ¾–1″ (19–25 mm). *Compact, short
wings.* Above, orange and brown with
broad, somewhat toothed border; male
stigma small and bent. Female lighter
than most females of genus, FW and
HW above fairly extensively patched
with tawny-orange. *Below,* FW orange
and black near base, olive outwardly,
with yellow patches. *HW* reddish to
olive-tan, coarsely scaled, with
*prominent yellowish, crescent-shaped spot,
spots often diffuse or lengthened into bars
between veins;* spots paler and more
distinct in female, yellower on male.

Similar Species: Sonora skipper has smaller HW spots
below on lighter background.

Life Cycle: Egg cream-colored, becoming pale
orange before hatching. Caterpillar
brown, turning gray, with dark
speckles on midline and head. Chrysalis
light gray with brown patches. Idaho
fescue (*Festuca idahoensis*) suspected as
host plant.

Flight: 1 brood; May–June in lowlands, June–
August in uplands.

Habitat: Gravel outwash prairies in Puget Sound
Basin; mountain meadows.

Range: Thurston and W. Yakima counties,
Washington.

This is one of the most narrowly
restricted North American butterflies,
and the only species to exist solely
within Washington State. A large
population of Mardon Skippers lies in
the stewardship of the Yakima Indians.

141 Draco Skipper
(*Polites draco*)

Description: ⅞–1″ (22–25 mm). *Above, male* dark brown near margin, with tawny-orange patch over most of FW toward base and on HW disk, and broad *stigma outwardly bordered by gray-brown patch on FW;* female mostly dark brown or blackish with yellow to tawny-orange smeared and streaked on both wings. Both sexes *below,* olive or greenish-gray; FW has orange near base and narrow yellow spot band; *HW has prominent chevron-shaped spot band of big, angular, interconnected yellow or whitish spots, the tip of chevron spot is conspicuously extended at margin and base well into cell.*

Similar Species: Montana Branded Skipper has tip spot of HW chevron band below not extended into cell at base. Sandhill Skipper has sawtooth border on FW above; male has narrower FW stigma above. Yellow Patch smaller, HW below more orange-brown, with broad, square yellow spots. Long Dash more orange above, HW band below crescent-shaped.

Life Cycle: Unreported.

Flight: 1 brood; June–August.

Habitat: Alpine and subalpine meadows, stream banks, slopes, and wet lowlands.

Range: Rocky Mountain states and provinces.

The Draco Skipper is alpine in its occurrence, barely reaching the high plains. In this genus, only the Sonora Skipper has such a montane orientation, but also flies in the coastal lowlands. The Draco is more common in the high Rockies than Morrison's Silver Spike.

211 Baracoa Skipper
(*Polites baracoa*)

Description: ¾–1″ (19–25 mm). *Male above dark olive-gray with conspicuous orange* on FW above stigma, in cell, and along costa; stigma thin, short, and straight. *Female mostly orange* on FW and HW. *Below, male FW 3-toned* from upper to lower margin in tawny-orange, dark brown, and buff; *HW olive-dusted, with yellow spots, vaguely defined outwardly, in continuous crescent band (sometimes diffused outward into rays).* Female similar below but with tawny overcast.

Similar Species: Tawny-edged Skipper larger with very poorly defined crescent band on HW below; male stigma larger.

Life Cycle: Egg pale pink. Mature caterpillar light brown with darker back stripes and side stripes; overwinters. Host plants are grasses (Poaceae), also sugar cane (*Sacchinarum officinarum*).

Flight: Several broods; year-round in Florida.

Habitat: Grassy flats and stream banks.

Range: S. Georgia, Florida, and Antilles.

Chiefly a Caribbean species, colonies of Baracoa Skippers in southern Georgia and Florida are local and ephemeral.

---

174 Tawny-edged Skipper
(*Polites themistocles*)

Description: ¾–1″ (19–25 mm). *Short, triangular wings.* Olive-brown above with bright tawny-orange patch all along costal FW margin, sometimes extending through much of wing in male. *Male stigma doubly curved;* female above has band of yellow spots through FW from below end of cell. *Below,* FW tawny-orange on costa, olive at margin, and black at base; *HW mustard-colored, often plain,* sometimes with *vague lighter spots in crescent row.*

Similar Species: Crossline Skipper duller tawny, with narrower stigma; female browner. Prairie *Hesperia* species are brighter, with less olive overall. Sachem is larger, yellower below.

Life Cycle: Caterpillar maroon or tan. Chrysalis

dull white, green, and brown; overwinters. Host plants are grasses, including panic grass (*Panicum*).

Flight: 1 brood in North and West; May–August. 2 broods start earlier and end later in South.

Habitat: Many kinds of grasslands.

Range: Throughout S. Canada and U.S., but rare in Pacific Northwest.

Many butterflies adopt different habitats in various parts of their range. This species inhabits grassy areas in the East; in California, it prefers forest glades, while in Washington it has been found only around boggy edges of high-elevation lakes.

147 Crossline Skipper
(*Polites origenes*)

Description: 1–1⅛″ (25–28 mm). Above, *male has pointed FW*, olive-brown with orange cell and costal area above stigma, pale orange at FW tip, with small FW cell spots outside *long, slender, single-curved, gray-brown stigma; female* olive-brown with subdued FW spots, *often lacking any tawny-orange*. Below, pale olive-brown to golden, with ill-defined buff-olive crescent band of spots; HW spots on female subtle or absent. Colorado populations tawnier, eastern population darker.

Similar Species: Tawny-edged Skipper male has S-shaped stigma, female has orange FW costa above; below on both sexes HW spots are less defined.

Life Cycle: Egg light green. Caterpillar dark brown with black head and whitish marbling. Host plant is desert bunchgrass (*Tridens flavus*).

Flight: 1 brood; June–early August.

Habitat: Grasslands, serpentine or sandy barrens, and canyon openings near plains.

Range: Montana east to New England, south to Georgia, and west to New Mexico.

Generally less abundant and more ecologically restricted than the Tawny-edged Skipper, the Crossline Skipper is nonetheless no rarity in its range. The 2 species cannot be separated without dissection of male genitalia.

162 Long Dash
(*Polites mystic*)

Description: 1–1¼" (25–32 mm). *Above, male* tawny-orange heavily marked with contrasting brown border on FW and HW; HW has brown loop near base; *stigma connects with black mark running toward FW tip.* Female tawnier overall, tawniness above limited to FW bases and bands of yellow spots on FW and HW. *Below* both sexes, FW and HW rust-gold; *HW crossed by sharp crescent of bright yellow spots enclosing spot at base.* In some populations east of Rockies, entire HW below may be yellow.

Similar Species: Sonora Skipper darker, less tawny-orange above; has HW crescent below narrower, less contrasting.

Life Cycle: Caterpillar dark brown, with short black hair among white mottling; while feeding, constructs protective cylinder from leaf blades, overwinters after 2nd or 3rd molt. Various grasses (Poaceae) serve as host plants, including bluegrass (*Poa*).

Flight: 2 broods in South; May and September. 1 brood in North; May–July.

Habitat: Wet meadows, streamsides, marshes, and watercourses.

Range: S. Canadian provinces, Washington east to New England, south to New Jersey, Virginia, Illinois, and west to Colorado.

Paler, vaguely marked midwestern prairie populations are thought by some to be a separate species, the Dakotah Dash (*P. dakotah*); but the 2 breed in Montana.

165, 176 Sonora Skipper
(*Polites sonora*)

Description: 1–1¼" (25–32 mm). Above, brown and orange; brown with ring of broad, dark tawny-orange spots vaguely contrasting with border and variable golden overscaling. *Below,* FW and HW range from mustard through *olive and cinnamon to chocolate; HW has narrow, distinct crescent of vague yellowish spots* enclosing single spot at base.

Similar Species: Long Dash paler tawny-orange, with broader crescent band on HW below and contrasting dark border on FW above. Mardon Skipper smaller, has darker HW below, with broader spots.

Life Cycle: Caterpillar dusky green with black head. Host plant is Idaho fescue (*Festuca idahoensis*).

Flight: 1 brood; May–September.

Habitat: Open, moist, flowery meadows, forest lanes, stream banks, and roadsides.

Range: British Columbia to Montana, south to Colorado, Arizona, California, and Baja California.

Montane in Colorado yet maritime in Washington, the Sonora Skipper seems to be limited to moist areas.

151, 155 Whirlabout
(*Polites vibex*)

Description: 1–1¼" (25–32 mm). *Male tawny-yellow above* with strong stigma and dark patch continuous with stigma; margin very contrasting but irregular; *below yellow with 2 broken rows of brown dots on HW Female dusky brown* above with light spots on FW; *below, gray-brown or grayish-yellow with broad, double curved, violet-tan band on HW;* sometimes band is slightly paler than ground color, with dark greenish spots on sides.

Similar Species: Dun Skipper, Northern Broken Dash, and Little Glassywing females do not have large, squarish tan spots of crescent band on HW below; Little Glassywing has a large, squarish, glassy spot below cell on FW above. Fiery Skipper has smaller dots below than male Whirlabout, narrower border above, and lacks dark patch beyond stigma. Sachem Skipper male also lacks dark patch above as well as dots below.

Life Cycle: Egg white. Caterpillar green with black head. Chrysalis light green with paler abdomen. Host plants are hairy paspalum (*Paspalum ciliatifolium*) and weedy lawn grasses (Poaceae).

Flight: 3 broods in Florida, fewer to north; midsummer peak period.

Habitat: Many damp or dry grassy places, including scrub and rolling hill country, as well as urban habitats.

Range: Arkansas and Virginia (occasionally to Connecticut) south to Texas and west to Arizona; also West Indies to Argentina.

As the name suggests, the Whirlabout has a rapid and often circular flight. It is an avid visitor of flowers.

150 Broken Dash
(*Wallengrenia otho*)

Description: 1–1¼" (25–32 mm). Brown above with tawny-orange patches on male FW and smaller, tawny-orange spots on female FW. *Male stigma broken into 2 segments,* interrupted by square, copper-brown patch. *Below,* reddish-tan, *HW has band of vague, pale spots.*

Similar Species: Northern Broken Dash has more brown, less tawny, and is darker.

Life Cycle: Caterpillar pale green with dark mottling; cuts circular pieces of grass blades to construct protective case and overwinters in it. St. Augustine grass (*Stenotaphrum secundatum*) is possible host plant as well as other grasses (Poaceae).

Flight: 1 or 2 broods in North; summer.

3 broods in South; April–October.
Habitat: Swamps, woods, pastures, hayfields.
Range: Chiefly East and Gulf coasts; Maryland to Texas south to Costa Rica.

The Broken Dash thrives in an array of habitats throughout its extensive Atlantic range, so its choice of grasses must be extremely wide.

173 Northern Broken Dash
(*Wallengrenia egeremet*)

Description: 1–1¼" (25–32 mm). *Sooty warm brown above with small apricot markings* near male divided stigma and on female FW cell. Below, FW similar to above; HW has buff spots vaguely in crescent.
Similar Species: Broken Dash lighter and tawnier.
Life Cycle: Egg greenish or yellowish. Caterpillar apple-green with darker green mottling, yellow-edged back stripe, and brown head. Chrysalis green with yellow abdomen and brown head. Host plants include panic grass (*Panicum*).
Flight: 1 brood in North; June–July. 2 broods southward; April–September.
Habitat: Grasslands, shrub rows, and fields.
Range: Ontario and Quebec south to Florida and west to N. Texas.

The 2 broken dash skippers were previously considered subspecies of the same butterfly. However, since they fly together without interbreeding in some places, it became apparent that they were similar species with overlapping ranges. This species is the rarer of the 2 *Wallengrenia* skippers.

206 Little Glassywing
(*Pompeius verna*)

Description: 1–1¼" (26–32 mm). Very dark brown with series of *glassy white spots on both* surfaces of FW of both sexes, the largest squarish spot below end of FW cell is especially prominent. *Below,* FW similar to above; *HW deep purplish-brown in male, blacker in female,* with vague yellowish spots. Fringes buff.

Similar Species: Dun, Broken Dash, and Long Dash lack glassy white spots.

Life Cycle: Egg white becoming greenish before hatching. Mature caterpillar tan or green with dark speckles and stripes, and rust-colored head. Host plant is desert bunchgrass (*Tridens flavus*).

Flight: 1 or more broods; April–August, depending on latitude.

Habitat: Damp woods, clearings, old fields, lanes, and hedgerows.

Range: Michigan east to New England, south to Georgia, Texas, and Nebraska.

The Little Glassywing is easily distinguishable from other dark skipper males. The female Dun, Northern Broken Dash, and Little Glassywing often occur together, and are known as "The Three Witches."

223 Beard-grass Skipper
(*Atrytone arogos*)

Description: ⅞–1¼" (22–32 mm). Male tawny-orange above with fairly broad *brown margins;* female has broader dark margins, with tawny-orange restricted to upper border on FW above and small patch on HW above. *Below, clear* tawny to bright yellow. Male lacks stigma.

Similar Species: Most other tawny skippers larger, with dark markings on male FW disk, and with tawny less restricted to costa. Delaware Skipper larger, with more discrete border, and black veins above.

Life Cycle: Egg cream-colored; laid on beard grass (*Andropogon gerardi*); panic grass (*Panicum*) may be host in Georgia. Caterpillar pale chartreuse with rust-colored spots and stripes; pupates in loose cocoon spun between 2 grass blades about a yard above ground. Chrysalis yellowish with paler thorax, abdomen, wing pads, and tongue case.

Flight: 1 brood in North; June–July. 2 broods in South; March–May and August–September.

Habitat: Beard grass fields and barrens.

Range: Minnesota east to New York, south to Florida, west to Texas, and north to Nebraska, Colorado, and Wyoming.

Beard-grass Skippers live in local colonies, never in large numbers. Their virgin prairie habitats have been largely eliminated; however, they can live near civilization, as demonstrated by the colony on one of the last serpentine barrens of Staten Island, New York.

135, 138 Delaware Skipper
(Atrytone delaware)

Description: 1–1⅜″ (25–35 mm). *Above, both sexes bright tawny-orange with small brown bars at end of FW cell;* male has *brown border and veins;* female has darker veins, broader border, and dark FW base. Below, FW black at base and along lower margin; otherwise bright yellow-gold, including HW.

Similar Species: Bunchgrass Skipper larger, with dark border, more brown above, and darker HW below. *Polites* skippers have darker or patterned HW below and dark male stigma. Beard-grass Skipper smaller, with more diffuse brown borders above, and lacking black veins.

Life Cycle: Caterpillar white with black crescent on hind segments and black and white head. Chrysalis greenish-white with black tips and bristly, blunt head. Host plants are woolly beardgrass (*Erianthus diverticatus*), bluestem (*Andropogon*), and switch grass (*Panicum virgatum*).

Flight: Early summer in most places, March in Florida, June in Massachusetts.

Habitat: Open woods and edges, flood plains, grassy lowlands, pond margins, bogs, and plains watercourses.

Range: Minnesota east to Massachusetts, south to Florida and Texas; eastern foothills of Rockies from Alberta to Mexico.

Although widespread, the Delaware Skipper seldom appears to be common. Populations may differ in color as well as in behavior at different times of day.

167, 175 Bunchgrass Skipper
(*Problema byssus*)

Description: 1¼–1½" (32–38 mm). *Relatively large.* Tawny-orange above with dark borders and veins; *male has narrow, 2-part black streak in upper FW cell,* female tawniness above limited to bands of tawny-yellow spots broken by dark veins on brown background. *Below,* FW has large black base; *HW yellow to dull tan to maroon* (darkest in female) with sickle of vague yellow spots, broader on male.

Similar Species: Sachem female has whitish spots on FW above. Delaware Skipper and Ottoe Skipper males are smaller with narrower black borders; Ottoe male has stigma.

Life Cycle: Egg white. Caterpillar dingy turquoise with rust-colored head; overwinters more than half grown. Chrysalis long and narrow, ivory dotted with brown; pupates in loose cocoon in tussock of bunchgrass; adult emerges 2 weeks later. In Missouri host plant is eastern grama grass (*Tripsacum dactyloides*).

Flight: 1 brood; June–July.

Habitat: Moister prairie remnants; strictly undisturbed prairies in Kansas, perhaps secondary grasslands elsewhere.

Range: Iowa south through Illinois to Alabama and Georgia, west to Kansas and Texas; probably scattered in intervening states.

Unlike the Rare Skipper, the Bunchgrass Skipper occupies an

extensive range and can be abundant, but like its close relative, it is very locally distributed within that range.

Rare Skipper
(*Problema bulenta*)

Description: 1¼–1½" (32–38 mm). *Above, male tawny-orange with dark border and narrow, dark veins* without long lines in FW cell; *female has broadly orange cell color equal to width of dark border. Below, both sexes have immaculate yellow wings* except for black base of FW.

Similar Species: Yehl Skipper male has stigma; female has bright spots on FW above.

Life Cycle: Unknown

Flight: May–August.

Habitat: Fresh and brackish estuarine marshes.

Range: Virginia, Carolinas, and Georgia; restricted to Chickahominy, Cape Fear, Santee, and Savannah rivers.

The Rare Skipper was lost for over a century. It was named from a painting made in the early 19th century by John Abbot, the pioneer Georgia naturalist; thereafter it disappeared until it was rediscovered in the marshes near Wilmington, North Carolina, in 1925. It may be fairly numerous on the 4 flood plains it inhabits.

163 Woodland Skipper
(*Ochlodes sylvanoides*)

Description: ¾–1⅛" (19–28 mm). *Above,* FW tawny-orange and black; *male has strong, black stigma extending in bent, dark dash out to jagged, toothed, dark border; similar pattern repeated more broadly in female.* HW above broadly tawny on male, mostly brown with tawny cell band and veins, dark border on female. Below, FW black at base, with orange costa,

yellow middle and darker margin; *HW
of both sexes highly variable below,* from
plain yellowish to dark honey-tan, with
separate, angular yellow spots in
submarginal band. Maritime
populations often darker below.

Similar Species: *Polites* skippers darker, with more olive
on HW below, and crisper light spots.
Rural Skipper has white FW spot.

Life Cycle: Egg ivory. Caterpillar light green.
Chrysalis cocoa with gray frosting.
Host plants are grasses (Poaceae).

Flight: 1 staggered brood; June–October. July
to August in Rockies, earlier in South.

Habitat: Scrub, ridges, forest edges, roadsides,
lawns, gardens, stream banks; in Pacific
Northwest, from tidewater marshes to
mid-elevation mountains.

Range: British Columbia south to Baja
California and east to Alberta, South
Dakota, and Colorado.

Ecologically versatile and climatically
tolerant, the Woodland Skipper adapts
so readily to both natural and altered
habitats that it prospers equally on
Santa Cruz Island, California, and in
the vacant lots of Seattle, Washington.
It accepts nectar sources eschewed by
most species, such as oxeye daisy and
everlasting.

134, 144   Rural Skipper
(*Ochlodes agricola*)

Description: ¾–1″ (19–25 mm). *Above,* FW and
HW tawny-orange with diffuse dark
border; male has prominent FW
stigma, female has dark patch on FW
in place of stigma, *both sexes have white
glassy spot between dark patch and tip.*
Below, FW has glassy spot between
base of wing and disk; male HW tawny
with vague, light spots in submarginal
row; female spots more prominent on
dusky, purplish-brown HW.

Similar Species: Woodland Skipper lacks FW spot.

Life Cycle: Egg gray-green, turns white. Young caterpillar white with glossy black head. Host plants grasses (Poaceae).

Flight: 1 brood; May–June, rarely August.

Habitat: Forest openings such as stream banks, glades, and clearings.

Range: SW. Oregon south through California, into N. Baja California.

The Rural Skipper and the Woodland Skipper occupy many of the same lowland places in California, but usually the 2 species are separated seasonally. The Rural Skipper will be found in glades and clearings during the early summer, while the Woodland Skipper does not appear in such places until mid- to late summer or fall.

203 Snow's Skipper
(*Ochlodes snowi*)

Description: 1⅛–1⅜″ (28–35 mm). Above, warm brown frosted with orange; male has small, vague, yellow spots; spots more prominent and diffuse in female; *both sexes have light hourglass-shaped cell spot on FW above. Below,* similar; FW has whitish flash on lower margin, *HW deep auburn, with row of minute, light orange spots on disk* and 1 spot at base.

Similar Species: Branded skippers and Morrison's Silver Spike much tawnier, with silvery spots on HW below. Other *Ochlodes* paler.

Life Cycle: Not recorded. Host plants probably grasses (Poaceae).

Flight: 1 brood; June–September.

Habitat: Montane meadows and roadsides at 7260–9900′ (2214–3020 m).

Range: Colorado, New Mexico, Arizona, and Mexico.

Snow's Skipper flies in moist northern mountains of the Southwest, but not in surrounding arid lands. In mountain meadows, Snow's Skipper takes nectar from the purple flowers of horsemint.

Yuma Skipper
(*Ochlodes yuma*)

Description: 1⅛–1⅜″ (28–35 mm). *Above, male
broadly yellow-gold with narrow, dark
margins* and prominent black stigma on
FW; *female nearly clear golden-tawny with
vague yellowish-white spots* in convex
pattern on FW and narrow dark
margins. *Below, both sexes yellow-gold*
with narrow, dark margins; male has
black FW base and female slightly
darker at base; otherwise plain.

Similar Species: Saw-grass Skipper larger, darker.
Delaware Skipper lacks stigma, has
more distinct brown border. Other
Ochlodes skippers more heavily and
contrastingly marked.

Life Cycle: Egg greenish-white. Young caterpillar
pale orange with black head, becoming
light green with cream-colored head.
Young caterpillar bends edges of blades
into shelters; older caterpillar makes
protective tubes out of entire rolled leaf
blades. Chrysalis dark brown, with
tongue case free from body halfway to
tip. Host plant is common reed
(*Phragmites communis*).

Flight: 2 broods in California; June and
August–September, July–August in
Colorado.

Habitat: Colonies rare and strictly limited to
stands of common reed: desert seeps,
sloughs, canals, wet spots on alkaline
flats and marshes.

Range: Central California west to W. Colorado
and south toward Rio Grande.

Much more habitat-limited than the
majority of western skippers, the Yuma
Skipper is seldom found more than a
few yards from its host plant. The small
and isolated colonies are thought to be
relics from the late Pleistocene era,
when a broader, more continuous
distribution probably existed. In
westernmost Colorado, the Yuma
Skipper flies with the Nokomis
Fritillary.

158 Mulberry Wing
(*Poanes massasoit*)

Description: 1–1⅛" (25–28 mm). *Wings stubby.*
Dark black-brown above with or
without minute spotting, orange on
males, white on females. *Below,* FW
similar to upperside; *HW brown to
maroon with band of bright yellow, angular
spots, 1 elongated into crossbar.*

Similar Species: Other *Poanes* skippers have longer,
more pointed wings, and brighter,
larger yellowish markings above.

Life Cycle: Egg chalk-white. Caterpillar olive with
long yellow hairs; feeds on grasses
(Poaceae) or sedges (Cyperaceae),
including tussock sedge (*Carex stricta*).

Flight: 1 brood; midsummer.

Habitat: Extensive tallgrass meadows and bogs
or marshes with sedges or tussocks.

Range: Minnesota, Ontario, New Hampshire
south to Maryland and west to South
Dakota, Nebraska, and Illinois.

Although widespread and sometimes
numerous, the Mulberry Wing favors
vulnerable, specialized wetland
habitats, making it very local.

159, 233 Hobomok Skipper
(*Poanes hobomok*)

Description: 1–1⅜" (25–35 mm). *Above, male*
yellow-orange with dark borders, *black
cell spot; female tawny-orange restricted,*
brown FW base. *Below,* FW more
yellow than above in both sexes; *tips
and outer margins dark and may be
bordered with violet; HW violet-edged,*
broadly on female, narrowly on male,
with brown base and broad, curved,
yellow spot band or broad, *yellow disk,
not enclosing row of brownish-red spots;*
bright spots on male, dull on female.
Dark female form dark brown above,
with white spots and small FW cell
spot.

Similar Species: Zabulon Skipper male has brownish-red spot row in yellow disk on HW below; female blackish with FW glassy spots.

Life Cycle: Caterpillar dark green or brown with rows of tiny black spines; caterpillar or chrysalis may overwinter. Host plants are grasses (Poaceae).

Flight: 1 extended brood; May–September, although flight period in a single area usually more restricted.

Habitat: Deciduous woods and adjacent roads, valley bottoms, and moist gullies; hedgerows, fields, and meadows.

Range: Saskatchewan to Maritime Provinces, south to Georgia, Arkansas, New Mexico, north to Dakota Black Hills.

The Hobomok Skipper is a fairly common butterfly over much of its broad range, which overlaps extensively with that of the Zabulon Skipper. A few Hobomok colonies occur in the Sangre de Cristo mountains of south-central Colorado, where the butterflies find a suitable habitat at a far higher altitude and farther west than is usual.

157, 235 Zabulon Skipper
(*Poanes zabulon*)

Description: 1–1⅜" (25–35 mm). *Slim, triangular wings. Above, male yellow-orange; female always blackish with conspicuous, angular glassy spots in disk,* lacks cell spot on FW. *Below, male* yellow-orange on HW, with margin, base, and *row of disk spots light reddish-brown;* female has row of red-brown spots below and violet flush.

Similar Species: Hobomok Skipper male lacks spots in yellow disk of HW below; dark female has white spots in disk and FW cell. Yellow Patch male has dark stigma.

Life Cycle: Host plants grasses (Poaceae), including tumble grass (*Eragrostis*).

Flight: 2 broods; peaking in May and August. Perhaps 3 broods in South.

Habitat: Many grassy places around hardwoods, scrub, stream banks, and in old fields.

Range: Wisconsin east to Massachusetts, south to Georgia, Texas, and Panama.

This species was first described from Georgia by the French naturalists Boisduval and Le Conte. In the Northeast, the Zabulon Skipper can be seen taking nectar from violets.

160, 212 Golden Skipper
(*Poanes taxiles*)

Description: 1¼–1⅜" (32–35 mm). *Male above brilliant, light brass-gold with narrow, scalloped, dark margin;* female mixed brown and tawny-orange. *Below,* FW tawny-orange with violet tip (more pronounced in female); male has light yellow crossbars on brown HW; *female has vague, angled row of small gold spots across violet-brown HW disk.*

Similar Species: Yuma Skipper male yellower above, with darker, larger stigma, and plainer below; female uniformly colored.

Life Cycle: Unrecorded.

Flight: 1 brood; June–July.

Habitat: Shaded gullies and valley bottoms, grassy areas near waterways, and pine forest clearings.

Range: Nevada east to Wyoming, Dakotas, and Nebraska, south to New Mexico, Arizona, and NW. Mexico.

This big, bright skipper is active in cloudier weather than most butterflies will tolerate. The Golden Skipper is the only truly western member of its genus.

171 Saffron Skipper
(*Poanes aaroni*)

Description: 1–1½" (25–38 mm). Male above tawny-orange toward base and dark olive-brown toward margin, with dark veins against tawny-orange HW disk and dark marginal markings distinct and scalloped inwardly; male has very long, slender silver stigma. Female mostly tawny-orange above with slightly broader dark margins. *Both sexes have small, black bar at end of FW cell above. Below* both sexes, FW tawny-orange with black base, olive edge, and *dark veins confining barlike orange areas across disk; HW olive-tan with distinctive ray of tarnished, brass-colored scales.*

Similar Species: Other tawny skippers in range lack pale ray below; males have dark stigmas. Rare Skipper larger, has vague border.

Life Cycle: Egg white. Caterpillar greenish; may feed on a marsh grass (Poaceae).

Flight: 2 broods in most of range; June and August peak periods. Perhaps 3 broods in Florida.

Habitat: Salt marshes, brackish wetlands; freshwater marshes and ponds, wet meadows, roadside ditches, and sedgy pine-cypress swamps.

Range: New Jersey south to Florida.

The orange bars on the hind wings above, particularly on the female, resemble the anthers of saffron crocuses, inspiring the species' common name.

148 Yehl Skipper
(*Poanes yehl*)

Description: 1⅛–1½" (28–38 mm). *Relatively large; striking sexual dimorphism. Male deep golden-orange above* with dark margins, prominent FW stigma, and well-defined orange HW disk with dark veins; below, FW has black base and HW tawny-orange with small yellow spots in loose row. *Female almost all brown above* with pale spots on FW and orange disk on HW; *below FW 2-toned;* HW brown with 3–4 tiny white spots.

Similar Species: Bunchgrass Skipper lacks male stigma,

female much darker.

Life Cycle: Not recorded.

Flight: 2 broods; May–June, August–October.

Habitat: Open, deciduous woods and nearby roads, fields, meadows, and swamps.

Range: Kentucky and Virginia south to Florida and west to Texas.

This locally common skipper avidly takes nectar from pickerelweed, milkweed, sweet pepper, and other flowers. It may be seen in the summer in the Great Dismal Swamp.

169 Broad-winged Skipper
(*Poanes viator*)

Description: 1¼–1¾″ (32–44 mm). *Relatively large;* wings broad and full. Male *above* mostly dark brown on FW with orange spots extending in orange veins to margin; *HW disk orange* with dark veins against disk; *broad scalloped margin.* No male stigma. Female above similar with several white FW spots; FW tip more pointed. *Below,* FW similar to above; *HW rust-tan with curved, jagged tawny-orange to yellow band* on disk.

Similar Species: Yehl Skipper has less orange on HW above, darker HW below.

Life Cycle: Egg gray. Caterpillar gray becoming brown; rests in natural cleft between grass sheath and stem. Host plants are marsh millet (*Panicum*), wild rice (*Zizania aquatica*), and lake sedge (*Carex lacustris*).

Flight: 1 brood in Wisconsin, 2 in Georgia and South; April–August.

Habitat: Salt marshes near coast, shrubby marshes and sedge bogs inland; also sandy, disturbed sites, such as landfill.

Range: Minnesota east to Maine, south to Alabama, west to Texas, and north to Nebraska.

The Broad-winged Skipper displays a peculiar distribution pattern; it occurs very locally over most of its broad range, suggesting strict habitat needs. However, it has adapted well to disturbed sites on the Atlantic Coastal Plain. The advance or retreat of this large, attractive skipper will depend on its adaptive powers in inland habitats.

164 Umber Skipper
(*Paratrytone melane*)

Description: 1⅛–1⅜″ (28–35 mm). Dark brown with golden-yellow spots on FW above and below; *HW golden-brown above* with diffuse spots. *Below, reddish-brown and slightly purplish* with HW crossed by band of yellow spots, squarish and continuous. Female larger, has pale FW spots. Male lacks stigma.

Similar Species: *Euphyes* skippers in range lack strong yellowish markings on HW beneath.

Life Cycle: Egg light green. Caterpillar dirty yellow with black back line, light side stripes, and yellow-brown head covered with whitish pile. Chrysalis straw-colored to tan; has long tongue case. Host plants are grasses (Poaceae).

Flight: Several broods; January–November in S. California, mostly late summer.

Habitat: Usually moist, well-vegetated places: from lawns to canyons, shady washes, and overgrown ditches.

Range: California in San Francisco Bay area to Baja California and Texas.

This butterfly has colonized new areas in southern California within the past half-century, and seems to be adapting well to landscape changes.

207 Palmetto Skipper
(*Euphyes arpa*)

Description: 1⅜–1¾″ (35–44 mm). *Long, triangular wings.* Male tawny-orange on inner ⅔ above, crossed by dark FW stigma, brown beyond; *female dark brown with square apricot-colored spots* across FW disk above. Both sexes *orange beneath on FW costa and tip and HW;* basal third of FW black. Head and collar golden.

Similar Species: Florida Swamp Skipper brown on HW below. Sedge Skipper red-tan below. Saw-grass Skipper brighter above, red-tan below.

Life Cycle: Caterpillar light green with black collar, black and white head, and yellow stripes; lives in cylinder of leaves. Host plant is saw palmetto (*Serenoa repens*).

Flight: Successive broods; March–December.

Habitat: Coastal stands of saw palmetto.

Range: Coastal plain of Mississippi, Alabama, Florida, and Georgia.

The Palmetto Skipper was first described from a painting done in Georgia by John Abbot, an early 19th-century naturalist; however, it is now a very uncommon insect in Georgia.

---

219 Saw-grass Skipper
(*Euphyes pilatka*)

Description: 1½–1¾" (38–44 mm). *Wings triangular.* Male bright orange-tan, dark borders and small black stigma. Female pale tawny-orange above with distinct, broad, dark brown borders; dark patch in light area extends near costa. *Below, HW dark tan;* FW tawny with black base, tan margin, and tip.

Similar Species: Palmetto and Sedge skippers darker above. Florida Swamp Skipper smaller, yellower above.

Life Cycle: Caterpillar yellow-green with black dots and brown head; feeds on saw grass (*Mariscus jamaicensis*).

Flight: 2 broods in Virginia; spring and fall. Several broods in Florida; most of year.

Habitat: Wetlands with saw grass stands.

Range: Virginia to Florida and Mississippi.

The Saw-grass Skipper's caterpillars draw blades of saw grass together, constructing tubes in which they conceal themselves. Adults fly to pickerelweed for nectar.

214 Sedge Skipper
(Euphyes dion)

Description: 1¼–1⅜" (32–41 mm). Relatively rounded wings. Male FW above tawny-orange with broad brown margin, black stigma; female FW above brown with row of tawny-orange spots. *Both sexes have HW above brown with bright orange ray from cell toward margin.* Below, FW orange on costal half and tip, black lower edge with lighter spots; HW reddish-tan with 1 or 2 pale streaks.

Similar Species: Florida Swamp Skipper yellower above, brown below. Palmetto Skipper brighter above, darker below.

Life Cycle: Egg pale green. Caterpillar yellow-green with black head; feeds on lake sedge (*Carex lacustris*) or wool grass (*Scirpus cyperinus*), a rush.

Flight: 1 brood; July–August in New Jersey.

Habitat: Swamps, marshes, wet meadows, and bogs with dense stands of sedges and tall grasses; also woodland shorelines.

Range: Wisconsin and Ontario, south to Texas and N. Florida.

The Sedge Skipper occupies a wide range, but colonies occur very locally and it rarely seems common in one area. Although this agile butterfly is capable of powerful flight, it seldom wanders far from beds of the host plants. A particularly dark population inhabits the southeastern states and is often considered a separate species, the Alabama Skipper (*E. alabamae*). A similar-looking, recently described Texas species, McGuire's Skipper (*E. mcguirei*) may be closely related to the Alabama.

170 Scarce Swamp Skipper
(*Euphyes dukesi*)

Description: 1¼–1½" (32–38 mm). *Short, broad, rounded wings.* FW deep brown above; male has black stigma, female midwing band of 2 or 3 small yellow spots. HW above lighter brown with tawny-orange base. *Below, FW similar; HW yellow-tan with yellow rays between veins.*

Similar Species: Dun Skipper narrower, HW below brown.

Life Cycle: Early stages unrecorded. Lake sedge (*Carex lacustris*) may be host plant.

Flight: 2 broods in Virginia; early–late summer. Fall only in Deep South.

Habitat: Shaded wetlands, coastal swamps, and ditches with aquatic vegetation.

Range: Michigan and Ohio south to North Carolina, Alabama, and Louisiana.

The Scarce Swamp Skipper was discovered only 50 years ago, much later than most eastern butterflies. Known colonies are quite widely separated, implying a previously more uniform distribution.

172 Black Dash
(*Euphyes conspicua*)

Description: 1–1⅜" (25–35 mm). *Male* tawny on FW above with broad brown borders and *conspicuous stigma. Female almost black on FW above with curved band of glassy yellow spots.* HW *above* of both sexes *has distinctive patch of light spots* (tawny-orange on male, yellowish on female) *crossed by dark veins* on otherwise dark HW. Below, FW and HW rich rust-colored with light spots.

Similar Species: Other *Euphyes* lack light HW patch.

Life Cycle: Unknown. Host plants probably sedges (*Carex*).

Flight: 1 brood; July–August, sometimes earlier farther south.

Habitat: Marshes, marshy meadows, mildly acid

woodland bogs, wet tallgrass prairies.

Range: Ontario south through Minnesota to Nebraska, and east through Massachusetts, Ohio, and Virginia.

The unwary Black Dash perches on tall grasses and takes nectar from the blossoms of buttonbush and swamp milkweed. In June, this butterfly flies together with the Mulberry Wing in the Great Dismal Swamp of Virginia.

146 Florida Swamp Skipper
(*Euphyes berryi*)

Description: 1⅛–1½" (28–38 mm). Male dark brown above with tawny-orange to yellow patches on both wings and conspicuous stigma in patches. Female dark brown above, FW crossed by orange square spots; HW vaguely orange-patched beyond cell. FW below paler, *HW brown to dull orange* with light veins and no pale spots.

Similar Species: Black Dash has HW below rust-colored with light spots. Sedge Skipper brighter. Palmetto Skipper HW bright orange below. Saw-grass Skipper more orange.

Life Cycle: Unknown.

Flight: 2 broods; March–May and September–October, or 1 long, staggered period.

Habitat: Swamps, pond edges, drainage canals, and marsh margins.

Range: Central Florida, rarely Georgia; also Coastal Plain.

Like several other uncommon southeastern wetland skippers, the Florida Swamp Skipper alights on pickerelweed flowers—this prolific swamp blossom attracts many skippers. However, land development and the pace of habitat alteration in central Florida threaten this and other rare skippers that should be monitored closely and perhaps protected.

224 Two-spotted Skipper
(*Euphyes bimacula*)

Description: 1⅛–1¼″ (28–32 mm). Pointed wings.
Dark above; male tawny-orange on
broad patch around stigma; *female
usually has 2 tiny yellowish FW spots
above.* Below, both sexes have tawny-
orange *FW with 2 bright yellow spots on
disk,* male stigma occasionally shows
through; HW dull yellow-tan with
bright yellow veins. Wing fringes and
palpi white; head and collar orange.

Similar Species: Dun Skipper dark below. Tawny-edged
Skipper lacks 2 yellow spots.

Life Cycle: Egg pale green. Caterpillar feeds on
sedges (*Carex*). Otherwise unrecorded.

Flight: 1 brood; late June–July.

Habitat: Bogs, marshes, pond edges and nearby
fields, and sedgy meadows in acid areas.

Range: Ontario to Nebraska; E. Colorado to
Maine, south along coast to Carolinas.

Although it enters the South, the Two-
spotted Skipper is mainly a northern
resident. Members of the genus survive
in a wide array of wetlands, but this
species prefers cool climates.

216, 240 Dun Skipper
(*Euphyes ruricola*)

Description: 1–1¼″ (25–32 mm). Triangular
wings. Dusky brown *above, virtually
unmarked* except for FW stigma on male
and tiny translucent spots on female
FW. Dull tan below with black at base;
female may have faint yellow HW
patch. *Head and palpi orange in some
populations.*

Similar Species: Red-headed Roadside Skipper smaller,
with light fringe. Orange-edged
Roadside Skipper smaller, with
reddish-orange fringe. Northern Broken
Dash has orange bars on FW above.

Life Cycle: Egg pale green becoming reddish on
top before hatching. Mature caterpillar

shiny green with overlay of silvery striations. Chrysalis blunt, ridged; various shades of green, yellow, and brown. Host plants are yellow nutgrass or chufa (*Cyperus esculentus*) in Missouri and sedge (*Carex heliophila*) in Colorado.

Flight: 1 brood in North; midsummer. 2 broods in South; spring, late summer.

Habitat: Deciduous woods, fields, roadsides, pastures, mountains to 9000′ (2745 m), swales in coastal California, and old logging roads near bogs in Northwest.

Range: S. Canada to Baja California and N. Mexico.

The Dun Skipper flies most of the summer in many natural and cultivated habitats throughout North America, commonly alighting on damp sand or leaves. Favored nectar flowers are mints, fireweed, dogbane, and lotus. Formerly called *E. vestris*.

218 Monk
(*Asbolis capucinus*)

Description: 1⅝–2″ (42–51 mm). *Large; robust, broad-bodied* with pointed wings. *Male rich brown-black above* with prominent orange-edged silver stigma; *female golden-brown* with diffuse light patch on FW. Both sexes chestnut below; FW has tawny patch. *Male has tawny fringes, female fringes buff.*

Life Cycle: Host plants in Florida include cabbage palmetto (*Sabal palmetto*), coconut palm (*Cocos nucifera*), date palm (*Phoenix*), and saw cabbage palm (*Paurotis*).

Flight: Successive broods; most of year.

Habitat: Where palms occur.

Range: S. Florida and Cuba.

Named for its warm brown-clad wings, the Monk was introduced from Cuba to Miami in 1947. Since then it has become established in southern Florida.

234 Dusted Skipper
(*Atrytonopsis hianna*)

Description: 1¼–1½" (32–35 mm). Pointed wings.
Dark brown with several tiny but
distinct translucent spots near FW tip
above and below. *Below, violet to pale
gray towards edge, brown base;* HW has 1
*translucent spot near base, diffuse brown
disk spots. Antennal club gently curved.*

Similar Species: Cloudywings have distinctly bent
antennal club.

Life Cycle: Caterpillar has 7 instars; makes tube
tents above bases of host plant clumps;
overwinters in tents until spring, then
resumes feeding. Probably beard grass
(*Andropogon scoparius*) is host plant.

Flight: 2 broods in Southeast; March–April
and October. 1 brood northward and
westward; May–June.

Habitat: Dry, open fields and acid, sandy barrens
of oak and oak-pine scrub in East; wet,
narrow canyons farther west.

Range: Saskatchewan to New England, south
to New Mexico, Arkansas, Georgia.

The Dusted Skipper shifts about and
colonizes fresh burns very efficiently.
This species is fond of taking nectar at
blackberry, strawberry, and clover.
Western populations are lighter and
more grayish both above and below.
The Southern Dusted Skipper (*A.
loammi*) lives in the southeast from the
Carolinas to Florida and Mississippi.
Sometimes considered the same as the
Dusted Skipper, it is redder, has more
whitish spotting above on the fore
wing, and bears 2 arcs of white spots
across the hind wing below.

Deva Skipper
(*Atrytonopsis deva*)

Description: 1½–1⅝" (38–41 mm). Long,
triangular wings. *Brown above with 3–5
glassy white spots around end of FW cell*

and in a bar from FW costa. Similar below but *HW violet-brown or cocoa-brown with vague darker shades* across disk. Buff, *uncheckered fringes.*

Similar Species: Mexican Cloudywing rounder, bands below more distinct, fringes checkered.

Life Cycle: Unreported.

Flight: 1 brood; late May–July.

Habitat: Gullies, arroyos, and wooded slopes.

Range: SE. Arizona; SW. New Mexico, and Sonora, Mexico.

The genus *Atrytonopsis* is very well represented in the American Southwest. A similar relative is the Moon-marked Skipper (*A. lunus*), which has a crescent-shaped light spot in the cell of the fore wing above, and white fringes on the hind wing. Relatively rare in the United States, it flies in June and July. Viereck's Skipper (*A. vierecki*) also has a crescent-shaped cell mark but is smaller, paler, with checkered fringes, and the male bears a prominent stigma on the fore wing. This species appears in April and May from Colorado and Utah south to Arizona and west Texas.

Sheep Skipper
(*Atrytonopsis ovinia*)

Description: 1⅜–1⅝″ (35–41 mm). Wings short. Light brown *above, heavily spotted with white in irregular, unconnected spot bands across both wings,* and with 1 small spot in HW cell and 1 large spot in FW cell. Female paler than male, with all spots larger and yellower. FW below similar to above but dusted with lighter scales; *HW below frosted purplish-brown with striations;* spots more cream-colored and numerous than above. Fringes checkered brown and white.

Similar Species: Moon-marked Skipper lacks white spots on HW.

Life Cycle: Unrecorded.

Flight: 1 brood in Texas; April–June. 2 broods

in Arizona; March–August, possibly longer.

Habitat: Gullies among well flowered, hilly country.

Range: Arizona east to W. Texas, south to Nicaragua.

The Sheep Skipper inhabits the same parched landscapes that support the Navajo's sheep, possibly inspiring the species name. Several similar skippers also occur in the desert Southwest. The Parchment Skipper (*A. pittacus*) is a small, pale, grayish species with a band of glassy white spots across the hind wing above, irregular white spots on the fore wing above, and a whitish, papery look to the hind wing below. It flies in spring and early summer from Arizona to west Texas and Mexico. The Belted Skipper (*A. cestus*) is similar, but is darker brown with a broad, white belt across the hind wing and a white spot in the hind wing cell above. It is known only in south-central Arizona. The Python Skipper (*A. python*) is warmer brown but lacks the very conspicuous hind wing spots; those on the fore wing are irregular and may be either glassy white (in New Mexico and Texas) or yellow (in Arizona). Its fringes are checkered and the hind wing beneath is frosted with grayish-violet.

Orange Roadside Skipper
(*Amblyscirtes simius*)

Description: 1–1⅛″ (25–28 mm). Usually quite *orange above* with dark brown area within white spot band on FW, and *apricot-colored cell spot;* sometimes very dark above, even blackish. Both wings *below frosty-gray* with submarginal row of white spots; FW has bright orange outlined area. *Fringes white.*

Life Cycle: Caterpillar feeds on blue grama grass (*Bouteloua gracilis*) in Colorado.

Flight: Possibly 2 broods in south, 1 brood
          northward; May–June.
Habitat: Small prairie prominences; roadsides
          with wildflowers across high plains.
Range: S. Saskatchewan south through
          Colorado to Arizona and Texas.

The habitat—unusual for the genus—
and certain structural details make
specialists wonder whether the Orange
Roadside Skipper is a true *Amblyscirtes*.
A more typical member is Oslar's
Roadside Skipper (*A. oslari*), which
covers almost exactly the same range as
the Orange Roadside Skipper. Flying
from April to August in dry canyons,
foothill slopes, and sandy gullies, it has
subtler coloring, and rather moderately
checkered fringes. The Cassus Roadside
Skipper (*A. cassus*) is brownish-orange
above, with large orange marks; below,
the hind wing is marbled with violet-
gray and grayish. This species inhabits
arroyos and rocky streambeds in the
Southwest, and flies from May to
August.

Large Roadside Skipper
(*Amblyscirtes exoteria*)

Description: 1–1¼" (25–32 mm). *Larger, more
               pointed wings than most in genus.*
               Brownish or rust-colored with *tiny white
               or light dots in curved line on outer FW
               above.* Orange dusting above. Below,
               gray dusting on brown ground color,
               with distinct white spots on FW and
               HW; FW cell below sometimes dark
               orange. *Fringes checkered.*
Similar Species: Other roadside skippers in range are
               smaller and paler below.
Life Cycle: Unknown.
Flight: 1 brood; June–July.
Habitat: Dry drainage systems, canyons, and
               wooded glades.
Range: SE. Arizona, SW. New Mexico, and
               Mexico.

One of the most challenging genera of butterflies to identify, *Amblyscirtes* includes 21 species in our area. Most are rather alike and may require a specialist to differentiate them. Some even have very similar genitalia. The Dotted Roadside Skipper (*A. eos*) flies in the same range as the Large Roadside Skipper. It has white, rather than cream-colored or buff spots below. One brood flies from April to July in Texas and from August to September in Arizona and Mexico. The Roadside Rambler (*A. celia*) is larger, with more rounded wings and cool gray below. It flies year-round in the lower Rio Grande Valley. Another southwestern member of this genus, the Prenda Roadside Skipper (*A. prenda*), is blackish-brown above, and has dispersed clusters of cream-colored spots on both wings above and below. The Prenda Roadside Skipper flies from June to August in arid and semiarid terrain.

191 Bronze Roadside Skipper
(*Amblyscirtes aenus*)

Description: ⅞–1⅛" (22–28 mm). Brown *above* with *brassy luster. FW has reddish cell below;* HW gray below, disk has transverse row of powdery, pale spots. Fringes checkered.

Similar Species: Cassus Roadside Skipper brighter orange, with stronger spot pattern above. Oslar's Roadside Skipper lighter below, lacks FW spots above; male has prominent stigma.

Life Cycle: Unknown.

Flight: 1 brood; May–June, Colorado; April–July, Texas; June–September, Arizona.

Habitat: Small, stony canyons, gullies on plains, rocky slopes of dry foothills and mountain ranges.

Range: Colorado and Utah, south through SE. Arizona, New Mexico, and W. Texas.

Despite its name, the Bronze Roadside Skipper is less brilliant than the Cassus Roadside Skipper. In west Texas, New Mexico, and southern Arizona, this species may be confused with the Texas Roadside Skipper (*A. texanae*), but the latter is paler overall, and has more extensive spots above than the Bronze Roadside Skipper. The similar Slaty Roadside Skipper (*A. nereus*), an uncommon and local species, is much grayer overall than both the Bronze and Texas Roadside skippers, and is greenish below, especially on the hind wing. The Erna Roadside Skipper (*A. erna*) is mostly unmarked. It flies in west Texas, Oklahoma, and Kansas. The Arkansas Roadside Skipper (*A. linda*) is darker; its fore wing below is nearly plain, but the hind wing is 2-toned with a contrasting brown cell on a buff background. It flies in Oklahoma, Arkansas, and Tennessee.

190 Pepper-and-salt Skipper
(*Amblyscirtes hegon*)

Description: ⅛–1″ (22–25 mm). Black above with vaguely *greenish cast;* faint, light FW spots and moderately developed *white, curved FW spot band. Below, putty-gray with understated light spots;* greenish cast over HW. Checkered fringes.

Similar Species: Slaty Roadside Skipper has more prominent spots above, paler below.

Life Cycle: Caterpillar light green with green back stripes, yellow side stripes, and brown head. Chrysalis straw-colored with greenish tinge. Host plants are Indian grass (*Sorghastrum nutans, S. secundum*) and Kentucky bluegrass (*Poa pratensis*).

Flight: 1 brood, perhaps 2; May–August.

Habitat: Forest edges, clearings and glades, coniferous and mixed woods, boggy stream banks, and hayfields.

Range: S. central and SE. Canada to Iowa and Georgia; maybe Texas and Mississippi.

Unlike most roadside skippers, this species prefers northern and Appalachian woodlands. Formerly called *A. samoset.*

187 Lace-winged Roadside Skipper
(*Amblyscirtes aesculapius*)

Description: 1–1¼" (25–32 mm). Dark gray-brown above and below, with *distinctive network of whitish spot bands and connecting white veins on HW below,* showing through to HW above. FW spots buff above, whiter below, in full curved band.

Similar Species: Cobweb and Uncas skippers are both tawny-orange on FW above and below.

Life Cycle: Unknown.

Flight: Successive broods in South, 1 brood in North; January–September.

Habitat: Wooded areas, edges of woods, forest paths, and canebrakes.

Range: Connecticut south to Florida and west to Missouri and New Mexico.

This skipper flies in the Great Dismal Swamp, on the edges of the vast wetland. The similar Carolina Roadside Skipper (*A. carolina*) has buff, not white, veins on the hind wing below, and its forewing below is 2-toned, with black near the base and yellow outwardly. It flies in swamps and marshes of the coastal plain, and in damp woods with maiden cane (*Arundinaria tecta*), from Virginia south to Georgia, and west to Mississippi.

Mottled Roadside Skipper
(*Amblyscirtes nysa*)

Description: ⅞–1" (22–25 mm). Mat-black above with tiny white spots near FW tip. *Below,* mottled with purplish-gray and brown; HW has *chocolate and tan patches*

surrounded by violet-gray. Prominently
checkered fringes.

Similar Species: Roadside Skipper is 2-toned violet and
brown below.

Life Cycle: Egg shiny white. Young caterpillar
white, maturing to green with dark
mid-back stripe, deeper green blotches,
and white, rust-colored striped head.
Host plants are crab grass (*Digitaria
sanguinalis*), St. Augustine grass
(*Stenotaphrum secundatum*), pungent barn
grass (*Echinochloa pungens*), and yellow
foxtail (*Setaria glauca*). Chrysalis ivory
with tawny-orange head.

Flight: 2 broods in northern part of range;
May–June and July–August. Successive
broods southward; March–November.

Habitat: Suburban and city lawns and gardens;
narrow, rocky gullies far from town.

Range: Kansas and Missouri to Arizona, Texas,
and Mexico.

Unlike most roadside skippers, the
Mottled Roadside Skipper can inhabit
both urban and wild areas. Along with
a few others of its family, this species
readily became a lawn skipper when
lawns replaced the prairies.

232 Roadside Skipper
(*Amblyscirtes vialis*)

Description: ⅞–1″ (22–25 mm). Rounded wings.
Dark brown to almost black, lacking
luster above and below; only tiny white
spots in minor cluster near FW tip.
Below, 2-toned: brown with prominent
*violet-gray shading on outer portion of FW
and HW*. Fringes brown and buff.

Similar Species: Mottled Skipper has mottled violet and
brown on HW below. Blue-dusted
Skipper has longer wings, dark in FW
cell and base; bluer, less 2-toned on
HW below, black and white fringes.

Life Cycle: Egg pale green; hemispherical.
Caterpillar light green with green dots
and soft hair; frosty head has rust-

colored stripes. Chrysalis green, reddish at both ends; overwinters. Host plants include Kentucky bluegrass (*Poa pratensis*), striped oats (*Avena striata*), bent grass (*Agrostis*), and Bermuda grass (*Cynodon dactylon*).

Flight: 1 brood in North, 2 in South; March–September, varying locally.

Habitat: Moist woods, glades, watersides, ravines; acid soil heaths and pine barrens; canyons and roadsides.

Range: British Columbia to Quebec and entire contiguous U.S. except S. Florida and S. California; more common in North.

The Roadside Skipper's caterpillar produces a waxy powder covering, then abandons its leaf shelter and rests in the open. Adults take nectar often, and particularly favor ground ivy.

199 Bell's Roadside Skipper
(*Amblyscirtes belli*)

Description: ⅞–1¼" (22–32 mm). Overall impression *black*. *FW above has largest spot in white FW band triangular or chevron-shaped. Beneath dark blue-gray,* with dirty gray spots in curved band on HW. Fringes black checkered.

Similar Species: Roadside Rambler grayer; its largest FW spot above is squarish.

Life Cycle: Egg white. Caterpillar pale, whitish-green with darker green lines and white and orange head. Chrysalis has rust-colored thorax, cream-white head, and yellowish wing cases. Host plant in Missouri is a grass (*Uniola latifolia*).

Flight: 2 or more broods; March–October in Texas, shorter period northward.

Habitat: Densely wooded ravines and woods; clearings, creek beds, fields, gardens.

Range: Florida through Georgia and Arkansas to Missouri, Oklahoma, and Texas.

A versatile species of the southeastern states, Bell's Roadside Skipper is one of

a minority of its genus that prefers
moist woodland conditions to dry,
stony canyonlands.

229 Blue-dusted Roadside Skipper
(*Amblyscirtes alternata*)

Description: ¾–1″ (19–25 mm). *Small; wings
narrow and noticeably pointed.*
Blackish gray-brown above, darker in
cell and at base, lighter outside curved
row of white spots across wing. *Below,
frosty blue scaling smeared across wings,*
brown at base. *Fringes white and black.*

Similar Species: Roadside Skipper has broader, less
pointed wings, buff checkered fringes;
HW below 2-toned, not smeared.

Life Cycle: Unknown.

Flight: Perhaps several broods; February–
April, July–August, and September–
November in southern part of range.

Habitat: Pine woods and moist forests.

Range: North Carolina south through
Alabama, Florida, and Texas.

Originally discovered in the Great
Dismal Swamp of Virginia and North
Carolina, the Blue-dusted Roadside
Skipper tends to fly close to the
ground.

Red-headed Roadside Skipper
(*Amblyscirtes phylace*)

Description: 1–1⅛″ (25–28 mm). *Pointed wings.
Above and below,* slate-gray or charcoal-
black *without spots. Unbroken buff-white
fringes. Rust-orange palpi, head, and
collar. Female palpi white below.*

Similar Species: Golden-headed Skipper less orange on
head, rounder, with subtle markings.

Life Cycle: Unreported.

Flight: 1 brood northward, probably 2
southward; late May–August.

Habitat: Semiarid mountains; bare hollows,

canyons, mouths of gullies, roadcuts;
moist mountain meadows in Colorado.

Range: Colorado, New Mexico, Arizona, and
W. Texas south into Mexico.

Throughout much of its southwestern
range, the Red-headed Roadside
Skipper dwells in arid draws and
gullies. Yet in the northern parts of its
territory, it climbs the mountain
shoulders into moist, cool locations.
The Orange-edged Roadside Skipper
(*A. fimbriata*) flies in southeast Arizona
and Mexico during June and July. It
has rust-colored fringes and is plain
mahogany above and slate below.

225 Eufala Skipper
(*Lerodea eufala*)

Description: ⅛–1¼" (22–32 mm). Triangular
wings. *Plain gray-brown above;* 3–5
small, white, glassy spots on middle of
FW above and below. *Below, FW cool
tan;* HW brown with gray dusting,
often with arrow of very obscure, pale
spots. Female somewhat larger and
lighter; male lacks FW stigma.

Similar Species: Swarthy Skipper lacks glassy spots.
Three-spotted Skipper larger, darker.

Life Cycle: Egg pale green. Caterpillar bright
green with green lines and yellow
blotches and white and rust-colored
head; feeds on grasses (Poaceae).
Chrysalis long and slim, light green
with green and yellow stripes and
frontal horn.

Flight: 1 brood in northern range; early
summer. 2 broods in Georgia; March–
April, August–September. Continuous
broods farther south; most of year.

Habitat: Grassy flats along canals, and pine
woods in keys; hill country in Texas;
desert valleys and coastal lowlands of
California; oak openings in Midwest.

Range: Most of southern U.S. from Central
Valley of California to mid-Atlantic,

north to Nebraska, and east to
Virginia; also south to Argentina.

This versatile skipper takes nectar from
goatweed, asters, and other flowers.
Two other small species enter our area
from the south. The Violet-clouded
Skipper (*L. arabus*) is darker brown
above with a pale purplish hind wing
below, enclosing a brown patch. This
little-known species flies in the early
spring in grassy arroyos of Arizona,
Baja California, and mainland Mexico.
The Olive-clouded Skipper (*L. dysaules*)
is still darker above, and has a 2-toned
olive-brown underside. It flies in south
Texas and Mexico in the fall.

202, 205 | Twin-spot Skipper
(*Oligoria maculata*)

Description: | 1¼–1⅜″ (32–35 mm). Broad wings.
Overall *dark brown above;* slightly
lighter *auburn-brown below.* Small
cluster of *3 bright white spots* on middle
of FW above and below. HW above
usually unspotted; HW below has 3
distinct white spots, 2 together near
end of cell and 1 toward costa.
Similar Species: | Little Glassywing lacks bright white
spots on HW below.
Life Cycle: | Caterpillar and chrysalis both green and
finely furred; caterpillar has brown
head; chrysalis has blunt head. Host
plants probably grasses (Poaceae).
Flight: | 2 broods in most of range; May–June,
August–September. Successive broods
in Florida year-round.
Habitat: | Varied swamps and pine flats.
Range: | Gulf and coastal southeast; rarely to
New York and Massachusetts.

In common with many subtropical
skippers, the Twin-spot Skipper finds
its way north as climate permits. Its
northern limit is imprecise, but it
seldom wanders far from the coast.

204 Brazilian Skipper
(*Calpodes ethlius*)

Description: 1¾–2¼" (44–57 mm). *Large;* FW narrow and pointed. Warm brown, darker on outer part and lighter on inner part, with *prominent row of translucent spots above and below,* spots larger on FW. Below, reddish with reddish-buff margins. *Head broad.*

Similar Species: Hammock Skipper lacks HW spots. Chestnut-spotted Skipper has reddish patches below.

Life Cycle: Caterpillar grayish-green with black-rimmed spiracle within brownish blotches, whitish lines on back, orange and black head. Chrysalis pale greenish, slender, with head and tail extensions, and long tongue case. Host plants are cannas (*Canna flaccida* and *C. indica*).

Flight: 3 broods in South; most of year.

Habitat: Gardens and other areas with cannas.

Range: S. California east to Carolinas; strays to Missouri, Mississippi, New York; also Argentina and West Indies.

Brazil is merely a middle point-of-reference for this large, powerful flier. Fluctuations in population lead to distant flights, when Brazilian Skippers appear in large numbers overnight to feed and lay eggs, then vanish within the hour. Few other skippers have the ability to disperse so widely.

210 Salt Marsh Skipper
(*Panoquina panoquin*)

Description: 1¼" (32 mm). Light brown above and below, with small light spots near tip of FW above; FW below has cream-colored spots; *prominent white dash, pointed outward from cell of HW below;* veins of HW below appear yellowish.

Similar Species: Other brown skippers lack single white dash on HW below.

Life Cycle: Egg pale greenish. Young cream-colored caterpillar matures to yellow-green; feeds on grasses (Poaceae) or sedges (Cyperaceae), with salt grass (*Distichlis spicata*) strongly suspected.

Flight: 2 northern broods; June and September. Probably more broods in South; early spring–late fall.

Habitat: Salt marshes; also tidal marshes and meadows near cord grass marshes.

Range: Connecticut along coast to S. Florida and west to Mississippi along Gulf.

Along with a surprising variety of habitat-limited butterflies, the Salt Marsh Skipper can still be found on Staten Island, New York.

179 Obscure Skipper
(*Panoquina panoquinoides*)

Description: 1–1¼" (25–32 mm). *Small. Dull yellowish-brown* above and below, *usually with 2–6 small pale spots on FW*, 1–2 faint light spots on HW below, and paler vein-scaling on FW or HW; sometimes unmarked. Occasionally slightly reddish due to tawny dusting of scales.

Similar Species: Salt Marsh Skipper has white bar on HW below. Eufala Skipper lacks HW spotting.

Life Cycle: Egg white. Caterpillar green with whitish and yellowish stripes. Chrysalis green with yellow and white stripes. Host plant may be salt grass (*Distichlis spicata*).

Flight: 2 broods in Florida; spring and fall.

Habitat: Salt marshes and immediate vicinities.

Range: Gulf Coast of Texas and Florida; also Antilles, Central and South America.

The Obscure Skipper can be rather common along the Florida coast north to about Titusville, but is very local and seldom seen. Some skipper specialists consider this the same species as the Wandering Skipper.

209 Wandering Skipper
(*Panoquina errans*)

Description: 1–1¼" (25–32 mm). *Dark yellowish-brown above and below,* with prominent yellow-orange spots on FW above; HW below has suggested or obvious yellow veins, sometimes yellow spot band.

Similar Species: Umber Skipper is chestnut-brown. Eufala Skipper lacks yellow veins.

Life Cycle: Egg white; caterpillar green with whitish and yellowish lines and stripes. Chrysalis translucent green. Host plant is salt grass (*Distichlis spicata*).

Flight: 1 brood, perhaps 2; July–September, later in Baja California.

Habitat: Edges of tidal marshes or beach areas; also ocean bluffs.

Range: S. California coast; Gulf of California.

A colony of these skippers has been found near the canals in Venice, California, a habitat enclave now undergoing rapid development.

217 Long-winged Skipper
(*Panoquina ocola*)

Description: 1¼–1⅜" (32–35 mm). *FW very slender and pointed.* Highly variable. *Dark yellowish-brown* above and below, with variably developed yellow FW spots but without spotting on HW above, and usually unmarked beneath.

Similar Species: Obscure Skipper and Salt Marsh Skipper lighter, smaller.

Life Cycle: Egg white. Caterpillar yellow, becomes blue-green on first 2 segments and gray-green on rest of body, with green spots and lines. Chrysalis greenish with short tongue case. Host plants grasses (Poaceae); also rice (*Oryza sativa*) and sugar cane (*Sacchinarum officinarum*).

Flight: 3 broods in Florida, fewer northward; most of temperate year.

Habitat: Salt marshes; also pine woods, low damp areas, and forest edges.

Range: Texas east to Florida; fairly regular in Kentucky, Arkansas, and New Jersey; rarely to Indiana, Ohio, New York.

The Long-winged Skipper can be abundant and sometimes has been seen in mass movements. A more spotted, redder species, the Russet Skipper (*P. hecebolus*), enters Texas from Mexico. It has a round, yellowish cell spot on the fore wing above.

231 Purple-washed Skipper
(*Panoquina sylvicola*)

Description: 1⅜" (35 mm). Slender, long *wings concavely curved along edges.* Brown *above, with prominent yellow-white spots* on FW, including elongated FW cell spot. *Below, FW brown;* HW brown in male; in *female, HW often splashed with purple-blue;* both sexes have *prominent row of light spots across most of HW below.*

Similar Species: Russet Skipper redder above, with rounder, yellower FW cell spot above. Others in genus lack purple wash, except Evans' Skipper, which is larger, has yellow spots, more purple.

Life Cycle: Egg cream-colored. Caterpillar yellow, maturing to green, has dark line down back, lighter stripes along sides, and dark green flecks overall. Chrysalis slim, green with yellowish body stripes, head projection, and short tongue case. Host plants include many grasses (Poaceae), as well as sugar cane (*Sacchinarum officinarum*).

Flight: August–December.

Habitat: Varied tropical and subtropical grassy areas, including city gardens in Texas.

Range: Lower Rio Grande Valley of Texas (and probably Florida) to Argentina and West Indies.

It is not unusual for tropical species of butterflies that wander into our range to adapt to subtropical gardens. The

Purple-washed Skipper, originally a denizen of the Brazilian forest, has found a suitable habitat in many Texas yards. A related tropical butterfly, Evans' Skipper (*P. evansi*) occasionally appears during late fall in southern Texas from Costa Rica. This skipper is warm brown above, with amber-yellow spots in the fore wing disk; below, the hind wing has a lavender-blue sheen, a dark border, and a blue disk.

198 Violet-banded Skipper
(*Nyctelius nyctelius*)

Description: 1⅜–1⅝" (35–41 mm). *Pointed, triangular wings.* Brown above; FW has large square spot and smaller white spots; HW has vague yellowish spot band. More white *below* on FW, *HW pale brown with violet and dark brown* bands, and small, subcostal dark spot near HW base. Head very wide.

Similar Species: Hammock Skipper larger, darker, with lobed HW. Long-winged Skipper smaller, with plain HW below. Purple-washed and Evans' skippers lavender-blue below, lack violet banding.

Life Cycle: Egg yellow-green. Caterpillar pale green with black head, becoming blue-gray as it feeds on grasses (Poaceae) as well as rice (*Oryza sativa*) and sugar cane (*Sacchinarum officinarum*). Chrysalis pale tawny with brown-mottled head, long tongue case.

Flight: Probably several broods; May and September–December.

Habitat: Grassy tropical and semitropical areas.

Range: Texas to Argentina; perhaps S. California; also West Indies.

Most hesperiine skippers use grasses and related plants as host plants for their caterpillars. This species and others adapt to sugar cane, rice, and other crops, although they seldom constitute significant agricultural pests.

GIANT SKIPPERS
(Megathymidae)

Approximately 20 species worldwide occurring only in the New World, most of them in the American Southwest and Mexico. Deserving of their name, the largest giant skipper species exceed 3" (76 mm) in wingspan. Unlike that of true skippers, the large, robust body is cigar-shaped; the head is narrower than the thorax, and the antennae clubbed rather than hooked. The wing venation, however, is similar to that of true skippers, and the large, triangular wings are usually black or brown with yellow or gold markings and frosty scaling. Giant skippers are reported to exceed 60 mph (96 kph) in flight, making them extremely difficult to observe except when they alight at damp earth to drink.

The caterpillars feed exclusively on yuccas, agaves, and manfredas; often a single butterfly species uses only one plant species. The caterpillars bore into the fleshy leaves, stems, or rootstocks, hollowing out feeding chambers in which they later pupate. In most species, the caterpillars build trapdoors through which the emerging adult escapes. The chrysalises, unlike those of any other butterfly family, move up and down inside their chambers, perhaps taking advantage of changes in temperature from day to night.

Much remains to be learned about the biology and relationships of giant skippers, the second smallest family of butterflies. Specialists differ widely about the number of species they recognize. A conservative approach is used in this guide, treating an array of butterflies as relatively few species.

195 Orange Giant Skipper
(*Agathymus neumoegeni*)

Description: 2–2⅜" (51–60 mm). *Bright orange above with blackish markings* on FW cell, costa, and margin. Below, FW gray-brown with frosty-gray margin, crossed by pale orange band; HW charcoal-gray to brownish, purplish or frosty-gray, with scalloped, paler submarginal band. *Thorax and abdomen conspicuously furred with orange.*

Life Cycle: Egg darkens from cream-colored to orange, hatching in 18–19 days. Caterpillar burrows in root, then back into leaf base of Parry's agave (*Agave parryi*) and related species, leaving trapdoor over burrow entrance before pupating. Feeds until July; rests several weeks before pupating.

Flight: September–October.

Habitat: Mesas and mountains at 7100–9150' (2165–2790 m).

Range: Central Arizona, Texas, N. Mexico.

Like most giant skippers, the Orange Giant Skipper adults are often hard to find; the caterpillars may be more easily located in the host plants. This bright insect can sometimes be spotted flying around agaves in Carlsbad Caverns National Park, New Mexico. All giant skippers in the genus *Agathymus* are very similar, differing only in the amount and color of the pale orange spots or markings above, and sometimes in the shape of the fore wing. They are best identified by locality and association with particular host plants.

196 Aryxna Giant Skipper
(*Agathymus aryxna*)

Description: 2–2⅜" (51–60 mm). *Dark brown above and below, with orange band from FW tip through HW disk* and 1 broad orange spot in FW cell; tawny-orange at wing bases. Below, FW dark brown with grayish tip and yellow band; HW frosty-gray with a few white spots.

Life Cycle: Egg dull greenish; laid on host plants, agaves (*Agave palmeri, A. chrysantha*); hatches 6 weeks later. Caterpillar feeds until May; pupates in late summer.

Flight: 1 brood; late August–early November.

Habitat: Arid but well-vegetated desert country, especially canyons.

Range: SE. Arizona, SW. New Mexico, and NE. Sonora, Mexico.

Long confused with the Orange Giant Skipper, Aryxna is biologically distinct and flies in different regions. It may be seen in moister areas of Ramsey Canyon Nature Preserve in Arizona. This species recycles its own excreted fluids, a habit which may relate to the rigorous physiological demands upon a large insect in an arid area. Another Arizona giant skipper, the Yavapai Giant Skipper (*A. baueri*), resembles the Aryxna but is darker muddy-brown with less tawny-orange above and darker gray on the hind wing below. Its caterpillars feed on agave (*Agave chrysantha*). Strictly limited to Arizona canyons, adults fly from September through November. In the same range, the smaller Amole Giant Skipper (*A. polingi*) is mostly orange above, with brown patches on the fore wing disk and tip; the hind wing is powdery gray with buff submarginal spots and basal marks. Its caterpillars feed on amole (*Agave schottii*). Adults fly in some numbers during October. The Molina Basin of the Santa Catalina Mountains near Tucson, Arizona supports a large colony.

197 Brigadier
(*Agathymus evansi*)

Description: 2″ (51 mm). *Dark brown above with bands of gold-orange* submarginal spots, some orange at base of FW, and gold spot in FW cell. Below, FW brown with orange band and grayish border; HW gray with white irregular submarginal band. Body brown, rarely with orange scaling.

Life Cycle: Host plant is Parry's agave (*Agave parryi*). Caterpillar makes tunnel near center of thick, succulent leaf and builds rough, black trapdoor on lower surface; feeds until midsummer.

Flight: 1 brood; late August–early October.

Habitat: Arid canyonlands and mountains.

Range: Arizona; restricted to Huachuca Mountains and environs in Cochise and Santa Cruz counties.

The habitat from which the species was described lies in Ramsey Canyon, a preserve belonging to The Nature Conservancy. The common name may refer to the golden epaulettes on the wings, or to the scientific name, which refers to Brigadier W. H. Evans, a 20th-century skipper specialist.

Pecos Giant Skipper
(*Agathymus mariae*)

Description: 1¾–2″ (44–51 mm). *Small;* wings blunt. Black-brown to tan-brown *above with varying number of yellow-gold spot bands of various widths, often with black patch across FW cell.* Warm brown-gray below, FW brown at base with yellow spots; white spots on HW below, or HW all gray below. Tan-gray body fur.

Life Cycle: Caterpillar feeds on lecheguilla (*Agave lecheguilla*). Mature caterpillar is bluish.

Flight: Late September–November.

Habitat: Rocky, often limestone hill-country, especially at bases of slopes, among agave, mesquite, juniper, and yucca.

Range: S. New Mexico, Trans-Pecos Texas, and Mexico.

Like most giant skippers, this species has been divided into a number of subspecies, some representing geographically isolated races. The West Texas Giant Skipper (*A. gilberti*) uses the same host plant but flies only in Brewster, Terrell, Val Verde, and Kinney counties of Texas.

---

194 California Giant Skipper
(*Agathymus stephensi*)

Description: 2–2⅛″ (51–54 mm). Umber-brown *above*, tan at bases with *buff spots at base and small, squarish white spots in submarginal band.* Below, FW dark, slate-brown *with white spots;* HW gray with row of small white spots shadowed by black spots. Body pale tan to grayish.

Life Cycle: Host plant is desert agave (*Agave deserti*).

Flight: Late September–early November.

Habitat: Gullies, washes, juniper flats, slopes.

Range: San Diego, Imperial, and Riverside counties, California; also N. Baja California.

Common in parts of southern California, this giant skipper may be looked for in Anza Borrego Desert State Park in autumn, when adults roost at night in junipers and other vegetation. In the Mojave Desert area is the similar Mojave Giant Skipper (*A. alliae*). Above, males of the 2 species resemble each other, but the female Mojave is more golden; below, both sexes have yellow spots. Its caterpillars feed on Utah agave (*Agave utahensis*). The Brown Bullet (*A. estelleae*) occasionally strays to southwestern Texas from Mexico. It is similar above but usually has fewer spots on the hind wing and is frosty-violet below with a straight row of pearly-white spots across the hind wing disk. Its caterpillars feed on lecheguilla (*Agave lecheguilla*). Adults fly from September to October.

237 Yucca Giant Skipper
(*Megathymus yuccae*)

Description: 2–2⅞" (51–73 mm). Black above with yellow rays outward from bases and cloudy-yellow HW borders; *FW has bright yellow rounded cell spot* (narrow or pointed in western populations) *and submarginal band of squarish yellow spots above;* sometimes HW has yellow spotting above. Below, dark HW frosty-violet with some black dusting and white spotting. *Big, rounded body,* generally black to dark brown or gray.

Similar Species: Cofaqui Giant Skipper has yellow spot band running into FW cell above; male has tuft of long hair on HW above.

Life Cycle: Caterpillars feed on all U.S. yuccas except Whipple's yucca (*Yucca whipplei*), sometimes devouring leaves before boring into the crown, making silk shelter of 2 or more leaves strapped together. Habits within plant vary with species of host plant, condition of plant, and population of skipper. Caterpillar overwinters, pupates in spring before feeding.

Flight: Late January–June, only about 4 weeks in any one locale except central Florida.

Habitat: Forest edges, granite outcroppings, old fields, and bottomlands with yucca.

Range: Utah and Great Basin east through Arkansas and Carolinas, south to Florida and Gulf, west to Nevada, S. California, Baja California, and Mexico.

By far the most widely distributed giant skipper, the Yucca Giant Skipper may be seen as readily in Georgia's pine woods as in Utah's deserts, and from the bottom of Black Canyon of the

Gunnison to high elevations in North Park, Colorado. Regional races may be quite distinctive. The western subspecies are often grouped in a separate species under the geographic name Colorado Giant Skipper (*M. coloradensis*). Adult Yucca Giants do not appear to feed as adults. The somewhat rare Ursine Giant Skipper (*M. ursus*) is the largest giant skipper in North America with a wingspan to 3⅛" (79 mm). This distinctive black skipper has a broad, yellow-orange spot patch across the fore wing tip above and below, and white antennae. It flies in summer and early fall in southeastern Arizona, southern New Mexico, southwestern Texas, and Mexico. Its caterpillars feed on yuccas (*Yucca*).

Cofaqui Giant Skipper
(*Megathymus cofaqui*)

Description:	2–2⅜" (54–60 mm). *Short, broad wings.* Tan to brown above with *broad submarginal band of yellow spots running into cell,* yellow spots near tip on FW above, and buff margin on HW above. Below, FW similar to upperside, but darker; HW nearly unmarked lilac-gray. *Male has conspicuous tuft of long dark hair near base of HW above.*
Similar Species:	Yucca Giant Skipper lacks tuft of hair on male HW above, and both sexes have yellow spot in FW cell above that is not connected to spot band.
Life Cycle:	Caterpillar burrows quickly into root of yucca (*Yucca filamentosa, Y. aliofolia*). Pupal tent location varies, sometimes projecting from root system up through soil or from surface of plant, but usually down among bayonets of yucca.
Flight:	2 broods; February–May and August–November.
Habitat:	Hardwood hammocks with yuccas; pine woods with yucca stands.
Range:	Florida; Georgia, and possibly

Tennessee; different races in upland Georgia and Coastal Plain.

The Cofaqui Giant Skipper occurs with the Yucca Giant Skipper in the Southeast; they are the only giants east of the Mississippi. But the Yucca Giant Skipper has no fall brood, whereas the Cofaqui extends its second flight well into autumn.

Strecker's Giant Skipper
(*Megathymus streckeri*)

Description: 2⅛–3″ (54–76 mm). *Large; broad wings.* Dark black to umber-brown *above with yellow-orange spots,* white marks at FW tips; *and broad* buff-white HW margin. Male sometimes nearly all black and white; female tawny. Below, FW similar to upperside; HW dusky gray-brown to black with small to medium light buff or frosty-white spots across disk. HW looks scalloped on inside edge of light fringe.

Similar Species: Yucca Giant Skipper has narrower, more pointed wings, darker orange on FW, and usually pale spots on HW above, with smooth, much narrower HW marginal band.

Life Cycle: Green egg glued to underside of leaves of yucca (*Yucca baileyi, Y. angustissima, Y. glauca, Y. constricta*). Caterpillars feed in main root system and build hidden, camouflaged tents when about to pupate.

Flight: Early April to mid-July, varying with elevation and latitude; usually 2–8 weeks after Yucca in any locality.

Habitat: Gravelly flats, yucca plains, semiarid hilly country.

Range: Florida, Georgia, and possibly Tennessee; different races in upland Georgia and Coastal Plain.

As with other giant skippers, colonies tend to shift from year to year as

females drift with the wind. Strecker's
Giant Skipper is named after a
pioneering 19th-century lepidopterist.
It was first described from the Petrified
Forest National Monument, Arizona,
where it may still be found.

Manfreda Giant Skipper
(*Stallingsia maculosus*)

Description: 1⅝–2½" (41–64 mm). *Broad, rounded
wings.* Dark brown above and below
with *narrow submarginal band of small
gold spots on FW and HW above,* and
small gold spot in cell. FW below has
band of large gold spots; HW below
has spots similar to upperside but less
distinct. Borders of both FW and HW
below are violet-tinged. Fringes buff
checkered.

Life Cycle: Eggs mature from chartreuse to white.
Caterpillar burrows into rootstocks,
leaves, and caudices of host plant,
manfreda (*Manfreda maculosa*). Develops
small tent, expanding it prior to
pupating; no trapdoor constructed.

Flight: 2 broods; April–May and September–
October.

Habitat: Semiarid landscapes with host plant.

Range: S. Texas south into Mexico.

The chiefly Mexican genus *Stallingsia*
differs from close relatives in the genera
Agathymus and *Megathymus* because its
caterpillars feed on manfreda instead of
agaves or yuccas. The first colony of
Manfreda Giant Skippers found in
Texas has been destroyed by
development and farming. Although
other colonies are still present in the
state, Texas lepidopterists consider the
species to be seriously endangered.

Part III
Appendices

GLOSSARY

Aestivation Dormancy during a hot, dry season.

Alpine Pertaining to or inhabiting mountains.

Antenna One of a pair of long, slender sensory appendages on either side of a butterfly's head. (*pl.* antennae)

Arroyo A gully or dry stream bed in arid country.

Arctic The region around and north of the Arctic Circle (23°30' from the North Pole); dominated by tundra.

Arctic-alpine Pertaining to higher elevations of mountains with arctic conditions, usually above timberline.

Anal fold The portion of the hind wing that folds against the abdomen when a butterfly is at rest.

Androconium A specialized scale that produces sex-attractant scent molecules. (*pl.* androconia)

Bald A shrubby or grassy area without trees on a mountain summit.

Bog Wetland characterized by saturated, acidic soil and often by the thick growth of sphagnum moss; usually treeless except for clumps of spruce, larch, or cedar.

Boreal Northern, pertaining to the cool, moist coniferous forest region of North

America that stretches from Alaska to Newfoundland.

Broadleaf Having wide leaves (as distinct from coniferous); pertaining to deciduous trees.

Brood A generation of butterflies hatched from the eggs laid by females of the previous generation; members of a brood fly during the same general period.

Cell The area of each wing that is entirely enclosed by veins; also called discal cell.

Chaparral Low, thick, scrubby growth consisting of evergreen shrubs or low trees; common in semiarid climates, especially in California.

Circumpolar Distributed around the world in polar regions.

Clubbed antennae Antennae that are thickened toward the end, not tapered to a point.

Cocoon A silken case that envelops the chrysalises of a few true butterflies, many skippers, and most moths.

Colony A distinct, resident group of butterflies with more or less determinable boundaries.

Composite A plant belonging to the family Asteraceae (formerly Compositae).

Compound eye An eye made up of thousands of light-sensitive units, or ommatidia.

Concave Curved inward.

Continental Divide The watershed or ridge running north and south that separates the Atlantic and Pacific drainages, and largely follows the crest of the Rocky Mountains.

Convex Curved outward.

Costa The upper edge of both the fore wing and the hind wing. (*adj.* costal)

Costal fold The costal margin in some butterflies, which contains pockets of scent-scales.

Cremaster Hook-bearing structure at the tail end of the chrysalis, frequently used for attachment to a support.

Desert An arid land having fewer than 15″ of annual precipitation.

Diapause A period of inactivity and reduced physiological function induced by environmental factors; more commonly occurs in caterpillars or chrysalises.

Dimorphism The occurrence of 2 distinct forms within a species.

Disjunct Widely spaced; usually refers to the range of a species.

Disk The central portion of a butterfly wing, tangential to the costal and trailing margins. (*adj.* discal)

Disturbed habitat Changed or altered landscape caused by human or natural disruption, such as plowing, overgrazing, or burning.

Dorsal surface The upperside of a wing.

Duff Humus; decomposed ground litter.

Emigrate To leave an area in large numbers, usually periodically and in a recognizable direction, but sometimes unpredictably.

Endemic Limited in natural distribution to a specific area.

Extinct Having no surviving populations or individuals.

Eyespot A scale-pattern on a butterfly's wing resembling an eye, usually with rims and pupils of contrasting colors.

Falcate Hooked; having the tip of the fore wing curved into a hook.

Fauna The collection of animal species inhabiting a given region.

Fell-field An arctic or arctic-alpine habitat consisting of lichens, mosses, grasses, sedges, dwarf wildflowers, and cushion plants growing among stones.

Fulvous Tawny-brown or dull orange.

Genetic swamping The absorption of one species by another through hybridization, which results in the loss of the weaker species' distinctive genes.

Girdle A silken band spun by some caterpillars prior to pupating, which encircles the thorax of the chrysalis and attaches it upright to a vertical surface.

Ground litter The partially decomposed, organic material on top of the soil.

Gynandromorph An individual expressing both male and female characteristics.

Hammock A raised clump or island of broadleaf trees, isolated by water or by other vegetation such as grasslands, everglades, or pine scrub; restricted to S. Florida in North America.

Hibernation Dormancy during the harsh winter months; also called overwintering.

Holarctic Designating the entire Northern Hemisphere north of the Tropics; used to indicate distribution in northern temperate regions.

Host plant The food plant of a caterpillar.

Host-specific Restricted to a single plant or group of plants as a source of food.

Hyaline Glassy or shiny; translucent.

Instar One of the stages between molts in the growth of a caterpillar.

Introduced Brought into an area; not native.

Legume A plant in the pea or bean family (Fabaceae). (*adj.* leguminous)

Lepidoptera The order of insects containing moths and butterflies.

Local Restricted to small, sparsely distributed colonies or geographical areas.

Lunule A crescent-shaped mark.

Maculation The spots, bars, chevrons, and other markings on a butterfly's wings.

Margin The edge of the wing.

Maritimes The southeastern coastal provinces of Canada, including New Brunswick, Nova Scotia, and St. Edward Island.

Marsh Wetland with standing water and emergent growth, such as grasses, reeds, rushes, and sedges, but with few shrubs or trees.

Melanic Particularly dark or blackish.

Mesa A tablelike, flat-topped land formation consisting of horizontal layers of sedimentary rock.

Migrate To move regularly from one region to another in large numbers.

Molt The shedding of skin by the caterpillar to permit growth.

Montane Pertaining to mountainous regions between foothills and subalpine areas.

Muskeg A mossy acid bog.

Naturalized Adapted to an area not originally within a species' range.

Nearctic The northern temperate and arctic regions of the New World.

Neotropics The tropical portions of the New World.

Osmeterium A pair of fleshy horns, located behind the head of all swallowtail caterpillars, that emits a strong odor believed to deter predators.

Over-scaling A loose, sparse coat of scales on top of the denser background layer of scales; often hairlike.

Overwinter To hibernate.

Palpus One of 2 furry facial appendages of butterflies, projecting on either side of the proboscis, which they protect. (*pl.* palpi)

Pheromones Sex attractant scent molecules produced by scent-scales, or androconia.

Piedmont A plateau between the sea and the mountains, specifically along the base of the Appalachians.

Pine barrens Sandy, acidic areas dominated by pines, scrub oaks, and other acid-tolerant plants.

Plains A broad expanse of level landscape, usually grasslands.

Pleistocene The geological epoch which began approximately a million years ago, and was characterized by the periodic advance and retreat of continental and alpine glaciers.

Polymorphism The occurrence of several distinct forms within a species.

Population An interbreeding group of individuals belonging to a species.

Postbasal Just beyond the base of the wing.

Postmedian Just beyond the middle of the wing, toward the margin.

Prairie A broad, level, or rolling grassland containing arid and moist areas.

Pupa The chrysalis.

Pupation The transformation from caterpillar to chrysalis.

Pupil A spot near the center of an eyespot.

Quaking bog A bog consisting of soil and vegetation that floats on ground water.

Race A distinctive geographical population, equivalent to a subspecies or variety.

Relict A population left behind or stranded far away from the main range of a species.

Resident A butterfly species that lives and breeds in a given area.

Riparian Pertaining to the banks of a river or stream.

Roost A place where butterflies gather to spend the night or to winter.

Salt marsh Tidal wetland, often with brackish water and salt-loving plants.

Savannah An open grassland punctuated by scattered trees and shrubs.

Scale One of millions of shinglelike plates covering the wings of butterflies.

Scarp The steep face of a slope or cliff.

Scent scales Specialized scales that produce and disperse sex attractants; also called androconia.

Scree A loose, unstable rockslide, or the rock pile at the base of a cliff or slope.

Seep A boggy spring or line of springs along a hillside in otherwise arid country.

Sex patch A thick pad of sex scales or scent scales on a butterfly's wing, known as stigmata or brands in skippers and hairstreaks.

Sibling species Two very closely related species, possessing subtle physical differences but often profound ecological distinctions.

Sphagnum A genus of mosses commonly found in bogs.

Sphragis A waxy pouch formed on the abdominal tip of a mated female parnassian that prevents further mating.

Steppe A dry, short-grass prairie, sometimes also with shrubs such as sagebrush.

Stigma A bold, sharply defined patch of scent scales on the fore wings of many male skippers and hairstreaks. (*pl.* stigmata)

Stray A butterfly that wanders far beyond its usual range but which is not a regular emigrant.

Striations Fine lines, close together and often parallel.

Submarginal Just inside the outer margin of the wing.

Subspecies A more or less distinct geographic population of a species that is able to interbreed with other members of the species.

Subtropical Pertaining to regions bordering on the tropical zone.

Swamp   A wetland area with standing water and woody vegetation.

Taiga   A swampy, coniferous forest of boreal regions.

Talus   A formation of rock debris at the base of a cliff, usually the result of a rockslide.

Tarsus   The foot section of the butterfly leg, which has hooks at the end for clinging.

Temperate Zone   The region between the Arctic and the Tropics.

Tibia   The fourth segment of the leg, linking femur and tarsus.

Timberline   The northern limit of forests in high altitudes and latitudes above which trees do not grow.

Tornus   The outer angle of the margin of a butterfly wing.

Tropics   The torrid zones, generally between the Tropic of Cancer and the Tropic of Capricorn (23°27′ north and south).

Truncate   Having the tip of the wing angled, appearing clipped.

Tubercle   A raised, wartlike knob, often with spines.

Tundra   Treeless vegetation above timberline, dominated by grasses, sedges, lichens, and mosses, and often having boggy or frozen soil.

Venation   The pattern of veins on a wing.

Ventral surface   The underside of a wing.

Virgin forest   Woodland that exists in its primeval state.

BUTTERFLY-WATCHING TIPS

Wherever you live—in city or suburb, near mountains, or at the shore—you will find butterfly-watching a pleasurable and engrossing experience. Although traditionally butterfly enthusiasts collected specimens, today most conservation-minded naturalists prefer to observe butterflies in their natural habitats.

Equipment: Your 2 essential tools for butterfly-watching are a field guide and a pocket notebook to record butterfly species, host plants, and observations of behavior. Many species are difficult to identify positively without viewing them at close range for several minutes with your guide in hand. An easy way to do this for all butterflies except skippers, many of which may be identified only by specialists, is by using a net. With practice you will be able to catch a butterfly without harming it. Experienced lepidopterists pick up the insect by its fore wings with blunt-ended tweezers. If you place the butterfly inside a clear glass jar or plastic box, you can examine it closely with a magnifying lens. Binoculars may be helpful if you wish to follow butterflies in flight or observe them from a distance. The binoculars should be capable of focusing quickly, and you should be able to view objects at a distance of 5–6′. A pair of 7 × 35 binoculars is convenient for most field observation.

Time: Butterflies usually fly during the time of year when maximum daily temperatures are 60° F or higher. Species have individual habitat preferences, and the frequency of their flights varies. For example, the Falcate Orangetip has only a single flight period each spring. The Gulf Fritillary flies year round in southern Florida, while the Pearly Eye in Virginia may produce 3 generations—in spring, early summer, and fall.
Butterflies and their host plants have different geographic limits and optimum climates, and generally no more than 100 species will be found in a local area. Subtropical regions are rich in butterflies, whereas Arctic areas are relatively poor. The length of the butterfly season is shorter as one proceeds northward or to higher elevations. Butterflies may be found throughout the year in southern Florida, but only for 6 to 8 weeks in some high western ranges.

Place: The best places to find butterflies are open, sunny areas with flowers. In low-lying, weedy fields or near river valleys, you will find the widest variety of butterfly species. Power lines, forest edges, mountain meadows, open ridge tops, or clover fields are all excellent places to look. Butterflies are not usually plentiful in shaded forests, although some satyrs and skippers may be found only there. Wetlands are often inhabited by regional colonies. An acid bog in Minnesota might have Bog Coppers, Jutta Arctics, and several northern fritillaries. Many kinds of butterflies congregate at mud puddles —especially swallowtails, sulphurs, blues, and duskywings.

Certain plants and other substances attract butterflies. In a backyard garden with zinnias, marigolds, privet, and butterfly bush, look for Tiger Swallowtails, Silver-spotted Skippers, and American Painted Ladies. Anglewings, Pearly Eyes, and the Red-spotted Purple like sap. Still other butterflies, such as the Viceroy and the Goatweed Butterfly are attracted to scat. Try to find butterflies in all habitats and visit each habitat at different times during the year. Once you have learned about the local species, explore other areas.

If you plan to observe or photograph butterflies in national wildlife refuges or private reserves, you will probably have to make arrangements in advance. To collect specimens in national or state parklands, you must apply for a permit usually several months prior to your visit. Also check with the local U.S. Fish and Wildlife Service or state conservation office for a list of protected species.

Keeping Records:

By taking notes on the butterflies you see, you will soon learn their habits and preferences, and this information may aid in identification. Always record the date, locality, weather conditions, type of habitat, and especially the plants where a butterfly seeks nectar or lays eggs. If you are unfamiliar with the local plants, take along a field guide or bring small samples home for identification.

Watch how each butterfly flies. The flap-and-glide flight of the Red-spotted Purple or other admirals is distinctive, just as the slow, meandering flight of the Cabbage White helps distinguish it from the white females of the Common Sulphur. Note also the time of day. At midday you might find a female Cabbage White laying slender, pale eggs under cabbage leaves; in late afternoon Tiger Swallowtails engage in their tremulous courtship flight; and at about dusk a male Red Admiral alternately perches and flies after other butterflies while seeking a mate.

Raising Butterflies:
By watching butterflies grow from eggs or caterpillars to the adult stage, we can observe their complete life cycles. To obtain eggs of non-threatened species, look for female butterflies laying eggs or check the species' host plants at appropriate times. Place the host plants with eggs attached in a small jar or vial until the eggs hatch. Caterpillars can be reared easily in wide-mouthed jars, plastic boxes, or transparent plastic bags. But never place caterpillars in a tightly closed container without first providing for air circulation. For jars with metal lids, punch holes in the lid that are smaller in diameter than the caterpillars, or cover a wide-mouthed jar with a piece of nylon stocking. Also prevent very dry or wet conditions.

Different types of caterpillars should be kept in separate containers. Crowding too many caterpillars into a single container is asking for trouble; they may devour each other, or disease may lead to mass mortality. Put sticks or other supports inside for attachment of chrysalises. Overwintering chrysalises of some butterflies, such as swallowtails, may be kept outdoors or refrigerated until spring.

Lepidopterists' Organizations:
If you join a Lepidoptera society or entomological group, you can meet other people interested in studying butterflies, and they are likely to know the haunts of local species. Every summer the Xerxes Society holds a Fourth of July butterfly count, and local Audubon Society chapters as well as other conservation groups often sponsor outdoor nature classes.

Paul Opler

HOW TO PHOTOGRAPH BUTTERFLIES

With their incredible colors and intricate patterns, butterflies are among the most beautiful photographic subjects. Naturally photogenic, these small creatures are also amazingly elusive, challenging even the most experienced photographers. While many species have been captured on film, several have never been photographed. Some of the butterflies in this book appear for the first time. Although photographing a living butterfly often requires more patience than catching and pinning it down for a collection, photography can be far more rewarding. Your picture provides a permanent record of both the butterfly and its natural habitat without harming the insect in the slightest. Moreover, a series of photographs can show various aspects of a single butterfly's behavior as well as important stages in its life—egg, caterpillar, and chrysalis.

Camera and Lens:
The centerpiece of your photographic equipment should be a 35mm single-lens reflex (SLR) camera. It should be able to accommodate a lens capable of stopping down to an aperture of at least f/16. The smaller the aperture (and greater the f/stop number), the more detail you can obtain.

Most lenses have an automatic diaphragm: the lens focuses wide open, but can shut down to a prearranged setting when the shutter release is pressed. This feature is absolutely essential when using a flash.

For taking close-up pictures, the most satisfactory lens is probably the micro, also called macro, lens. This lens can focus from infinity to close up, and you need not approach your subject closer than about 10'. Nor does it require changing lenses, adapter rings, and extension tubes.

There are 3 standard sizes from which you can choose: 55mm, 90mm, and 105mm. Consider both the size of the subject you want to photograph and how close you will be able to approach without causing it to fly away. At a distance of 10', a 55mm lens will not produce a large enough image of most butterflies, and if you get closer you may lose your subject. The lens is good, however, for subjects that are stationary or do not move quickly, such as a chrysalis or slow-moving bug, or for photographing very tiny things, such as butterfly eggs.

Although they cannot focus as closely as the 55mm lens, the 90mm and 105mm lenses are both excellent for obtaining a large image. You can keep a greater distance between you and the butterfly and still photograph relatively small subjects. Telephoto/macro lenses and zoom lenses are also available for some cameras. They enable you to take close-up pictures from a distance of 4–5'. However, these lenses are not necessary for photographing most butterflies.

Flash: An inherent problem of close-up photography is the narrow depth of field. As you move the camera closer to the subject or as the amount of available light decreases at higher magnifications, the lens aperture must be opened wider or the exposure time increased. Either step may result in a photograph that is out of focus. Electronic flash virtually eliminates this problem because it guarantees a reliable source of strong light. Also, a narrow lens opening may be used, thereby expanding the depth of field. At 5′ use either a fairly weak flash with a guide number of about 8 and ASA 64, or a slightly stronger flash with a guide number of 20 and ASA 25. Some cameras now have a built-in flash.

Bellows and Tripod: If you wish to photograph tiny objects, such as butterfly eggs and very young caterpillars, you may need to use a bellows attachment, which makes it easier to adjust the distance between the lens and the camera body. The bellows must be used with a tripod.

Film: If you choose not to use a flash, high-speed film is recommended to help eliminate the depth-of-field problem; however, you will have clearer detail with slower film. Select high-speed film, such as ASA 400 or ASA 1000, for color slides, and Tri-X pan for black-and-white prints. With flash photography, use slower, moderate speed film (ASA 25, 64, or 100) for color slides and Panatomic-X (ASA 32) or another fine-grained film for black-and-white prints.

Camera Settings: How you adjust the lens opening and shutter speed to the available light determines whether your photograph will be dark and muddy or clear and beautiful. For close-up flash photography of most butterflies, set the lens at f/16 and the exposure time at 1/60th of a second, or the setting on your camera that indicates flash synchronization. When photographing white or very light-colored butterflies or blossoms, use f/22. For shooting in natural light without a flash, take an exposure reading before each set of pictures. Set the lens opening as small as possible, with the exposure time fast enough to prevent blurring (usually no slower than 1/30th of a second). When photographing without flash in a deep forest with dense shade, try taking pictures with your high-speed film "pushed" to ASA 800 or even ASA 1000. Be sure to instruct the photo lab to develop these rolls accordingly.

Approaching a Butterfly: Probably the best time for photographing butterflies is in the early morning when temperatures are fairly low and butterflies are just warming up. Look for a place where butterflies are common and at rest: when on flowers, when taking moisture from mud and sand, during mating, or when basking in the sun.

When following a butterfly, the 2 guiding principles are low and slow. Try to keep your body even with or not much higher than the butterfly and avoid rapid or jerky movements. Kneel a few feet from the butterfly and slowly inch forward. With your eye to the viewfinder, flash in place, and finger on the shutter release, slowly lean forward until the butterfly is in focus and snap the picture.

Framing: Try to aim so that the butterfly is at the center of the frame. Position the camera lens at a right angle to the butterfly's wings. If a butterfly is resting on a flower with its wings spread, shoot from above; if the wings are folded vertically over the back, approach it from the side. Always aim the flash at the butterfly's head, because it is aesthetically more pleasing to have the head and wings well lit, even if the underside is in shadow.

Paul Opler

PICTURE CREDITS

The numbers in parentheses are plate numbers. Some photographers have pictures under agency names as well as their own. Agency names appear in boldface. Photographers hold copyrights to their works.

Donald B. Adelberg (7, 15 left and right, 38, 41 left and right, 232 left, 515, 526 right)

David H. Ahrenholz (54 right, 63 left, 108, 122, 127, 129, 142 left and right, 159 right, 174 left, 232 right, 269, 274, 277, 283 left, 312 right, 338, 361 right, 362 left and right, 378 left and right, 390, 392, 393, 395, 429, 452, 463, 472 left and right, 473 left and right, 474 left and right, 478 right, 496, 497, 530 left and right, 543, 549, 563 left and right, 579 left and right, 630 left, 650 right, 659 right, 666 left, 704, 721, 733, 736, 739, 744, 749, 752)

John Alcock (484 left)

Ardea London
Elizabeth S. Burgess (4, 368 right)

Karölis Bagdonas (266, 437, 562 right, 587 left and right, 617, 618, 620, 622) left and right, 633 left, 745)

Gregory R. Ballmer (39, 42 left and right, 59, 76, 79, 83, 106, 176 left and right, 194, 295 right, 367 right, 370 right, 405, 424, 425, 426, 428, 445, 447, 465, 471, 487 left, 493 left, 532 left and right, 537, 553, 559 left, 569 right, 581 left, 602 left, 603 left, 605 right, 621, 711, 717)

Ben M. Burns (117)
Robert P. Carr (78, 137 left, 501, 509, 713)
R. D. Coggeshall (253)

Bruce Coleman, Inc.
Jen and Des Bartlett (658 right) Robert P. Carr (150 left) John Markham (357) Gary Meszares (564 right, 646 left) David T. Overcash (85, 228 left, 374 right) Kjell B. Sandved (70 right) Larry West (366 left, 500)

Bruce Coleman, Ltd.
Stephen Dalton (381) B. Davies (354) John Shaw (25 right, 35 right, 682 right) Peter Ward (309)

Charles V. Covell, Jr. (214 right, 361 left)
Robert Dana (133, 149 left and right, 177 left and right, 223 left and right)
Douglas W. Danforth (145, 184, 188, 189, 191, 196, 197, 200, 201, 203 left and right, 246, 249, 289, 291, 296 left, 297 left, 305, 432, 460, 484 right, 487 right, 567, 583, 606 left, 607 right, 662 left, 663 left, 697, 707)
Harry N. Darrow (52 left, 55 left, 55 right, 63 right, 65, 67, 72 left and right, 73 left and right, 74 left and right, 101, 104, 105, 118, 119, 123, 125, 130, 147, 157 right, 158 right, 161 left, 162 right, 168, 171, 172 left and right, 173 right, 202 right, 206, 207, 210 right, 211, 218, 219, 225 left, 233 left and right, 234 left, 235, 236, 241 left and right, 244, 250 left and right, 255, 258, 265, 268, 270, 271, 280, 281, 282, 285 right, 287, 294, 297 left, 298, 299, 300, 307, 310 left, 311 right, 312 right, 314, 315, 324, 327, 341, 344, 363 right, 364 left and right, 365 left and right, 373 right, 375 right, 380, 383, 387, 388, 389, 396, 398, 408, 409, 412, 413, 418, 419, 420, 442, 443, 449, 453, 455, 462, 468 left, 480 left and

right, 481, 505 right, 513 left, 525
left, 527 right, 542, 544, 573, 574
right, 575 left and right, 584, 592
right, 595 left, 601 left and right, 609
left, 614 left and right, 625 right, 626
right, 629 left and right, 631 left, 647
left and right, 648 left and right, 649
right, 653 left and right, 654 left and
right, 655, 663 right, 664 left, 669
left, 677, 678, 679 left and right, 680
left and right, 681 left and right, 682
left, 683 left and right, 684 left, 686,
687, 689, 692, 705, 714, 716, 725,
758, 759)
Thomas W. Davies (14 left and right,
30, 75 right, 102, 152, 185, 208
right, 212, 251 left, 330, 345, 379,
394, 450, 468 right, 475, 489 left,
557, 565, 571, 576 right, 607 left,
610 right, 611 right, 649 left, 651 left
and right, 659 left, 668 left, 708,
719, 753)
E. R. Degginger (3, 5, 6, 9, 12, 16
left and right, 17 left and right, 19 left
and right, 21 right, 23 right, 24 right,
26 right, 27 left, 34 left, 48 left and
right, 49 left and right, 50 left and
right, 51 left, 272, 316, 328, 335,
340, 402, 609 right, 644 right, 665
left and right, 685, 709)

Design Photographers International
John H. Gerard (175 left) Richard
Parker (20 right, 97)

John DiLiberti (349, 352, 355, 367
left, 370 left, 538, 540)
Robert Duncan (204 left)
Peter Luis Eades (96)
David L. Eiler (242 left)
Jane Evans (593 left)
Chuck Farber (93, 112, 141, 159 left,
160 left and right, 182 left and right,
186, 216 left, 227 left and right, 369
right, 441, 456, 459, 522, 748)
William E. Ferguson (18 left and
right, 36 left, 40 right, 726)
Clifford D. Ferris (612 left, 624 right)
Richard S. Funk (371 right, 632 left

Robert J. Long (747)

Marcon Photo/Marion Latch (628 left)

Peter May (56 right, 139)
Sturgis McKeever (318)
James W. Mertins (20 left)
C. Allan Morgan (1)
James R. Mori (469, 738)
William P. Mull (485 left and right, 671 left and right)

National Audubon Society Collection/Photo Researchers, Inc.
N. E. Beck, Jr. (375 left) Danny Brass (308) Ken Brate (52 right, 53 left, 84, 107, 124, 136 left, 155 right, 202 left, 204 right, 210 left, 239, 263, 276, 284 left, 285 left, 457, 488 right, 600) Michael P. Gadomski (628 right) Gilbert Grant (755) Sturgis McKeever (23 left) Richard Parker (115) Louis Quitt (627 right) William Ray (53 right) Kjell B. Sandved (98, 99, 116, 323, 656) John Serrao (596 right) Virginia P. Weinland (114)

Natural History Photographic Agency
Stephen Dalton (594 right)

John and Vikki Neyhart (70 left, 377 left and right, 446, 482, 483, 494 left, 524 right, 568 right, 570 left and right, 581 right, 590, 611 left, 612 right, 639 right, 731)
Peter G. Nice (144 left, 154 left, 164 left, 183 left and right, 225 right, 247, 252 left, 287, 399, 433, 435, 448, 458, 467 left, 529 left, 559 right, 560 left and right, 561 right, 569 left, 572, 586 right, 605 left, 746)
Philip Nordin (61 left, 140, 209 left and right, 248, 275, 347, 467 right, 499, 673, 674)
Paul A. Opler (88, 89, 90, 157 left, 198, 224 left and right, 385, 416, 438, 479 left, 639 left)

Larry J. Orsak (44 left, 427, 470, 486 right, 492, 520, 534 right, 604 left)
Keith R. Palmer (730)
Kenelm W. Philip (94, 368 left, 634, 635, 636, 637 left and right, 638 left and right, 718, 722, 729, 732, 751)
Robert M. Pyle (757)
Betty Randall (296 right, 652 left and right)
R. D. Richard (66)
Stephen Roman (68, 146, 167, 175 right, 178, 179, 215 right, 217, 222, 421, 461)
Edward S. Ross (22 left, 25 left, 31, 32, 40 left, 44 right, 46 left, 71 right, 86, 100, 126, 128, 134, 144 right, 163 right, 164 right, 195, 254, 267, 278, 342, 360, 516, 524 left, 531 right, 550, 554, 556, 558, 566, 577 left, 586 left, 597 left, 603 right, 606 right, 615 left, 616, 627 left, 645 left, 668 right, 675, 723, 724, 750)
Thomas Ruckstuhl (28, 46 right, 143 left, 156 left, 479 right, 507 left and right, 613 left and right, 619)
Clark Schaack (662 right)
Rosemary Scott (720)
John Shaw (138 left and right, 143 right, 161 right, 169, 190, 228 right, 259, 264, 293, 317, 320, 351, 369 left, 376 left and right, 404, 410, 439, 440, 454, 488 left, 505 left, 513 right, 514, 526 left, 527 left, 555, 588 left, 594 left, 596 left, 608 left and right, 624 left, 625 left, 630 right, 658 left, 669 right, 670 left and right, 690, 693, 701, 703, 712)
Richard Singer (321, 366 right, 382, 477, 518 left, 528 right, 631 right, 650 left, 667 left, 672 left)
Paul A. Spade (329, 332, 353, 356, 358, 531 left)
Keith M. Spencer (640)

Tom Stack and Associates
Harry Ellis (111)

Ray E. Stanford (208 left)
Gayle T. Strickland (150 right, 151,

170, 187, 205, 213, 215 left, 220,
229, 237, 240, 251 right, 273, 284
right, 286, 506 left and right, 552,
695, 696)
Edmund R. Taylor (57 right, 199,
397, 657, 676)
George Taylor (348, 597 right)
Paul M. Tuskes (21 left, 29, 33, 34
right, 36 right, 595 right, 646 right,
684 right)
John L. Tveten (322, 336)
M. W. F. Tweedie (301)

Valan Photos
Albert Kuhnigk (672 right)

Harold F. Webb (310 right)
Larry West (27 right, 60, 69, 71 left,
77, 80, 81, 95, 103, 120, 121, 136
right, 153, 154 right, 155 left, 156
right, 158 left, 162 left, 166, 173 left,
174 right, 192, 214 left, 234 right,
238, 243 left, 252 right, 262, 331,
363 left, 371 left, 374 left, 384, 400,
401, 403, 406, 430, 491, 508, 523
left, 534 left, 541, 551, 562 left, 588
right, 589, 592 left, 593 right, 610
left, 626 left, 644 left, 664 right, 688,
698, 702, 728, 735, 742)
Douglas Whitman (43 left and right,
165 left and right, 502, 535)
John Wilkie (56 left, 226 left and
right, 283 right, 373 left, 386, 503,
504, 523 right, 574 left, 613 left and
right, 623 left and right, 633 right,
641, 700, 743)
E. N. Woodbury (13 left and right, 45
left, 51 right, 57 left, 319, 339, 343,
518 right, 599, 661 left and right,
666 right, 706)
David M. Wright (490, 525 right)
Maria Zorn (2, 10, 11, 26 left, 35 left)

HOST PLANT INDEX
This index contains the common and scientific names of the host plants eaten by caterpillars. Also included are two genera of aphids that are consumed by the Harvester's caterpillars.

A

Abies, 351

Abutilon, 743, 765, 768
incanum, 482

Acacia, 487, 491, 738

Acacia, 487
angustissima, 738
koa, 487

Acamptopappus sphaerocephalus, 586

Acanthaceae, 600

Acanthus, 592, 600

Acanthus family, 570, 571

Actinomeris alternifolia, 582

Aeschynomene viscidula, 396

Agave, 848
desert, 850
Parry's, 847, 849
Utah, 850

Agave
chrysantha, 848
deserti, 850
lecheguilla, 849, 851
palmeri, 848
parryi, 847, 849
schottii, 848
utahensis, 850

Agrostis, 784, 791, 837

Albizzia, 491

Alder, 342, 345, 613

Alfalfa, 372, 486, 506

Alnus, 342, 345, 613

Alternanthera, 741

Althaea, 627, 765, 766, 768

Amaranth, 741

Amaranthus, 769
retroflexus, 740

Ambrosia trifida, 581, 591

Amelanchier alnifolia, 483

Amole, 848

Amorpha, 486
californica, 384, 385
fruticosa, 384

Amphicarpa, 393, 730

Amyris elemifera, 339

Anacardium occidentale, 650

Anaphalis, 624

Andropogon, 777, 796, 811
gerardi, 810
scoparius, 774, 794, 829

Androsace, 517

Angelica, 330
seaside, 335

Angelica, 330
lucida, 335

Anisacanthus wrightii, 592

Annonaceae, 349

Annona globiflora, 423

Antelope brush, 448

Antennaria, 624

Aphid, woolly, *401*

Apiaceae, 328, 329, 335, 336

Apocynum, 712

Apple, 425, 636, 637

Aquilegia
canadensis, 757
vulgaris, 757

Arabis, 355, 358, 363, 365,
369, 370

Arbutus, 425

Arceuthobium, 434
campylopodum, 435

Arctostaphylos uva-ursi, 425,
427, 566

Aristida, 796

Aristolochia, 326
californica, 325
longiflora, 325
macrophylla, 325
serpentaria, 325

Artemisia
arctica, 333
dracunculus, 331, 332

Artocarpus integrifolia, 650

Aruncus dioicus, 495

Arundinaria
gigantea, 663
tecta, 663, 665

Asclepiadaceae, 713

Asclepias, 712, 714
amplexicaulis, 714
curassavica, 715

Ash, 341, 343, 454
prickly, 347, 644
white, 603

Asimina, 348
triloba, 348

Aspen, 342, 636, 637, 638

Aster, 575, 577, 582, 583,
584, 585
blue wood, 574
golden, 589
mojave, 586
showy, 589

Aster, 573, 575, 577, 582
conspicuus, 584, 589
occidentalis, 584
umbellatus, 583
undulatus, 574

Asteraceae, 600, 625

Astragalus, 373, 489, 493,
503, 513
alpinus, 377
calycosus, 511
spatulatus, 511

Atriplex, 483
canescens, 509, 770, 771
lentiformis, 771

Avena striata, 837

Avens
arctic, 561, 570
mountain, 565, 567

Azalea, 425, 613
flame, 455
western, 614

B

Baccharis glutinosa, 520

Bacopa monniera, 633

Balloon vine, 465

Balm, mountain, 345

Baptisia, 731
laevicollis, 758
tinctoria, 428, 758

Barbarea, 358, 369

Batis, 484
maritima, 361

Bay, 346
red, 347
sweet, 347

Bean, 489, 492, 720, 722, 727, 730, 737
lima, 485

Bear grass, 432

Bearberry, 425, 427, 566

Beardgrass
bluestem, 794, 811
prairie, 774
woolly, 811

Beardtongue, 602, 605, 606

Bebbia juncea, 522

Bedstraw family, 641

Beech, 466

Bee plant, 356, 357

Beet, wild, 740

Beggar's tick, 720

Beggar-weed, 727

*Beloperone
californica*, 599
guttata, 571

Bernardia myricaefolia, 476

Betula, 613, 619, 635, 748

Betulaceae, 341

Bidens pilosa, 398

Bilberry
arctic, 379

dwarf, 379

Birch, 341, 613, 619, 635, 748

Bird's beak, 593

Bistort, 569
alpine, 558, 559

Blackberry, 760

Bladder-pod, 354

Blechum brownei, 634

Bleeding heart, 323

Bloodberry, 447

Blueberry, 382, 424, 429, 456, 494, 517, 570
alpine, 566
dwarf, 566
velvet-leaf, 381

Bluegrass, 778, 806
Kentucky, 834, 837
pine, 789

Bluestem, 811

Boehmeria, 628
cylindrica, 610

Borage family, 606

Boraginaceae, 606

Bouteloua, 796
gracilis, 785, 787, 792, 795, 831

Brachyelytrum erectum, 664

Brassica nigra, 354

Brassicaceae, 357, 360, 368, 726

Breadfruit, 650

Bromus, 789

Buckbrush, 532, 754, 755

Buckthorn, 407, 620, 755

Buckthorn family, 344

Buckwheat, *442, 474, 499, 529*
alpine hoary, *442*
California, *441*
kidney-leaved, *501*
wild, *406, 410, 443, 445, 446, 497, 498, 499, 500, 513, 514, 515*

Buckwheat family, *415*

Bunchgrass, *792*
chaparral, *793*
desert, *792, 805, 810*

Butterfly-pea, *727*

C

Cabbage, wild, *366*

Caesalpina mexicana, 528

Calamagrostis purpurascens, 771

Calliandra, 391

Cane
giant, *663*
maiden, *663, 665*
sugar, *776, 778, 784, 804, 843, 844, 845*

Canna, *726, 841*

Canna, 726
flaccida, 841
indica, 841

Caper, *353, 356, 361*

Capparaceae, *357*

Capparis, 353

Capsella bursa-pastoris, 370

Cardamine, 369

Cardiospermum
corindum, 465
halicacabum, 465

Carduus pycnocephalus, 580

Carex, 666, 667, 674, 825, 827
heliophila, 828

lacustris, 821, 824, 825
stricta, 817

Carica, 715

Carpinus, 636

Carrot family, *328, 329, 335, 336*

Carya, 453, 454
ovata, 453

Cashew, *650*

Cassia, 387, 388, 390, 393, 395, 397, 719, 747

Castanea dentata, 749

Castilleja, 593, 606, 607
integra, 595
lanata, 595, 596

Cat's claw, *489*

Caulanthus, 355, 366

Ceanothus, *451, 494*

Ceanothus, 345, 425, 451, 494, 620, 755
americanus, 754
cuneatus, 459
fendleri, 532, 754, 755
macrocarpus, 459
sanguineus, 459
velutinus, 459

Cedar
Atlantic white-, *440*
eastern red-, *439*
incense, *436*
southern red-, *439*
western red-, *436*

Celtis, 535, 609, 621, 654, 655, 657, 659
laerigata, 658
pallida, 525, 656, 660

Cenizo, *596*

Cercis canadensis, 429

Cercocarpus betuloides, 458, 459

Chamaecrista, 490

cinerea, 387
fasciculata, 393

Chamaecyparis thyoides, 440

Chaste-tree, 729

Cheeseweed, 626, 769

Chelone glabra, 603

Chenopodium, 484, 741
album, 740, 769
ambrosioides, 740

Cherry, 341, 343, 636, 637
Barbados, 747
holly-leaf, *345*
wild, *483, 533*

Chestnut, American, 749

Chickweed, common, *398*

Chinkapin, *418*
giant, 642

Chokecherry, 640

Chrysolepis chrysophylla, 418,
642

Chrysopsis breweri, 589

Chrysothamnus
nauseosus, 584
paniculatus, 584
viscidiflorus, 584, 585

Chufa, 828

Chuparosa, 599

Cimicifuga racemosa, 494

Cinquefoil, 409, 415, 416,
761
shrubby, *416*

Cirsium, 578, 579, 580, 625
horridulum, 519
muticum, 523

Citrus, 335, 338, 340, 744

Citrus family, *328*

Cleome, 356, 357
spinosa, 361

Clitoria, 727

Clover, 371, 376, 384, 385,
387, 393, 395, 492, 508,
511, 735
bush, 731
owl's, 607
Parry's, 376
slender bush, 492
whiproot, 376
white, 372
white sweet, 375

Cocos nucifera, 828

Coffeeberry, 345

Columbine
garden, 757
wild, 757

Composite, *398, 600, 624,*
625

Composite family, *524, 591*

Coneflower, *582*
purple, 790

Conifer, 352, 436

Cordylanthus pilosus, 593

Corn, 473
Indian, 777

Cornus, 494

Coronilla varia, 758

Corydalis, 322
gigantea, 322

Corylus, 466

Cotton, 473

Cottonwood, 341, 621, 638

Cowania mexicana
var. *stansburiana, 425*

Crabgrass, 797, 836

Cranberry, wild, 414

Crassulaceae, 631

Crataegus, 456, 461, 635, 636

Crazyweed, 506

Cress, *358, 369, 370*
rock, *355, 363, 365, 370*

Crotalaria, 485

Croton, *420, 468, 475*
myrtle, *476*
wild, *480*
woolly, *651*

Croton, 468, 475
capitatum, 652
linearis, 480, 651, 652
monanthogynus, 652
niveus, 420

Crowberry, *505*
black, *566*

Crownbeard, *582, 583*

Crucifer, *354, 355, 356, 357,
358, 360, 361, 364, 368,
726*

Cupressaceae, *436*

Cupressus
forbesii, 438
sargenti, 436

Currant, *613, 615, 617, 618*
squaw, *615*
wild, *402*

Cuscata, 425

Cycad, *419, 420*

*Cynodon dactylon, 670, 676,
728, 781, 782, 784, 799,
837*

Cyperaceae, *669, 675, 678,
689, 692, 694, 696, 705,
708, 817, 842*

Cyperus esculentus, 828

Cypress
Sargent, *436*
tecate, *438*

D

Daisy
cowpen, *591*

showy, *584*
wild, *588*

Dalbergia, 391

Dalea, 385

Dalechampia, *644, 647, 648*

Dalechampia, 644, 647, 648

Danthonia, 789

Daucus carota, 328, 329

Deer weed, *441, 445, 503,
504, 513*

Dentaria, 358
diphylla, 359

Descurainia, 367
pinnata, 366
richardsonii, 363

Desert candle, *355, 366*

Desmodium, 492, 720, 731
tortuosum, 727

Desmonicus, 778

Diapensia, *517*

Diapensia lapponica, 517

Dicentra formosa, 323

Dicliptera, 571
brachiata, 570

Digitaria, 797
sanguinalis, 836

Diospyros texana, 429

Distichlis
spicata, 842, 843
stricta, 786

Dock, *404, 405, 408, 409,
411, 415*
curly, *403, 411*

Dodder, *425*

Dodecatheon, 517

Dogbane, *712*

Dogwood, *494*
Jamaica, *721*

Douglas-fir, *351*

Draba, *364*

Draba, 364

Dropseed, *664*

Dryas
integrifolia, 561, 565, 570
octopetala, 565, 567

Drypetes lateriflora, 353

Dudleya, *496*

Dutchman's pipe, *325*

E

Ebony, Texas, *733*

Echeveria, 433

Echinacea, 790

Echinochloa, 777
pungens, 836

Elm, *609, 610, 615, 621*

Empetrum, 505
nigrum, 566

Encelia californica, 520

Eragrostis, 791, 818

Erianthus diverticatus, 811

Erigeron
leiomeris, 588
speciosus, 584

Eriogonum, 442, 443, 446,
474, 497, 498, 513, 514,
529
effusum, 500
elongatum, 410
fasciculatum, 406, 441, 514
flavum, 500
incanum, 442
kearneyi, 499
latifolium, 406, 445
leptocladon, 500

microthecum, 406, 499
nudum, 410, 445
nudum complex, *406*
ovalifolium, 514
plumatella, 499
pusillum, 499
racemosum, 500
reniforme, 501
umbellatum, 406, 444, 446,
514
umbellatum subaridum, 442
wrightii, 500, 515

Eriophorum spissum, 707

Eupatorium
betonicifolium, 521
greggii, 524
havanense, 524
odoratum, 521
serotinum, 521

Everlasting, *624*

F

Fabaceae, *387, 503, 722,*
724, 726, 733, 734, 756

Fagus, 466

Fennel, *335*

Fescue, *704, 706, 789, 797,*
799
Idaho, *702, 802, 807*
red, *789*

Festuca, 704, 706, 789, 797
idahoensis, 702, 802, 807
ovina, 799

Ficus, 630, 650, 715

Fig, *630, 650, 715*

Figwort family, *595, 604,*
606, 607, 631

Fir
Douglas-, *351*
true, *351*

Flax, *543*

Foeniculum vulgare, 335

Fog fruit, *572*

Foxglove, false, *603*

Foxtail, yellow, *836*

Fragaria, *473*

Fraxinus, *341, 344, 454*
americana, *603*

G

Galactia, *485*

Gaultheria, *425*
hemifusa, *382*

Gaylussacia, *429*

Gerardia
grandiflora, *603*
pedicularia, *603*

Glasswort, *484*

Gleditsia, *720*

Globemallow, *627*

Glycyrrhiza, *506, 720*

Gnaphalium, *624*

Goat's beard, *495*

Goatweed, *652*

Golden banner, *373, 759*

Goldeneye, *581, 590*

Goldenhead, *586*

Goldenrod, *584, 588*

Golden weed, *587*

Gooseberry, *402*
straggly, *617*

Goosefoot, *740, 741*

Goosefoot family, *483*

Gossypium, *473*

Grass, *664, 668, 669, 670,*
671, 672, 673, 675, 678,
679, 680, 681, 682, 684,
685, 686, 687, 688, 689,
690, 691, 692, 693, 694,
695, 696, 698, 699, 700,
701, 702, 703, 704, 705,
706, 708, 728, 780, 784,
788, 789, 790, 792, 796,
799, 800, 801, 804, 805,
806, 808, 814, 815, 817,
818, 822, 837, 839, 840,
842, 843, 844, 845
alpine, *691, 709*
arctic, *709*
bear, *432*
beard, *796, 810, 829*
bent, *784, 791, 837*
Bermuda, *670, 676, 728,*
781, 782, 784, 799, 837
blue-, *778, 806*
blue grama, *785, 787, 792,*
795, 831
bluestem beard-, *794, 811*
bunch-, *792*
chaparral bunch-, *793*
crab-, *797, 836*
desert bunch-, *792, 805, 810*
desert salt-, *786*
eastern grama, *812*
fall witch-, *790*
grama, *796*
Indian, *673, 834*
Kentucky blue-, *834, 837*
long-awned wood, *664*
marsh, *820*
needle, *787, 796*
oat, *789*
panic, *776, 778, 791, 805,*
809, 810
pepper, *361*
pine blue-, *789*
prairie, *797*
prairie beard-, *774*
pungent barn, *836*
purple reed-, *771*
St. Augustine, *676, 775,*
777, 784, 808, 836
salt, *842, 843*
saw, *823*
switch, *811*
tumble, *791, 818*
water, *777*

western needle-, *799*
wool, *824*
woolly beard-, *777, 811*
yellow nut-, *828*

Gray nicker, *489*

Grouseberry, *516*

Guamuchil, *526*

Guava, *718*
common, *718*

Guilandina, 489

H

Hackberry, *525, 535, 609, 621, 654, 655, 657, 658, 659, 660*
Spiny, *656*

Haplopappus squarrosus, 587

Hawthorn, *456, 461, 635, 636*

Hazelnut, *466*

Helenium autumnale, 398

Helianthus, 581, 582
annuus, 590, 591

Heracleum, 328, 330
lanatum, 328, 335

Heterotheca grandiflora, 587

Hibiscus, *476*

Hibiscus tubiflorus, 476

Hickory, *453, 454*
shagbark, *453*

Holly, *456*

Hollyhock, *627, 765, 766*

Honeysuckle, *604, 608*

Hops, *609, 610, 628*

Hoptree, *338*
common, *344*

Horkelia, *409, 760*

Horkelia, 761

fusca, 409
tenuiloba, 409

Hornbeam, *636*

Horse sugar, *455*

Hosackia purshiana, 509

Huckleberry, *429*

Humulus, 609, 628
lapulus, 610

Hyptis pectinata, 480

Hyssop, water, *633*

I

Ichthyomethia, 721

Ilex, 456

Indian paintbrush, *593, 595, 596, 606, 607*

Indigo, *491, 747*
false, *384, 385, 428, 486*
wild, *731, 758*

Indigofera, 491, 747

Inga edulis, 737

Isomeris arborea, 354

Iva xanthifolia, 581

J

Jacobinia carnea, 570

Jewel flower, *355*
Arizona, *366*
mountain, *366*

Juglans, 453

Juniper, *437, 439*
California, *438*
western, *436*

Juniperus
californica, 437, 438
occidentalis, 436
osteosperma, 437
scopulorum, 437
silicicola, 439

K

Kalmia, 505

Knotweed, *411, 412, 413, 415, 513*

Koa, Hawaiian, *487*

Koeleria cristata, 793

L

Lamb's quarters, *769*

Lamiacea, 473

Lantana, *471, 480*

Lantana, 471
camara, 480

Lathyrus, 374, 492, 493, 503, 735
odoratus, 486

Lauraceae, *344*

Laurel, *344, 505*

Lead plant, *384*

Leadwort, *485, 486*

Lecheguilla, *849, 851*

Ledum palustre, 505

Legume, *371, 372, 373, 375, 377, 378, 386, 387, 393, 395, 396, 473, 485, 489, 491, 492, 509, 513, 528, 720, 722, 726, 727, 733, 734, 747, 755, 756, 759*

Legume family, *389, 503, 526, 724*

Lenscale, *771*

Lepidium virginicum, 361

Leptoloma cognatum, 790

Lespedeza, 492, 731

Leucophyllum
frutescens, 596
minus, 597

Libocedrus decurrens, 436

Licorice, *720*
wild, *506*

Lignumvitae, Texas, *391*

Ligusticum scothicum, 330

Lilac
California, *425*
mountain, *345*
wild, *459*

Lily, canna, *726, 841*

Lime, wild, *339*

Lindera benzoin, 346

Linum, 543

Lippia, *480*

Lippia, 632
alba, 480
graviolens, 480
lanceolata, 572
nodiflora, 572

Liriodendron tulipifera, 341

Lithocarpus densiflorus, 418

Live-forever, *496*

Locoweed, *374, 489, 493, 503, 511, 513*

Locust, *720, 748, 755*

Lolium, 789

Lonicera, 604
involucrata, 608

Lotus, *474, 759*

Lotus, 513, 759
argophyllus, 474
scoparius, 441, 445, 474, 503, 504

Lousewort, *607*

Lupine, *374, 378, 428, 449, 502, 503, 504, 505, 506, 510, 511, 513, 759*

Lupinus, 374, 428, 449, 502, 503, 505, 506, 510, 513, 759
albifrons, 502
arboreus, 504
arcticus, 378
excubitus, 502
lyalli, 511

M

Machaeranthera, 575
canescens, 585
tortifolia, 586
viscosa, 585

Madrone, 425

Magnolia virginiana, 347

Mahogany, mountain, 458, 459

Malachra fasciata, 743

Mallow, 473, 477, 478, 625, 626, 743, 763, 764, 765, 766, 768
alkali, 477, 762
globe, 762, 766
Indian, 482

Mallow family, 630

Malpighia glabra, 747

Malpighiaceae, 746

Malus, 425, 636, 637

Malva, 765
parviflora, 627
rotundifolia, 769

Malvaceae, 625, 626, 630, 763, 764, 766, 768

Malvastrum, 768
coromandelianum, 478

Mamake, 487, 629

Manfreda, 854

Manfreda maculosa, 854

Mangrove
black, 391, 632

red, 718

Marigold
bur, 398
garden, 398

Mariscus jamaicensis, 823

Maytenus phyllanthoides, 472

Meadowsweet, 494

Medicago sativa, 372, 486, 506

Melilotis alba, 375

Menispermum, 543

Mesquite, 470, 489, 491
honey, 530

Milkvetch, 373
alpine, 377

Milkweed, 712, 713, 714
blunt-leaved, 714
rambling, 714
red-flowered, 715

Millet, 778
marsh, 778, 821

Mimosa, 738

Mimosa, 722
padica, 394
pigra, 738

Mimosa family, 386, 737

Mint, 473

Mint family, 480

Mistletoe, 422
dwarf, 434
western dwarf, 435

Moonseed, 543

Mountain mahogany, 458, 459

Muelleria, 722

Muelleria moniliformis, 722

Muhlenbergia, 664

Mule fat, *520*

Mustard, *363, 365*
black, *354*
hedge, *365, 369*
mountain tansy, *363*
tansy, *367*
western tansy, *366*

Myrica, 468

N

Nasturtium, *360*

Needlegrass, western, *799*

Nettle, *609, 610, 623, 627*
false, *628*
stinging, *611*

Nettle family, *629, 648*

Nolina microcarpa, 432

Nutgrass, yellow, *828*

O

Oak, *423, 451, 452, 453,
456, 460, 473, 717, 724,
749, 753*
Arizona, *717, 724*
black, *750*
California scrub, *457*
canyon, *418*
canyon live, *642*
coast live, *642, 751*
Emory, *724*
Gambel, *417, 751, 752*
huckleberry, *418*
interior live, *457*
live, *423, 461*
Muehlenberg's, *752*
northern red, *750*
northwest Oregon white, *751*
scrub, *452*
tanbark, *418*
Texas, *752*
white, *750*

Oats, striped, *837*

Ocimum basilicum, 480

Odontonema callistachus, 592

Olana, *629*

Opahe, *629*

Orthocarpus, 607

Oryza sativa, 778, 843, 845

Oxyria digyna, 403, 404

*Oxytheca
perfoliata, 501
trilobata, 501*

Oxytropis, 374, 506

P

Palm, *778*
coconut, *828*
date, *828*
saw cabbage, *828*

Palmetto
cabbage, *828*
saw, *823*

Palo verde, Mexican, *446*

*Panicum, 778, 791, 797, 805,
809, 810, 821*
maximum, 776, 778
virgatum, 811

Pansy, *543*

Papaya, *715*

Parietaria, 628

Parkinsonia aculeata, 446

Parsley family, *335, 336*

Parsnip
cow, *328, 330, 335*
meadow, *328*

Paspalum, *777*
hairy, *808*

*Paspalum, 777
ciliatifolium, 808*

Passion flower, *538, 540,
541, 542, 543, 544*

Passiflora, 540, 541, 542, 543
foetida, 544
incarnata, 538

Paurotis, 828

Pawpaw, 348, 349, 423

Pea, 493, 735
milk, 485
partridge, 387, 393, 490
sweet, 486
wild, 374, 492, 503

Pea family, 385, 486

Peanut, hog, 393, 730

Pedicularis, 607

Pellitory, 628

Pemphigus, 401

Pencil flower, 396

Penstemon, 601

Penstemon, 601, 602, 606
antirrhinoides, 605
subserratus, 605

Pentaclethra macroloba, 389

Pepper, 337
Brazilian, 463

Persea, 346
borbonia, 347

Persimmon, Texas, 429

Phaseolus, 490, 492, 720,
722, 727, 730, 737
limensis, 485

Phleum arvense, 783

Phoenix, 828

Phoradendron, 422

Phragmites communis, 816

Picea mariana, 430

Pickleweed, 483

Pigweed, 484, 769

Pine, 351, 431
hard, 430
jack, 430
lodgepole, 431
pitch, 430
ponderosa, 352, 431
scrub, 430

Pink family, 398

Pinus, 351
banksiana, 430
contorta, 431
ponderosa, 352, 431
rigida, 430
virginiana, 430

Piperaceae, 337

Pipevine, 325, 326

Pipturus albidus, 487, 629

Pithecellobium, 386, 489
dulce, 526
flexicaule, 733
guadelupense, 390

Plantago, 543, 604, 607
lanceolata, 603
major, 605

Plantaginaceae, 606, 631

Plantain, 543, 603, 604, 607
common, 605

Plantain family, 606, 631

Platanaceae, 342

Plum, 456, 483, 637
Guiana, 353
wild, 429

Plumbago, 485, 486

Poa, 778, 789, 806
pratensis, 834, 837

Poaceae, 666, 668, 669, 670,
672, 673, 675, 678, 679,
680, 681, 682, 684, 685,
686, 687, 688, 689, 690,
691, 692, 693, 694, 695,
696, 698, 699, 700, 701,

702, 703, 704, 705, 706,
708, 709, 728, 780, 784,
788, 790, 792, 796, 799,
800, 801, 804, 806, 808,
814, 815, 817, 818, 820,
822, 839, 840, 842, 843,
844, 845

Polygonaceae, 415

Polygonum, 411, 415
aviculare, 513
bistortoides, 569
douglasii, 412, 413
viviparum, 558, 559

Pongam, 721

Pongamia, 721

Poplar, 342, 619, 635, 636,
637, 640, 748, 759
tulip-, 341

Populus, 619, 621, 635, 636,
637, 638, 640, 748, 759

Porliere angustifolia, 391

Portulacaceae, 630

Potato, 424

Potentilla, 409, 415, 416, 761
fruticosa, 416

Powder puff, 390

Prickly-ash
common, 338
lime, 744

Prince's plume, golden, 354

Prosopis, 489, 491
juliflora, 470, 530

Prunus, 341, 343, 429, 456,
483, 636, 637
harvardii, 533
ilicifolia, 345
virginiana, 640

Pseudabutilon, 743

Pseudotsuga, 351

Psidium, 718
guajava, 718

Ptelea trifoliata, 338, 344

Punctured bract, 501

Purshia
glandulosa, 448
tridentata, 448

Purslane family, 630

Q

Queen Anne's lace, 328, 329

Quercus, 423, 451, 453, 456,
460, 473, 717, 724, 749,
753
agrifolia, 642, 751
alba, 750
arizonica, 717, 724
chrysolepis, 418, 642
dumosa, 457
emoryi, 724
gambelii, 417, 751, 752
garryana, 751
ilicifolia, 452, 461
laurifolia, 461
muehlenbergii, 752
rubra, 750
texana, 752
vaccinifolia, 418
velutina, 750
virginiana, 423, 461
wislizenii, 457

R

Rabbit brush, 584, 585

Ragweed, 581
giant, 591

Ragwort, 520

Ram's horn, 390

Raspberry, 561

Rattlebox, 485

Redbud, 429

Redcedar
eastern, *439*
southern, *439*
western, *436*

Red root, *754*

Reed, common, *816*

Reedgrass, purple, *771*

Rhamnus
californicus, 345
crocea, 425, 407

Rhizophora mangle, 718

Rhododendron, *613, 615,*
617

Rhododendron, 613, 615, 617,
425
calendulaceum, 455
occidentale, 614

Rhubarb, wild, *405*

Rhus copallina, 468

Ribes, 402, 613, 615, 617,
618
cereum, 615
divaricatum, 617

Rice, *778, 843, 845*
wild, *821*

Rivinia humilis, 447

Robinia, 720, 748, 755

Rock-jasmine, *517*

Rondia, 641

Rosaceae, *456*

Rose, cliff, *425*

Rose family, *456*

Rubiaceae, *641*

Rubus, 561
chamaemorus, 760

Rudbeckia laciniata, 582

Rue, *328, 338*

Ruellia, *633*

Ruellia, 570, 571
occidentalis, 633

Rumex, 409, 415
acetosella, 403
conglomeratus, 408
crispus, 403, 408, 411
hymenosepalus, 405, 408
obtusifolia, 408
paucifolius, 404
triangularis, 405

Rutaceae, *328, 335, 338,*
340, 744

Ruta graveolens, 328, 338

S

Sabal palmetto, 828

Sacchinarum officinarum, 776,
778, 784, 804, 843, 844,
845

Sagebrush, arctic, *333*

Salal, *425*

Salicaceae, *341, 342*

Salicornia, 484
ambigua, 483

Salix, 383, 450, 451, 456,
559, 561, 562, 569, 613,
619, 621, 635, 636, 637,
638, 640, 748, 759
nivalis, 563
reticulata, 382

Saltbush, *483, 770, 771*
hoary, *509*

Saltgrass, desert, *786*

Salt-loving shrub, *472*

Saltwort, *361, 484*

Sapindus saponaria, 462

Sarcostemma hirtellum, 714

Sassafras, *346, 347*

Sassafras albidum, 346, 347

Saxifraga bronchialis, 568

Saxifrage, spotted, 568

Schinus terebinthifolius, 463

Schizoneura, 401

Schrankia, 727

Schrankia, 727

Scirpus cyperinus, 824

Scotch lovage, 330

Scrophulariaceae, 595, 604, 606, 607, 631

Sedella, 426

Sedge, 666, 667, 669, 674, 675, 678, 689, 692, 694, 696, 705, 707, 708, 817, 825, 827, 828, 842
lake, 821, 824, 825
tussock, 817

Sedum, 324, 426
allantoides, 433
lanceolatum, 324, 426, 543
obtusatum, 324
purpureum, 496
spathulifolium, 426
texana, 433

Senecio obovatus, 520

Senna, 386, 387, 388, 390, 393, 395, 397, 719, 747

Senna spectabilis, 386

Sensitive plant, 394, 722

Serenoa repens, 823

Serviceberry, western, 483

Setaria glauca, 836

Seymeria tenuisecta, 595

Shepherd's purse, 370

Shooting star, 517

Shrimp plant, 571

Sida, 768

Sida, 477, 478, 765, 768
filipes, 768
hederacea, 477, 762

Sidalcea, 627

Sidalcea, 627

Silver-leaf, Big Bend, 597

Silybum marianum, 580

Siphonoglossa pilosella, 577, 599

Sisymbrium, 369
officinale, 365

Snakeroot, 521
black, 494
Virginia, 325

Sneezeweed, 398

Snowberry, 604, 605, 608

Soapberry, 462

Solanum umbellatum, 424

Solidago
californica, 584
multiradiata, 588

Sorghastrum nutans, 673, 834
secundum, 834

Sorrel, 415
mountain, 403, 404
sheep, 403

Sphaeralcea, 627, 743, 766, 768
coccinea, 762

Spicebush, 346

Spider flower, 361

Spider plant, 357

Spiraea, 494

Spruce, black, 430

Stanleya pinnata, 354

Stellaria media, 398

Stemodia, 632

Stenardrium barbatum, 592

Stenotaphrum secundatum, 676, 775, 777, 784, 808, 836

Stipa, 789
nevadensis, 787
occidentalis, 799

Stonecrop, 324, 426, 496, 543

Stonecrop family, 631

Strawberry, 473

Streptanthella longirostris, 367

Streptanthus, 355, 366

Stylocanthes biflora, 396

Sulphur flower, 442, 444

Sumac, dwarf, 468

Sump-weed, 581

Sunflower, 581, 582, 590, 591
bush, 520

Sweetbush, 522

Sycamore, 342

Symphoricarpos
albus, 604, 605, 608
vaccinoides, 605

Symplocos tinctoria, 455

T

Tagetes, 398

Tea, Hudson Bay, 505

Telegraph weed, 587

Thamnosma
montana, 329
texana, 328

Thermopsis, 373, 759

Thistle, 578, 579, 580, 625
mild, 580
plumeless, 580
swamp, 523
yellow, 519

Thlaspi, 358

Thuja plicata, 436

Timothy, 783

Toothwort, 358, 359

Torchwood, 339

Touchardia, 629

Tragia, 644, 646, 647, 648

Tragia, 646, 647, 648
ramosa, 644
volubilis, 646

Trefoil
bird's foot, 513
tick, 492, 731

Trema, 481, 645

Trema, 645
floridana, 481

Tridens, 792
flavus, 805, 810
pulchella, 792

Trifolium, 371, 384, 385, 387, 393, 395, 492, 735
dasyphyllum, 376, 511
longipes, 508
monanthum, 508
parryi, 376
repens, 372
wormskioldii, 508

Trilobia, 501

Tripsacum dactyloides, 812

Tropaeolaceae, 360

Tube-tongue, 577, 599

Tulip-poplar, 341

Turnera, 544

Turnera ulmifolia, 544

Turpentine broom, *329*
Texas, *328*

Turtlehead, *603*

Twist flower, long-beaked, *367*

U

Ulmus, 609, 610, 615, 621

Uniola latifolia, 837

Urera, 629, 648

Urtica, 611, 623, 627
dioica, 610

Urticaceae, *609, 627, 629, 648*

Urticastrum, 648

V

Vaccinium, 382, 425, 429, 456, 494, 517, 570
caespitosum, 379, 566
canadense, 381
macrocarpum, 414
myrtilloides, 381
myrtillus, 516
uliginosum, 379, 566

Verbena, 596

Verbenaceae, *631*

Verbesina
encelioides, 591
helianthoides, 582, 583
virginica, 582

Vervain, *596*

Vervain family, *631*

Vetch, *375, 493, 503, 735*
crown, *758*
joint, *396*

Viburnum, *494*

Viburnum, 494

Vicia, 493, 503, 735
angustifolia, 375

Viguiera
deltoidea var. *parishii, 590*
multiflora, 581

Viola, 543, 545, 548, 550, 556, 559, 561, 564, 569
adunca, 551, 554, 555, 557
canadensis, 555
cuneata, 551
fimbriatula, 547
glabella, 565
lanceolata, 547
lobata, 551
nephrophylla, 549
nuttallii, 547, 550, 552
ocellata, 565
pedunculata, 552
primulifolia, 547
purpurea, 555
rotundifolia, 546
sempervirens, 565

Violet, *543, 545, 546, 547, 548, 550, 551, 552, 554, 555, 556, 557, 559, 561, 564, 565, 569*
blue, *549*

Vitex, 729

W

Walnut, *453*

Waxmyrtle, *468*

White-cedar, Atlantic, *440*

Willow, *340, 342, 382, 383, 450, 451, 456, 559, 561, 562, 569, 613, 619, 621, 635, 636, 637, 638, 640, 748, 759*
netvein dwarf, *382*
snow, *563*

Wingstem, *582*

Wintergreen, creeping, *382*

Wissadula, 743, 768

Wisteria, 720

Wisteria, 720

Witchgrass, fall, 790

Wormwood, dragon, *331, 332*

Y

Yerba papagayo, *634*

Yucca, *851, 852, 853*
Whipple's, *851*

Yucca, 853
aliofolia, 852
angustissima, 853
baileyi, 853
constricta, 853
filamentosa, 852
glauca, 853
whipplei, 851

Z

Zamia, 419, 420
integrifolia, 419
loddigesii, 420

Zanthoxylum, 340, 644
americanum, 338
fagara, 339, 744

Zea mays, 473, 777

Zizania aquatica, 821

BUTTERFLY INDEX

Numbers in boldface type refer to color plates. Numbers in italics refer to pages. Alternate common names appear in quotation marks. Circles preceding common names of butterflies make it easy for you to keep a record of the butterflies you have seen.

A

Achalarus
casica, 732
jalapus, 733
lyciades, **294,** *731*
toxeus, **290,** *732*

Achylodes thraso, **258,** *744*

Adelpha
bredowii, **31,** **652,** *642*
fessonia, **654,** *641*

Admiral
Asian, *629*
○ Lorquin's, **651,** *639*
○ Red, **672,** *627*
○ Weidemeyer's, **649,** *638*
○ White, **650,** *635*

Adopaeoides prittwitzi, **145,** *782*

Agathymus
alliae, 850
aryxna, **196,** *847*
baueri, 848
estelleae, 850
evansi, **197,** *849*
gilberti, 850
mariae, 849
neumoegeni, **195,** *847*
polingi, 848
stephensi, **194,** *850*

Aglais
milberti, **658,** *622*
urticae, 623

Agraulis vanillae, **19,** **593,** *538*

Agriades franklinii, **483,** **502, 507, 516**

○ **Aguna, Gold-spot, 302,** *723*

Aguna asander, **302,** *723*

"Alderman," *627*

"Alfalfa Butterfly, *372*

Alpine
○ Arctic, *689*
○ Banded, **718,** *691*
  "Butler's," *695*
○ Colorado, *696*
○ Common, **750, 753,** *695*
○ Magdalena, **756,** *690*
○ Northwest, **757,** *688*
○ Red-disked, **748, 751,** *692*
  "Rockslide," *690*
  "Ross's," *689*
○ Spruce Bog, **749, 752,** *689*
○ Theano, **754,** *693*
  "Vidler's," *688*

"Young's," 694
○ Yukon, 694

Amblyscirtes
aenus, 191, 833
aesculapius, 187, 835
alternata, 229, 838
belli, 199, 837
carolina, 835
cassus, 832
celia, 833
eos, 833
erna, 834
exoteria, 832
fimbriata, 839
hegon, 190, 834
linda, 834
nereus, 834
nysa, 835
oslari, 832
phylace, 838
prenda, 833
samoset, 835
simius, 831
texanae, 834
vialis, 232, 836

○ **Amymone,** 73, 679, 645

Anaea
aidea, 653
andria, 362, 652
floridalis, 361, 651
glycerium, 653
pithyusa, 653
portia, 651

Anartia
fatima, 682, 633
jatrophae, 690, 693, 633

Ancyloxypha
arene, 779
numitor, 142, 778

Anglewing
○ Colorado, 613
○ Faunus, 378, 612
○ Oreas, 377, 617
○ Satyr, 372, 611
○ Sylvan, 14, 379, 614
○ Zephyr, 370, 615

Anteos
clorinde, 386
maerula, 105, 386

Anthanassa
ptolyca, 571
texana, 573, 570

Anthocharis
cethura, 76, 366
lanceolata, 83, 370
midea, 5, 37, 57, 74, 369
pima, 367
sara, 75, 368

Apaturines, 538

Aphrissa statira, 127, 390

○ **Aphrodite,** 608, 617, 623, 546

Apodemia
chisosensis, 533
hepburni, 530
mormo, 569, 570, 528
mormo langei, 529
multiplaga, 531
nais, 591, 532
palmerii, 577, 530
phyciodoides, 531
walkeri, 531

Appias drusilla, 69, 352

Arawacus jada, 423, 424

Arctic
○ Alberta, 737, 740, 704
○ California, 734, 738, 741, 699
○ Canada, 744, 747, 701
○ Chryxus, 743, 746, 702
○ Great, 727, 700
○ Jutta, 733, 707
"Katahdin," 709
"Labrador," 705
"Macoun's," 701
○ Melissa, 735, 708
"Nevada," 700
○ Polixenes, 729, 730, 709
Sentinel, 703
○ Uhler's, 728, 703
○ White-veined, 731, 705

Arctic Grayling, 732, 706

Argynnis, 538

Artogeia, 350
napi, 65, 54, 358
rapae, 46, 56, 360
virginiensis, 67, 359

Asbolis capucinus, 218, 828

Ascia monuste, 52, 53, 84, 361

Asterocampa
alicia, 656
antonia, 663, 655
celtis, 45, 664, 653
celtis "alicia," 654, 656
clyton, 12, 666, 657
flora, 665, 658
leilia, 662, 656
louisa, 661, 659
montis, 656
subpallida, 655, 658
texana, 658

Astraptes
○ Dull, 307, 729
○ Flashing, 309, 729

Astraptes
anaphus, 307, 729
fulgerator, 309, 729

○ Atala, 29, 461, 646, 419

Atalopedes campestris, 139, 168, 799

Atlides halesus, 460, 685, 421

Atrytone
arogos, 223, 810
delaware, 138, 811

Atrytonopsis
cestus, 831
deva, 829
hianna, 234, 829
loammi, 829
lunus, 830
ovinia, 830
pittacus, 831

python, 831
viereckii, 830

Autochton
cellus, 253, 676, 730
pseudocellus, 731

Azure
"Dusky," 495
○ Sooty, 464, 509, 495
○ Spring, 477, 482, 493

B

○ Baltimore, 11, 21, 563, 602

Basilarchia
archippus, 26, 597, 637
arthemis, 650, 635
astyanax, 317, 320, 636
lorquini, 651, 639
weidemeyerii, 649, 638

Battus, 321
philenor, 25, 318, 321, 325
polydamas, 329, 332, 326

Biblis hyperia, 680, 646

○ Black, Crimson-banded, 680, 646

Black Swallowtail
Eastern, 324, 335, 338, 327
Short-tailed, 330, 333, 335
Western, 326, 334, 337, 358, 331

Black Swallowtails, 321

○ Blomfild's Beauty, 686, 648

Blue
○ Acmon, 494, 531, 512
○ Antillean, 488, 504, 489
"Arctic," 516
○ Arrowhead, 476, 502
○ Cassius, 508, 485
○ Common, 44, 467, 510
○ Cyna, 488

○ Dotted, 470, *498*
○ Eastern Pygmy, 506, *484*
○ Eastern Tailed, 401, 402, 490, *491*
  "Emerald-studded," *512*
○ Greenish, 463, 478, 497, *507*
  "Hawaiian," *487*
○ High Mountain, 483, 502, 507, *516*
○ Karner, *507*
○ Lupine, 495, *514*
○ Marine, 487, *486*
  "Melissa," *506*
○ Miami, 480, *488*
○ Mojave, *499*
○ Northern, 473, 496, 538, *505*
○ Orange-bordered, 472, 500, 501, *506*
○ Orange-veined, 533, *515*
○ Pale, 469, *499*
○ Reakirt's, 474, 503, *490*
○ Rita, *500*
○ San Emigdio, 493, *508*
○ Shasta, 499, *511*
○ Silvery, 468, *503*
○ Small, 465, *501*
  "Solitary," *490*
○ Sonoran, 489, *496*
○ Square-spotted, 486, 532, *497*
○ Western Pygmy, 505, *483*
○ Western Tailed, 471, *492*
○ Xerces, 475, *504*
○ Yukon, 479, *516*

○ **Blue Wing**, 645, *643*
○ Cyananthe, *643*

Bluet, Blackburn's, 485, *487*

Boloria, *538*
napaea, 638, *557*

Brephidium
exilis, 505, *483*
isophthalma, 506, *484*

○ **Brigadier**, 197, *849*

Brown
○ Appalachian, 702, 705, *667*
○ Eyed, 701, 704, *666*
○ Red-bordered, 755, *697*
○ Smokey Eyed, *666*

Browns, *662*

Brush-footed Butterflies, *537*

○ **Buckeye**, 23, 688, 691, *630*
○ Dark, *631*
  "Florida," *632*
○ West Indian, 689, 692, *632*

○ **Bullet, Brown**, *850*

C

Cabares potrillo, **287**, *737*

Calephelis
arizonensis, 550, *524*
borealis, 542, *519*
driesbachi, *525*
freemani, *524*
guadeloupe, *524*
muticum, 551, *522*
nemesis, 545, *520*
perditalis, 544, *521*
rawsoni, 543, 546, *523*
virginiensis, 549, 552, *519*
wrighti, 39, 547, *522*

Calico
○ Ferentina, 758, *647*
○ Guatemalan, 759, *647*

Callophrys
affinis, 430, *444*
apama, 432, *443*
comstocki, 436, *441*
dumetorum, 431, *440*
lemberti, 433, *442*
sheridanii, 437, *445*
viridis, 435, *444*

Calpodes ethlius, 204, *841*

Calycopis
cecrops, **393**, **467**, *467*
isobeon, **392**, **468**, *468*

Caria ino, **541**, **548**, *525*

Carrhenes canescens, **256**, *742*

Carterocephalus palaemon, **252**, *771*

Catasticta, *351*

Celastrina
argiolus, *494*
ebenina, **464**, **509**, *495*
ladon, **477**, **482**, *493*

Celotes
limpia, *768*
nessus, **249**, *768*

Cercyonis
behrii, *686*
meadii, **726**, *684*
oetus, **699**, *686*
paula, *686*
pegala, **698**, **725**, *683*
silvestris, *686*
sthenele, **708**, **711**, *685*

Chalceria
ferrisi, *406*
heteronea, **43**, **466**, **492**, **516**, *406*
rubidus, **511**, **522**, *405*

Charidryas
acastus, **581**, **589**, *585*
damoetas, **587**, *587*
gabbii, **585**, *587*
gorgone, **632**, *580*
harrisii, **588**, *583*
hoffmanni, **586**, *589*
neumoegeni, **602**, *586*
nycteis, **631**, *581*
palla, **560**, **566**, *584*

Checkered Skipper
Alpine, **250**, *759*
Common, **251**, *762*
Desert, **246**, *765*
Small, **248**, *762*
Southern, *761*
Tropical, **243**, *764*
Two-banded, **247**, *760*
White, **245**, *764*

Checkerspot
○ Anicia, **556**, **562**, *606*
○ Arachne, **580**, *601*
○ Aster, **586**, *589*
○ Chalcedon, **22**, **557**, **571**, *603*
○ Chara, **603**, *599*
○ Chinati, *597*
○ Colon, **568**, *605*
○ Cyneas, **567**, *595*
○ Desert, **602**, *586*
○ Dotted, *602*
○ Dymas, *598*
○ Edith's, **561**, **565**, *607*
○ Elada, **579**, *600*
○ Fulvia, **558**, *594*
○ Gabb's, **585**, *587*
○ Gillette's, **564**, *608*
○ Harris', **588**, *583*
○ Leanira, **559**, **572**, *593*
○ Northern, **560**, **566**, *584*
    "Paintbrush," *606*
○ Rockslide, **587**, *587*
○ Sagebrush, **581**, **589**, *585*
    "Snowberry," *605*
○ Theona, **583**, *596*
    "Yellowstone," *608*

Chioides
catillus, **314**, *722*
zilpa, **313**, *723*

Chiomara asychis, **244**, *746*

Chlorostrymon
maesites, *464*
simaethis, **439**, *465*
telea, **438**, *464*

Chlosyne
californica, **660**, *589*
definita, **578**, *591*
janais, **683**, *592*
lacinia, **659**, **675**, *590*
rosita, *593*

Clossiana, 538
acrocnema, 563
alberta, 567
astarte, 635, 568
bellona, 614, 560
charidea, 639, 569
epithore, 615, 564
freija, 566
frigga, 637, 561
improba, 636, 562
kriemhild, 616, 564
polaris, 634, 565
selene, 3, 16, 625, 559
titania, 619, 613, 568

Cloudywing
○ Drusius, 736
○ Eastern, 284, 735
○ Mexican, 288, 735
○ Northern, 283, 286, 734
○ Southern, 285, 733
○ Valeriana, 735
○ Western, 734

Codatractus
arizonensis, 305, 725
melon, 725

Coenonympha
ampelos, 723, 681
california, 720, 682
haydenii, 715, 677
inornata, 721, 679
kodiak, 722, 678
nipisquit, 680
ochracea, 719, 724, 680
tullia, 679, 680, 681, 682

Cogia
caicus, 289, 739
calchas, 304, 738
hippalus, 291, 738

Colias
alexandra, 103, 110, 113, 373
behrii, 91, 92, 379
boothii, 378
cesonia, 385
eurydice, 386

eurytheme, 86, 97, 100, 130, 372
gigantea, 382
hecla, 89, 376
interior, 95, 380
meadii, 96, 375
nastes, 94, 377
occidentalis, 131, 374
palaeno, 90, 379
pelidne, 87, 132, 381
scudderii, 93, 383
thula, 88, 378

○ **Comma,** 371, 375, 610
○ Gray, 376, 618
  "Green," 612
○ Hoary, 368, 616

Copaeodes
aurantica, 137, 781
minima, 136, 782

Copper
○ American, 513, 403
○ Blue, 43, 466, 492, 516, 406
○ Bog, 525, 530, 414
○ Bronze, 514, 527, 411
○ Dorcas, 523, 416
○ Edith's, 521, 529, 409
○ Ferris', 406
  "Flame," 403
○ Gorgon, 520, 524, 410
○ Great Gray, 528, 536, 537, 408
○ Hermes, 425, 407
  "Lilac-bordered," 412
○ Lustrous, 512, 404
○ Mariposa, 519, 412
○ Nivalis, 540, 412
○ Purplish, 515, 526, 414
○ Ruddy, 511, 522, 405
○ Snow's, 405
○ Tailed, 422, 517, 539, 402

  "Cosmopolite," 625

Crescentspot
  "Bates'," 574
○ Cuban, 629, 571

○ Field, **576, 590,** *575*
○ Gorgone, **632,** *580*
○ Mylitta, **17, 627,** *579*
○ Orseis, **582,** *578*
○ Painted, **576**
○ Pallid, *579*
○ Pearly, **600, 628, 630,** *573*
○ Phaon, **574,** *572*
○ Ptolyca, *571*
○ Silvery, **631,** *581*
○ Tawny, **575,** *574*
○ Texan, **573,** *570*
○ Vesta, **584,** *577*

Cyanophrys
goodsoni, **429,** *447*
miserabilis, **434,** *446*

○ Cycad Butterfly, **462,** *419*

Cyllopsis
gemma, **716,** *670*
henshawi, *668*
pertepida, **717,** *669*
pertepida dorothea, *670*
pyracmon, *668*

Cymaenes
odilia, **180,** *776*
tripunctus, **215,** *775*

Cynthia, *624*

D

Daggerwing
○ Banded, **364,** *650*
○ Ruddy, **363,** *650*

Danaidae, *711*

Danaus
eresimus, **595,** *714*
gilippus, **1, 33, 594,** *713*
plexippus, **2, 35, 596,** *711*

Dash
Black, **172,** *825*
Broken, **150,** *808*
Dakota, *806*
Long, **162,** *806*

Northern Broken, **173,** *809*
See Skipper

Diaethria
anna, **687,** *645*
clymena, *645*

○ Diana, **323, 656, 657,** *544*

Dione moneta, **601,** *539*

○ Dogface, California, **99, 102,** *385*

○ Dogface Butterfly, **98, 101,** *384*

Dolymorpha, *424*

Doxocopa, *661*
laure, **380, 653,** *660*
pavon, **655,** *659*

Dryas iulia, **592,** *540*

Duskywing
○ Afranius, *759*
○ Columbine, *756*
○ Dreamy, **274,** *748*
○ Florida, **265, 280,** *747*
○ Funereal, **277,** *756*
○ Horace's, **254, 270,** *752*
○ Juvenal's, **269,** *749*
○ Mottled, **273,** *753*
○ Mournful, **278,** *753*
○ Pacuvius, **275,** *754*
○ Persius, *758*
○ Propertius, **267,** *751*
○ Rocky Mountain, **266,** *750*
○ Scudder's, *753*
○ Sleepy, **268, 271,** *748*
○ Southern, *751*
○ Wild Indigo, **272,** *757*
○ Zarucco, **276,** *755*

Dymasia
chara, **603,** *599*
dymas, *598*

Dynamine
dyonis, *644*
mylitta, **647,** *644*

E

○ **Eighty-eight, South America,** *645*

○ **Eighty-eight Butterfly,** **687,** *645*

Electrostrymon
angelia, **421,** *463*
endymion, **391,** *462*

○ **Elf, 677,** *598*

Elfin
○ Bog, **457,** *429*
○ Brown, **449, 452,** *424*
○ Early, **451,** *425*
○ Eastern Pine, **38, 459,** *430*
○ Frosted, **454,** *428*
○ Henry's, **455, 456,** *429*
○ Hoary, **453,** *427*
○ Moss, **30, 450,** *426*
○ Western Pine, **458,** *431*

Emesis
ares, **553, 554,** *528*
emesis, *527*
zela, **555,** *527*

Emperor
"Hackberry," *653*
○ Mountain, *656*
○ Pale, *655*
○ Tawny, **12, 666,** *657*
○ Texas, *658*

Empress
○ Alicia, **654,** *656*
○ Antonia, **663,** *655*
○ Flora, **665,** *658*
○ Leilia, **662,** *656*
○ Louisa, **661,** *659*

Enodia
anthedon, **700, 703,** *664*
creola, **696,** *665*
portlandia, **694, 695,** *663*

Epargyreus
clarus, **308,** *719*
zestos, *720*

Ephyriades brunnea, **265, 280,** *747*

Epidemia
dorcas, **523,** *416*
epixanthe, **525, 530,** *414*
helloides, **515, 526,** *414*
mariposa, **519,** *412*
nivalis, **540,** *412*

Erebia
callias, *696*
disa, **749, 752,** *689*
discoidalis, **748, 751,** *692*
epipsodea, **750, 753,** *695*
fasciata, **718,** *691*
magdalena, **756,** *690*
rossii, *689*
theano, **754,** *693*
vidleri, **757,** *688*
youngi, *694*

Eresia, *572*
frisia, **629,** *571*

Erora
laeta, **409, 481,** *466*
quaderna, **410, 484,** *467*

Erynnis
afranius, *759*
baptisiae, **272,** *757*
brizo, **268, 271,** *748*
funeralis, **277,** *756*
horatius, **254, 270,** *752*
icelus, **274,** *748*
juvenalis, **269,** *749*
lucilius, *756*
martialis, **273,** *753*
meridianus, *751*
pacuvius, **275,** *754*
persius, *758*
propertius, **267,** *751*
scudderi, *753*
telemachus, **266,** *750*
tristis, **278,** *753*
zarucco, **276,** *755*

Euchloe
ausonides, **40, 77,** *362*
creusa, *364*
hyantis, **58,** *365*

olympia, 78, **80,** *364*

Eumaeus
atala, **29, 461, 646,** *419*
minijas, **462,** *419*

Eunica
monima, **644**
tatila, **684,** *643*

Euphilotes
battoides, **486, 532,** *497*
enoptes, **470,** *498*
mojave, *499*
pallescens, **469,** *499*
rita, *500*

Euphydryas, **604,** *609*
anicia, **556, 562,** *606*
chalcedona, **22, 557, 571,** *603*
colon, **568,** *605*
editha, **561, 565,** *607*
gilletti, **564,** *608*
phaeton, **11, 21, 563,** *602*

Euphyes
alabamae, *824*
arpa, **207,** *822*
berryi, **146,** *826*
bimacula, **224,** *827*
conspicua, **172,** *825*
dion, **214,** *824*
dukesi, **170,** *825*
mcguirei, *824*
pilatka, **219,** *823*
ruricola, **216, 240,** *827*
vestris, *828*

Euptoieta
claudia, **20, 626,** *542*
hegesia, **599,** *543*

Euptychia, **668, 671, 673, 676**
dorothea, *670*

Eurema
boisduvaliana, **123,** *396*
chamberlainii, *394*
daira, **68, 124,** *395*
dina, *394*
lisa, **107, 129,** *392*

messalina, *398*
mexicana, **106,** *397*
nicippe, **117, 120,** *394*
nise, **122,** *393*
proterpia, **125, 126, 128,** *392*
salome, *397*

Euristrymon
favonius, **388,** *460*
ontario, **387,** *460*
polingi, *461*

Eurytides
marcellus, **48, 340, 343,** *347*
philolaus, **341, 344,** *349*

Everes
amyntula, **471,** *492*
comyntas, **401, 402, 490,** *491*

F

○ **Fatima, 682,** *633*

Feniseca tarquinius, **518,** *401*

"Flying Pansy," *385*

Fritillaries
Greater, *538*
Lesser, *538*

Fritillary
○ Adiaste, *554*
○ Alberta, *567*
○ Arctic, **639,** *569*
○ Astarte, **635,** *568*
○ Atlantis, **618, 624,** *554*
○ Bog, **633,** *558*
○ Callippe, **610, 621,** *552*
○ Coronis, **605,** *550*
○ Dingy Arctic, **636,** *562*
○ Edwards', **622,** *549*
○ Egleis, **612,** *553*
○ Freya's, *566*
○ Frigga's, **637,** *561*
○ Great Spangled, **609, 640, 641,** *545*
○ Gulf, **19, 593,** *538*

○ Hydaspe, **611**, *555*
○ Kriemhild, **616**, *564*
○ Leto, *546*
○ Meadow, **614**, *560*
○ Mexican, **599**, *543*
○ Mormon, **606**, **620**, *557*
○ Napaea, **638**, *557*
○ Nokomis, **604**, **642**, *548*
○ Polaris, **634**, *565*
○ Regal, **667**, *547*
○ Silver-bordered, **3**, **16**, **625**, *559*
○ Titania's, **613**, **619**, *568*
○ Uncompaghre, *563*
○ Variegated, **20**, **626**, *542*
○ Western Meadow, **615**, *564*
○ Zerene, **607**, *551*

G

Gaeides
editha, **521**, **529**, *409*
gorgon, **520**, **524**, *410*
xanthoides, **528**, **536**, **537**, *408*

Ganyra josephina, **55**, *362*

Gesta gesta, **279**, *746*

Giant Skipper
○ Amole, *848*
○ Aryxna, **196**, *847*
○ California, **194**, *850*
○ Cofaqui, *852*
○ Colorado, *852*
○ Manfreda, *854*
○ Mojave, *850*
○ Orange, **195**, *847*
○ Pecos, *849*
○ Strecker's, *853*
○ Ursine, *852*
○ West Texas, *850*
○ Yavapai, *848*
○ Yucca, **237**, *851*

Giant Skippers, *846*

Giant Sulphur
Argante, **108**, *389*

Cloudless, **32**, **115**, **118**, *387*
Orange, **121**, *390*
Orange-barred, **116**, **119**, *388*
Tailed, *388*

Giant Swallowtail, **27**, **328**, **331**, *337*

Giant Swallowtails, *321*

○ **Glassywing, Little**, **206**, *809*

Glaucopsyche
lygdamus, **468**, *503*
piasus, **476**, *502*
xerces, **475**, *504*

○ **Goatweed Butterfly**, **362**, *652*
"Florida," *651*

Goatweeds, *538*

"Gold Rim," *326*

Gorgythion begga, **261**, *741*

Gossamer Wings, *400*

Grais stigmaticus, **255**, *745*

Graphium
marcellus, *348*
philolaus, *349*

Grayling
○ Arctic, **732**, *706*
"Blue-eyed," *683*

Green Hairstreak
Alpine, **433**, *442*
Bluish, **435**, *444*
Bramble, **431**, *440*
Canyon, **432**, *443*
Desert, **436**, *441*
Immaculate, **430**, *444*
White-lined, **437**, *445*

○ **Greenwing**
Dyonis, *644*
Mylitta, **647**, *644*

Gyrocheilus patrobas, **755**, *697*

H

Habrodais grunus, 426, 427, *418*

Hackberries, 537

○ **Hackberry Butterfly,** 45, 664, *653, 654*

Hairstreak
○ Acadian, 404, *449*
○ Acis, 384, *480*
○ Alea, *475*
○ Alpine Green, 433, *442*
○ Amethyst, *464*
○ Angelic, 421, *463*
○ Aquamarine, 412, *423*
○ Arizona, 410, 484, *467*
○ Avalon, 414, *474*
○ Azia, *471*
○ Banded, 394, *453*
○ Barry's, *436*
○ Bartram's, *481*
○ Behr's, 400, 534, *448*
○ Bluish Green, 435, *444*
○ Bramble Green, 431, *440*
○ California, 399, *450*
○ Canyon Green, 432, *443*
○ Cestri, 416, 510, *479*
○ Clytie, *469*
○ Colorado, 382, 498, *417*
○ Columella, 406, *477*
  "Common," *472*
○ Coral, 403, *482*
○ Desert Green, 436, *441*
○ Dryope, *452*
○ Dusky Blue, 392, *468*
○ Early, 409, 481, *466*
○ Edwards', 398, *452*
○ Endymion, 391, *462*
○ Golden, 426, 427, *418*
○ Gold Hunter's, 428, *457*
○ Goodson's, 429, *447*
○ Gray, 386, *472*
  "Great Blue," *421*
○ Great Purple, 460, 685,' *421*
○ Hedgerow, 42, 424, *459*
○ Hessel's, 7, 41, 442, *439*
○ Hickory, 396, *454*

○ Immaculate Green, 430, *444*
  "Incense Cedar," *436*
○ Jade-blue, 423, *424*
○ Johnson's, *435*
○ Juniper, 441, *437*
○ King's, 397, *455*
○ Leda, 407, *470*
○ Limenia, *478*
○ Loki, 444, 445, *438*
  "Long-tailed," *481*
○ Marsyas, 413, *420*
○ Martial, 383, *481*
○ Mexican Gray, 390, *475*
○ Miserabilis, 434, *446*
  "Mistletoe," *435*
○ Mountain Mahogany, 415, *458*
○ Muir's, *436*
○ Nelson's, 447, 448, *436*
○ Northern, 387, *460*
○ Olive, 443, *438*
○ Palegon, *421*
○ Poling's, *461*
○ Red-banded, 393, *467*
○ Reddish, 420, *478*
○ Sandia, *432*
○ Silver-banded, 439, *465*
  "Skinner's," *438*
○ Soapberry, *462*
○ Sonoran, *472*
○ Sooty, 411, *449*
○ Southern, 388, *460*
○ Spurina, *421*
○ Striped, 395, *456*
○ Sylvan, 405, 535, *451*
  Tecate Cypress, *438*
○ Telea, 438, *464*
○ Thicket, 446, *434*
○ White, 418, *482*
○ White-lined Green, 437, *445*
○ White M, 389, *422*
○ Xami, 440, *433*
○ Yojoa, 419, *476*
○ Zebina, 385, *421*

Hamadryas
februa, 758, *647*
guatemalena, 759, *647*

Harkenclenus titus, **403,** *482*

○ **Harvester, 518,** *401*

○ **Heath, Large,** *679*

Heliconiidae, *539*

Heliconiines, *538*

Heliconius
charitonius, **4, 34, 644,** *541*
erato, **643,** *541*

Heliopetes
arsalte, **72, 241,** *767*
domicella, *766*
ericetorum, **59, 82, 242,** *766*
laviana, **293,** *767*

Hemiargus
ceraunus, **488, 504,** *489*
isola, **474, 503,** *490*
thomasi, **480,** *488*

Heraclides, **321, 322**
anchisiades, **340**
andraemon, **339**
aristodemus, **339**
cresphontes, **27, 328, 331,** *337*
thoas, **327, 336**

Hermelycaena hermes, **425,** *407*

Hermeuptychia
hermes, **736, 739,** *671*
sosybius, **712,** *672*

○ **Hermit, 255,** *745*

Hesperia
attalus, *794*
columbia, **152,** *793*
comma, **156, 183, 185,** *788*
dacotae, **177,** *796*
juba, *788*
leonardus, **153,** *791*
lindseyi, *789*
meskei, **178,** *795*

metea, **192,** *793*
miriamae, **193,** *797*
nevada, **182,** *798*
ottoe, **133, 149,** *790*
pahaska, **186,** *792*
pawnee, *791*
sassacus, **166,** *797*
uncas, **189,** *787*
viridis, **184,** *794*
woodgatei, **201,** *789*

Hesperiidae, *716*

Hesperiinae, *716*

○ **Hoary Edge, 294,** *731*
○ Mexican, *732*

"Hop Merchant," *610*

"Hunter's Butterfly," *623*

Hylephila phyleus, **154,** *784*

Hyllolycaena hyllus, **514, 527,** *411*

Hypaurotis crysalus, **382, 498,** *417*

Hypolimnus misippus, **673, 674,** *629*

Hypostrymon critola, *472*

I

Icaricia
acmon, **494, 531,** *512*
icarioides, **44, 467,** *510*
lupini, **495,** *514*
neurona, **533,** *515*
shasta, **499,** *511*

Incisalia
augustinus, **449, 452,** *424*
eryphon, **458,** *431*
fotis, **451,** *425*
henrici, **455, 456,** *429*
irus, **454,** *428*
lanoraieensis, **457,** *429*
mossii, **30, 450,** *426*
niphon, **38, 459,** *430*
polios, **453,** *427*

J

○ **Julia, 592,** *540*

Junonia
coenia, **23, 688, 691,** *630*
evarete, **689, 692,** *632*
nigrosuffusa, *631*

K

○ **Kamahameha, 671,** *628*

Karwinski's Beauty, 648

Kricogonia lyside, **104,** *391*

L

Lady
○ American Painted, **13, 669,** *623*
○ Painted, **18, 670,** *625*
○ "Virginia," *623*
○ West Coast, **668,** *626*

Lantana Butterfly
○ Larger, **408,** *471*
○ Smaller, **417,** *479*

Lasaia
narses, *526*
sula, **491,** *526*

○ **Laure, 380, 653,** *660*

Leafwing
○ Blue, *653*
○ Crinkled, *653*
○ Florida, **361,** *651*
○ Tropical, *653*

"Leopard-spot," *645*

Leptotes
cassius, **508,** *485*
marina, **487,** *486*

Lerema accius, **213, 239,** *777*

Lerodea
arabus, *840*
dysaules, *840*
eufala, **225,** *839*

Lethe, **664,** *667*

Libytheana
bachmanii, **366,** *534*
carinenta, **365,** *535*

Libytheidae, *534*

Limenitis, *636*

Longtail
○ Brown, **312,** *728*
○ Lilac-banded, **315,** *726*
○ Teleus, **311,** *727*
○ White-striped, **314,** *722*
○ Zilpa, **313,** *723*

Longwing
○ Crimson-patched, **643,** *541*
○ Zebra, **4, 34, 644,** *541*

Longwings, *538*

Lycaeides
argyrognomon, **473, 496,** *538, 505*
melissa, **472, 500, 501,** *506*

Lycaena, *401*
cupreus, **512,** *404*
cupreus snowii, *405*
phlaeas, **513,** *403*
thoe, *411*

Lycaenidae, *400*

Lycorea cleobaea, *715*

Lycorella ceres, *715*

○ **Lyside, 104,** *391*

M

○ **Malachite, 648,** *634*

○ **Marble, Gray, 83,** *370*

Marblewing
○ Creamy, **40, 77,** *362*
○ Northern, *364*
○ Olympia, **78, 80,** *364*
○ Pearly, **58,** *365*

Marpesia
chiron, **364,** *650*

coresia, **381,** *649*
petreus, **363,** *650*

Megathymidae, *846*

Megathymus
cofaqui, *852*
coloradensis, *852*
streckeri, *853*
ursus, *852*
yuccae, **237,** *851*

Megisto
cymela, **706, 709,** *675*
rubricata, **707, 710,** *676*

Melanis pixe, **678,** *526*

Mestra amymone, **73, 679,**
645

Metalmark
○ Ares, **553, 554,** *528*
○ Arizona, **550,** *524*
○ Blue, **491,** *526*
○ Chisos, *533*
○ Crescent, *531*
　"Dusky," *520*
○ Falcate, *527*
○ Fatal, **545,** *520*
○ Freeman's, *524*
○ Gray, **577,** *530*
○ Hepburn's, *530*
○ Lange's, *529*
○ Little, **549, 552,** *519*
○ Lost, **544, 546,** *521*
○ Mormon, **569, 570,** *528*
○ Nais, **591,** *532*
○ Narrow-winged, *531*
○ Narses, *526*
○ Nogales, *525*
○ Northern, **542,** *519*
○ Rawson's, **543,** *523*
○ Red-bordered, **541, 548,**
　525
　"Schaus'," *525*
○ Swamp, **551,** *522*
　"Virginia," *519*
○ Walker's, *531*
○ Wright's, **39, 547,** *522*
○ Zela, **555,** *527*

　Metalmarks, *518*

Microtia elva, **677,** *598*

Milkweed Butterflies, *711*

○ Milkweed, Tropical, *715*

○ Mimic, **673, 674,** *629*

Ministrymon
clytie, *469*
leda, **407,** *470*

Mitoura
barryi, *436*
gryneus, **443,** *438*
hesseli, **7, 41, 442,** *439*
johnsoni, *435*
loki, **444, 445,** *438*
nelsoni, **447, 448,** *436*
nelsoni muiri, *436*
siva, **441,** *437*
spinetorum, **446,** *434*
thornei, *438*

○ Monarch, **2, 35, 596,** *711*

Monca tyrtaeus, **238,** *774*

○ Monk, **218,** *828*

○ Mourning Cloak, **10, 24,**
681, *621*

○ Mulberry Wing, **158,**
817

Myscelia
cyananthe streckeri, *643*
ethusa, **645,** *643*

N

Nastra
julia, **221,** *775*
lherminier, **220, 222,** *774*
neamathla, *775*

Nathalis iole, **109, 112,**
398

Neominois ridingsii, **742,**
745, *698*

Neonympha
areolatus, **714,** *673*
mitchellii, **713,** *674*

Neophasia
menapia, 61, *351*
terlootii, 66, 598, *352*

Nisoniades rubescens, 281, 739

Nyctelius nyctelius, 198, 845

Nymphalidae, *537, 711*

Nymphalis
antiopa, 10, 24, 681, *621*
californica, 367, *620*
polychloros, 621
vau-album, 369, 619

O

Oarisma
edwardsii, 781
garita, 227, *780*
poweshiek, 226, *779*

Ochlodes
agricola, 134, 144, *814*
snowi, 203, *815*
sylvanoides, 163, *813*
yuma, 816

Oeneis
alberta, 737, 740, *704*
bore, 732, *706*
chryxus, 743, 746, *702*
excubitor, 703
ivallda, 734, 738, 741, *699*
jutta, 733, *707*
macounii, 744, 747, *701*
melissa, 735, *708*
nevadensis, 727, *700*
polixenes, 729,'730, *709*
taygete, 731, *705*
uhleri, 728, *703*

Oenomaus ortygnus, 412, *423*

Oligoria maculata, 202, 205, *840*

Orange
◯ Sleepy, 117, 120, *394*

◯ Tailed, 125, 126, 128, *392*

"Orange Dog," *337*

Orangetip
◯ Desert, 76, *366*
◯ Falcate, 5, 37, 57, 74, *369*
◯ Pima, *367*
◯ Sara, 75, *368*

P

Panoquina
errans, 209, *843*
evansi, 845
hecebolus, 844
ocola, 217, *843*
panoquin, 210, *841*
panoquinoides, 179, *842*
sylvicola, 231, *844*

Papilio, 321, *322*
anchisiades, 340
bairdii, 326, 334, 337, 358, *331*
brevicauda, 330
indra, 330, 333, *335*
joanae, 328
kahli, 328
machaon, 354, 357, *333*
nitra, 332
oregonius, 352, 355, *332*
polyxenes, 324, 335, 338, *327*
rudkini, 325, 359, *329*
zelicaon, 36, 353, 356, *334*

Papilionidae, *321*

Paramacera
allyni, 697, *677*
xicaque, 677

Paratrytone melane, 164, *822*

Parnassian
◯ Clodius, 70, *323*
◯ Eversmann's, *322*
◯ Phoebus, 28, 71, *324*

Parnassians, *321*

Parnassius
clodius, 70, *323*
eversmanni, *322*
phoebus, **28**, 71, *324*

Parrhasius m-album, 389, *422*

Patch
○ Bordered, 659, 675, *590*
○ California, 660, *589*
○ Definite, **578**, *591*
○ Janais, 683, *592*
○ Rosita, *593*
○ White, 244, *746*

○ **Pavon**, 655, *659*

○ **Peacock, White**, 690, 693, *633*

○ **Pearly Eye**, 694, 695, *663*
○ Creole, 696, *665*
○ Northern, 700, 703, *664*

Perichares philetes, 236, *777*

Phaeostrymon alcestis, *462*

Philotes sonorensis, 489, *496*

Philotiella speciosa, 465, *501*

Phocides
pigmalion, **300**, *717*
polybius, 299, *718*
urania, *718*

Phoebis
agarithe, **121**, *390*
argante, 108, *389*
neocypris, *388*
philea, 116, 119, *388*
sennae, **32**, 115, 118, *387*

Pholisora
alpheus, *770*
catullus, **282**, *769*
graceliae, *771*
libya, **295**, *769*
mejicana, *769*

Phyciodes
batesii, 575, *574*
campestris, **576**, 590, *575*
mylitta, **17**, 627, *579*
orseis, **582**, *578*
pallida, *579*
phaon, 574, *572*
picta, *576*
tharos, 600, 628, 630, *573*
vesta, **584**, *577*

Pieridae, *350*

Pieris, *350*

Pipevine Swallowtail, 25, 318, 321, *325*

Pipevine Swallowtails, *321*

Piruna
pirus, 208, *772*
polingii, 200, *773*

○ **Pixie**, 678, *526*

Plebejus saepiolus, 463, 478, 497, *507*

Plebulina emigdionis, 493, *508*

Poanes
aaroni, 171, *819*
hobomok, **159**, 233, *817*
massasoit, 158, *817*
taxiles, 160, 212, *819*
viator, 169, *821*
yehl, 148, *820*
zabulon, 157, 235, *818*

Poladryas
arachne, **580**, *601*
minuta, *602*

Polites
baracoa, **211**, *803*
coras, 161, *800*
dakotah, *806*
draco, **141**, *803*
mardon, *802*
mystic, 162, *806*
origenes, 147, *805*
peckius, *801*

sabuleti, 140, 181, *801*
sonora, 165, 176, *807*
themistocles, 174, *804*
vibex, 151, 155, *807*

Polygonia
comma, 371, 375, *610*
faunus, 378, *612*
gracilis, 368, *616*
hylas, *613*
interrogationis, 9, 15, 373, 374, *609*
oreas, 377, *617*
progne, 376, *618*
satyrus, 372, *611*
silvius, 14, 379, *614*
zephyrus, 370, *615*

Polygonus
leo, 298, *721*
manueli, *721*

Pompeius verna, 206, *809*

Pontia, *350*
beckerii, 79, *353*
occidentalis, 62, *357*
protodice, 60, 63, 81, *355*
sisymbrii, 64, *355*

"Portia," *651*

Precis lavinia, *631*

Problema
bulenta, *813*
byssus, 167, 175, *812*

Proclossiana, *538*
eunomia, 633, *558*

Proteides mercurius, 292, 301, *719*

Pseudocopaeodes eunus, *786*

Pseudolycaena marsyas, 413, *420*

Pterourus, *321, 322*
eurymedon, 342, 345, *344*
glaucus, 6, 49, 322, 348, 351, *340*
multicaudatus, 346, 349, 360, *343*

palamedes, 50, 336, 339, *346*
pilumnus, *344*
rutulus, 347, 350, *342*
troilus, 51, 316, 319, *345*

Purple
"Banded," *635*
○ Red-spotted, 317, 320, *636*

Purplewing
○ Dingy, *644*
○ Florida, **684**, *643*

Pyrginae, *716*

Pyrgus
albescens, 245, *764*
centaureae, 250, *759*
communis, 251, *762*
oileus, 243, *764*
philetas, 246, *765*
ruralis, 247, *760*
scriptura, 248, *762*
syrichtus, *765*
xanthus, *761*

Pyrrhopyge araxes, **296**, *717*

Pyrrhopyginae, *716*

Q
○ Queen, 1, 33, **594**, *713*
○ Tropic, **595**, *714*

○ **Question Mark**, 9, 15, 373, 374, *609*

R

Rambler, Roadside, *833*

Ringlet
○ California, **720**, *682*
○ Hayden's, **715**, *677*
○ Kodiak, **722**, *678*
○ Nipisquit, *680*
○ Northwest, **723**, *681*
○ Ocher, 719, **724**, *680*
○ Prairie, **721**, *679*
"Wyoming," *677*

Riodinidae, *518*

○ Roadside Rambler, *833*

Roadside Skipper, 232, *836*
Arkansas, *834*
Bell's, 199, *837*
Blue-dusted, 229, *838*
Bronze, 191, *833*
Carolina, *835*
Cassus, *832*
Dotted, *833*
Erna, *834*
Lace-winged, 187, *835*
Large, *832*
Mottled, *835*
Orange, *831*
Orange-edged, *839*
Oslar's, *832*
Prenda, *833*
Red-headed, *838*
Slaty, *834*
Texas, *834*

"Roller, Bean-leaf," *726*

S

○ Sachem, 139, 168, *799*

Sandia mcfarlandi, 432

Satyr
○ Canyonland, 717, *669*
○ Carolina, 712, *672*
○ Gemmed, 716, *670*
○ Georgia, 714, *673*
○ Hermes, 736, 739, *671*
○ Little Wood, 706, 709, *675*
○ Mitchell's Marsh, 713, *674*
○ Nabokov's, *668*
○ Pine, 697, *677*
○ Red, 707, 710, *676*
○ Riding's, 742, 745, *698*
○ Sonoran, *668*
○ Xicaque, *677*

Satyridae, *662*

Satyrium
acadica, 404, 449

adenostomatis, 458
auretorum, 428, 457
behrii, 400, 534, 448
calanus, 394, 453
californica, 399, 450
caryaevorus, 396, 454
dryope, 452
edwardsii, 398, 452
falacer, 454
fuliginosum, 411, 449
kingi, 397, 455
liparops, 395, 456
saepium, 42, 424, 459
sylvinus, 405, 535, 451
tetra, 415, 458

Satyrodes
appalachia, 702, 705, 667
eurydice, 701, 704, 666
fumosus, 666

Satyrs, *662*
Grass, *673*

Seminole Crescent, *571*

○ Silver Spike, Morrison's, *786*

○ Silverspot, Mexican, 601, *539*

Siproeta stelenes, 648, 634

Sister
○ California, 31, 652, *642*
○ Mexican, 654, *641*

Skipper
○ Acacia, 291, *738*
○ Alabama, *824*
○ Alpine Checkered, 250, *759*
Amole Giant, *848*
○ Apache, 201, *789*
○ Araxes, 296, *717*
○ Arctic, 252, *771*
○ Arizona, 305, *725*
○ Arizona Powdered, 259, *744*
○ Arkansas Roadside, *834*
○ Arsalte, 72, 241, *767*
Aryxna Giant, 196, *847*

○ Baracoa, 211, *803*
○ Beard-grass, 223, *810*
○ Bell's Roadside, 199, *837*
○ Belted; *831*
○ Black Dash, 172, *825*
○ Blue-banded, 279, *746*
○ Blue-dusted Roadside, 229, *838*
○ Brazilian, 204, *841*
○ Broad-winged, 169, *821*
○ Broken Dash, 150, *808*
○ Bronze Roadside, 191, *833*
○ Brown-banded, *745*
○ Bunchgrass, 167, 175, *812*
○ Caicus, 289, *739*
California Giant, 194, *850*
○ Carolina Roadside, *835*
○ Cassus Roadside, *832*
○ Clouded, 213, 239, *777*
○ Cobweb, 192, *793*
Cofaqui Giant, *852*
Colorado Giant, *852*
○ Columbian, 152, *793*
○ Common Branded, 156, 183, 185, *788*
○ Common Checkered, 251, *762*
○ Common Streaky, 249, *768*
○ Coyote, 290, *732*
○ Crossline, 147, *805*
○ Dakota, 177, *796*
○ Dakotah Dash, *806*
○ Delaware, 135, 138, *811*
○ Desert Checkered, 246, *765*
○ Desert Gray, 188, *785*
○ Deva, *829*
○ Dixie, 178, *795*
○ Dotted, *794*
○ Dotted Roadside, *833*
○ Draco, 141, *803*
○ Dun, 216, 240, *827*
○ Dusted, 234, *829*
○ Erna Roadside, *834*
○ Eufala, 225, *839*
○ European, 143, 228, *783*

○ Evans', *845*
○ Falcate, 303, *737*
○ False Golden-banded, *731*
○ Fawn-spotted, 180, *776*
○ Fiery, 154, *784*
○ Florida Swamp, 146, *826*
○ Golden, 160, 212, *819*
○ Golden-banded, 253, 676, *730*
○ Green, 184, *794*
○ Green-backed, 236, *777*
○ Guava, 299, *718*
○ Hammock, 298, *721*
○ Hoary, 256, *742*
○ Hoary Edge, 294, *731*
○ Hobomok, 159, 233, *817*
○ Indian, 166, *797*
○ Jalapas, *733*
○ Juba, *788*
○ Julia, 221, *775*
○ Lace-winged Roadside, 187, *835*
○ Large Roadside, *832*
○ Large White, 59, 82, 242, *766*
○ Laviana, 293, *767*
○ Leonardus, 153, *791*
○ Lindsey's Branded, *789*
○ Long Dash, 162, *806*
○ Long-tailed, 8, 47, 310, *725*
○ Long-winged, 217, *843*
○ Malicious, 230, *773*
Manfreda Giant, *854*
○ Mangrove, 300, *717*
○ Manuel's, *721*
○ Mardon, *802*
○ McGuire's, *824*
○ Mercurial, 292, 301, *719*
○ Mexican Melon, *725*
○ Mimosa, 304, *738*
Mojave Giant, *850*
○ Moon-marked, *830*
○ Mottled Roadside, *835*
○ Neamathla, *775*
○ Nevada, 182, *798*
○ Northern Broken Dash, 173, *809*
○ Obscure, 179, *842*

○ Olive-clouded, *840*
○ Orange-edged Roadside, *839*
　Orange Giant, **195**, *847*
○ Orange Roadside, *831*
○ Oslar's Roadside, *832*
○ Ottoe, **133, 149**, *790*
○ Pahaska, **186**, *792*
○ Palmetto, **207**, *822*
○ Parchment, *831*
○ Pawnee, *791*
○ Peck's, *801*
　Pecos Giant, *849*
○ Pepper-and-salt, **190**, *834*
○ Plains Gray, *784*
○ Potrillo, **287**, *737*
○ Prenda Roadside, *833*
○ Purple-washed, **231**, *844*
○ Purplish Black, **281**, *739*
○ Python, *831*
○ Rare, *813*
○ Red-headed Roadside, *838*
○ Roadside, **232**, *836*
○ Rural, **134, 144**, *814*
○ Russet, *844*
○ Saffron, **171**, *819*
○ Salenus, *773*
○ Salt-grass, *786*
○ Salt Marsh, **210**, *841*
○ Sandhill, **140, 181**, *801*
○ Saw-grass, **219**, *823*
○ Scarce Streaky, *768*
○ Scarce Swamp, **170**, *825*
○ Sedge, **214**, *824*
○ Sheep, *830*
○ Short-tailed Arizona, **297**, *724*
○ Sickle-winged, **258**, *744*
○ Sierra, **193**, *797*
○ Silver-spotted, **308**, *719*
○ Slaty Roadside, *834*
○ Small Checkered, **248**, *762*
○ Snow's, **203**, *815*
○ Sonora, **165, 176**, *807*
○ Southern Checkered, *761*
○ Southern Dusted, *829*
　Strecker's Giant, *853*
○ Sunrise, **145**, *782*

○ Swarthy, **220, 222**, *774*
○ Tawny-edged, **174**, *804*
○ Texas Powdered, **260**, *743*
○ Texas Roadside, *834*
○ Three-spotted, **215**, *775*
○ Tropical Checkered, **243**, *764*
○ Twin-spot, **202, 205**, *840*
○ Two-banded Checkered, **247**, *760*
○ Two-spotted, **224**, *827*
○ Umber, **164**, *822*
○ Uncas, **189**, *787*
○ Urania, *718*
　Ursine Giant, *852*
○ Variegated, **261**, *741*
○ Viereck's, *830*
○ Violet-banded, **198**, *845*
○ Violet-clouded, *840*
○ Violet-patch, **238**, *774*
○ Wandering, **209**, *843*
　West Texas Giant, *850*
○ White-banded, *766*
○ White Checkered, **245**, *764*
○ White Patch, **244**, *746*
○ White-tailed, **306**, *728*
○ Window-winged, **257**, *742*
○ Woodland, **163**, *813*
　Yavapai Giant, *848*
○ Yehl, **148**, *820*
○ Yellowpatch, **161**, *800*
　Yucca Giant, **237**, *851*
○ Yuma, *816*
○ Zabulon, **157, 235**, *818*
○ Zestos, *720*

Skipperling
○ Edwards', *781*
○ Garita, **227**, *780*
○ Least, **142**, *778*
○ Orange, **137**, *781*
　"Pirus," *772*
○ Poweshiek, **226**, *779*
○ Russet, **208**, *772*
○ Southern, **136**, *782*
○ Spotted, **200**, *773*
○ Tropical Least, *779*

Skippers
Giant, 846
True, 716

Smyrna blomfildia, **686,**
648

Snout Butterflies, 534

○ Snout, Southern, 365,
535

○ Snout Butterfly, 366,
534

Sootywing
○ Common, 282, 769
○ Golden-headed, 264, 740
○ Great Basin, 295, 769
○ MacNeill's, 771
○ Mexican, 769
○ Saltbrush, 770
○ Scalloped, 262, 741
○ Southern Scalloped, 263,
740

Spathilepia clonius, **303,**
737

Speyeria, 538
adiaste, 554
aphrodite, **608, 617, 623,**
546
atlantis, **618, 624,** 554
callippe, **610, 621,** 552
coronis, **605,** 550
cybele, **609, 640, 641,** 545
diana, **323, 656, 657,** 544
edwardsii, **622,** 549
egleis, **612,** 553
hydaspe, **611,** 555
idalia, **667,** 547
leto, 546
mormonia, **606, 620,** 557
nokomis, **604, 642,** 548
zerene, **607,** 551

Stallingsia maculosus, 854

Staphylus
ceos, **264,** 740
hayhurstii, **262,** 741
mazans, **263,** 740

○ *Statira,* **127,** *390*

Stinga morrisoni, 786

Strymon
acis, **384,** 480
acis bartrami, 481
albata, **418,** 482
alea, 475
avalona, **414,** 474
bazochii, **417,** 479
bebycia, **390,** 475
buchholzi, 475
cestri, **416, 510,** 479
columella, **406,** 477
facuna, 447
laceyi, 476
limenia, 478
martialis, **383,** 481
melinus, **386,** 472
pastor, 447
rufofusca, **420,** 478
yojoa, **419,** 476

Sulphur
○ Argante Giant, 108, 389
"Barred," 395
○ Behr's, 91, 92, 379
○ Blueberry, 87, 132, 381
○ Booth's, 378
"Clouded," 371
○ Cloudless Giant, 32, 115,
118, 387
○ Common, 85, 111, 114,
371
"Dainty," 398
○ Great Northern, 382
○ Greenland, 89, 376
○ Labrador, 94, 377
○ Mead's, 96, 375
○ Orange, 86, 97, 100, 130,
372
○ Orange-barred Giant, 116,
119, 388
○ Orange Giant, 121, 390
○ Palaeno, 90, 379
○ Pink-edged, 95, 380
○ Queen Alexandra's, 103,
110, 113, 373
○ Scudder's Willow, 93, 383

"Sierra," *379*
○ Tailed Giant, *388*
○ Thula, **88**, *378*
○ Western, **131**, *374*
○ White Angled, *386*
○ Yellow Angled, **105**, *386*

Sulphurs, *350*

Swallowtail
○ Anise, **36**, 353, 356, *334*
○ Bahaman, *339*
  "Baird's," *331*
  "Blue," *325*
○ Dark Zebra, 341, 344, *349*
○ Desert, 325, 359, *329*
○ Eastern Black, 324, 335, **338**, *327*
○ Giant, **27**, 328, 331, 337
  "Green-clouded," *345*
  "Indra," *335*
○ Kahli, *328*
  "Maritime," *330*
○ Old World, 354, 357, *333*
○ Oregon, 352, 355, *332*
○ Ozark, *328*
○ Palamedes, **50**, 336, 339, *346*
○ Pale Tiger, 342, 345, *344*
○ Pipevine, **25**, 318, 321, *325*
○ Polydamas, 329, 332, *326*
○ Ruby-spotted, *340*
○ Schaus', *339*
○ Short-tailed, 326, *330*
○ Short-tailed Black, 330, **333**, *335*
○ Spicebush, **51**, 316, 319, *345*
○ Thoas, 327, *336*
○ Three-tailed Tiger, *343*
○ Tiger, **6**, **49**, 322, 348, 351, *340*
○ Two-tailed Tiger, 346, 349, 360, *343*
○ Western Black, 334, 337, 358, *331*
○ Western Tiger, 347, 350, *342*
○ Zebra, **48**, 340, 343, 347

Swallowtails, *321*
Black, *321*
Giant, *321*
Pipevine, *321*
Tiger, *321*

Synapte
malitiosa, **230**, *773*
salenus, *773*

Systasea
pulverulenta, **260**, *743*
zampa, **259**, *744*

T

Texola elada, **579**, *600*

Tharsalea arota, **422**, 517, 539, *402*

Thereus
palegon, *421*
spurina, *421*
zebina, **385**, *421*

Thessalia
chinatiensis, *597*
cyneas, **567**, *595*
fulvia, **558**, *594*
leanira, 559, 572, *593*
theona, **583**, *596*

"Thistle Butterfly," *625*

Thorybes
bathyllus, **285**, *733*
confusis, **284**, *735*
diversus, *734*
drusius, *736*
mexicana, **288**, *735*
pylades, 283, 286, *734*
valeriana, *735*

Thymelicus lineola, **143**, 228, *783*

Tiger Swallowtail, **6**, **49**, 322, 348, 351, *340*
Pale, 342, 345, *344*

Three-tailed, *344*
Two-tailed, **346, 349, 360,** *343*
Western, **347, 350,** *342*

Tiger Swallowtails, *321*

Timochares ruptifasciatus, *745*

Tmolus
azia, *471*
echion, **408,** *471*

Tortoiseshell
○ California, **367,** *620*
○ Compton, **369,** *619*
○ European Large, *621*
○ European Small, *623*
○ Milbert's, **658,** *622*

True Skippers, *716*

U

Urbanus
dorantes, **315,** *726*
doryssus, **306,** *728*
procne, **312,** *728*
proteus, **8, 47, 310,** *725*
teleus, **311,** *727*

V

Vacciniina optilete, **479,** *516*

Vaga blackburnii, **485,** *487*

Vanessa
annabella, **668,** *626*
atalanta, **672,** *627*
cardui, **18, 670,** *625*
carye, *627*
huntera, *624*
indica, *629*
tameamea, **671,** *628*
virginiensis, **13, 669,** *623*

○ Viceroy, **26, 597,** *637*

W

○ Waiter, **381,** *649*

Wallengrenia
egeremet, **173,** *809*
otho, **150,** *808*

○ **Whirlabout,** **151, 155,** *807*

White
○ Becker's, **79,** *353*
○ Cabbage, **46, 56,** *360*
"California," *355*
○ Caribbean Dainty, *398*
○ Checkered, **60, 63, 81,** *355*
○ Chiricahua Pine, **66, 598,** *352*
"Common," *355*
"European Cabbage," *360*
○ Florida, **69,** *352*
○ Giant, **55,** *362*
○ Great Southern, **52, 53, 84,** *361*
"Mustard," *358*
○ Pine, **61,** *351*
"Sagebrush," *353*
"Small," *360*
○ Spring, **64,** *355*
○ Veined, **54, 65,** *358*
○ West Virginia, **65,** *359*
○ Western, **62,** *357*

"White Mountain Butterfly," *708*

White Patch, 244, *746*

Whites, *350*

Wood Nymph
○ Dark, **699,** *686*
○ Great Basin, **708, 711,** *685*
○ Large, **698, 725,** *683*
"Mead's," *684*
○ Red-eyed, **726,** *684*

X

Xamia xamia, **440,** *433*

Xenophanes trixis, **257,** *742*

Y

Yellow
- Boisduval's, **123**, *396*
- Chamberlain's, *394*
- Dina, *394*
- Dwarf, **109**, **112**, *398*
- Fairy, **68**, **124**, *395*
- Little, **107**, **129**, *392*
- Mexican, **106**, *397*
- Mimosa, **122**, *393*
- Salome, *397*

Yvretta
carus, **188**, *785*
rhesus, *784*

Z

Zebra Swallowtail, **48**, **340**, **343**, *347*
Dark, **341**, **344**, *349*

Zerene
cesonia, **98**, **101**, *384*
eurydice, **99**, **102**, *385*

Zestusa dorus, **297**, *724*

Zizula cyna, *488*

BUTTERFLY SOCIETIES

Lepidopterists' Society
Department of Biology
University of Louisville
Louisville, Kentucky 40208

Lepidoptera Research Foundation
Santa Barbara Museum of
Natural History
2559 Puesta del Sol Road
Santa Barbara, California 93105

Xerces Society
Department of Zoology and Physiology
University of Wyoming
Laramie, Wyoming 82071

STAFF

Prepared and produced by
Chanticleer Press, Inc.

Founding Publisher: Paul Steiner
Publisher: Andrew Stewart

Staff for this book:

Editor-in-Chief: Gudrun Buettner
Executive Editor: Susan Costello
Senior Editor: Jane Opper
Text Editor: Olivia Beuhl
Assistant Editor: Ann Whitman
Art Director: Carol Nehring
Production: Helga Lose, Amy Roche
Picture Library: Edward Douglas
Silhouettes and Maps: Paul Singer
Drawings: Vichai Malikul, M. J. Spring

Original series design by
Massimo Vignelli

All editorial inquiries should be
addressed to:
Chanticleer Press
665 Broadway, Suite 1001
New York, NY 10012

To purchase this book or other
National Audubon Society illustrated
nature books, please contact:
Alfred A. Knopf
1745 Broadway
New York, NY 10019
(800) 733-3000
www.randomhouse.com

NATIONAL AUDUBON SOCIETY
FIELD GUIDE SERIES

Also available in this unique all-color, all photographic format:

African Wildlife

Birds *(Eastern Region)*

Birds *(Western Region)*

Fishes

Fossils

Insects and Spiders

Mammals

Mushrooms

Night Sky

Reptiles and Amphibians

Rocks and Minerals

Seashells

Seashore Creatures

Trees *(Eastern Region)*

Trees *(Western Region)*

Tropical Marine Fishes

Weather

Wildflowers *(Eastern Region)*

Wildflowers *(Western Region)*

Part I
Color Plates

Key to the Color Plates

The color plates on the following pages are divided into 15 groups:

Eggs, Caterpillars, and Chrysalises
White Butterflies
Sulphurs
Folded-winged Skippers
Spread-winged Skippers
Swallowtails
Angled-winged Butterflies
Hairstreaks and Elfins
Blue Butterflies
Copper Butterflies
Metalmarks
Checkered Butterflies
Fritillaries and Orange Patterned Butterflies
Boldly Patterned Butterflies
Eyespot Patterned Butterflies

Thumb Tab Guide: To help you find the correct group, a table of silhouettes precedes the color plates. Each group is represented by a silhouette of a typical member of that group on the left side of the table. On the right, you will find the silhouettes of butterflies found within that group.

The representative silhouette for each group is repeated as a thumb tab at the left edge of each double page of color plates, providing a quick and convenient index to the color section.

Thumb Tab	Group	Plate Numbers
	Eggs	1–12
	Caterpillars and Chrysalises	13–51
	White Butterflies	52–84
	Sulphurs	85–132
	Folded-winged Skippers	133–240

Typical Shapes	Butterflies	Plate Numbers
	eggs	1–12
	caterpillars and chrysalises	13–51
	whites, orangetips, marblewings, Fairy Yellow, and Amymone	52–58; 60–69; 73–81; 83, 84
	parnassians	70, 71
	skippers	59, 72, 82
	sulphurs and dogface butterflies	85–132
	roadside skippers, skipperlings, and other hesperiine skippers	133–193, 198–236, 238–240
	giant skippers	194–197, 237

Thumb Tab	Group	Plate Numbers
	Spread-winged Skippers	241–315
	Swallowtails	316–360
	Angled-winged Butterflies	361–381

Typical Shapes	Butterflies	Plate Numbers
	checkered skippers, sootywings, duskywings, cloudywings, and other pyrgine skippers	241–305, 307–309
	tailed skippers	306, 310–315
	black, tiger, giant, and other swallowtails	316, 318, 319, 321, 322, 324–360
	Red-spotted Purple and Diana	317, 320, 323
	anglewings, leafwings, tortoiseshells, question marks, and commas	361, 362, 367–380
	daggerwings	363, 364, 381
	snouts	365, 366

Thumb Tab	Group	Plate Numbers
	Hairstreaks and Elfins	382–462
	Blue Butterflies	463–510
	Copper Butterflies	511–540
	Metalmarks	541–555
	Checkered Butterflies	556–591

Typical Shapes	Butterflies	Plate Numbers
	hairstreaks, Eastern Tailed Blue, Tailed Copper, and Hermes Copper	382–448, 460–462
	elfins	449–459
	blues, hairstreaks, and Blue Copper	463–490, 492–510
	Blue Metalmark	491
	coppers, blues, hairstreaks, and Harvester	511–540
	metalmarks	541–555
	checkerspots and crescentspots	556–568, 571–576, 578–590
	metalmarks	569, 570, 577, 591

Thumb Tab	Group	Plate Numbers
	Fritillaries and Orange Patterned Butterflies	592–642
	Boldly Patterned Butterflies	643–687
	Eyespot Patterned Butterflies	688–759

Typical Shapes	Butterflies	Plate Numbers
	fritillaries, crescentspots, and checkerspots	599–642
	Julia, Gulf Fritillary, Queen, Monarch, Viceroy, and Chiricahua Pine White	592–598
	admirals, sisters, patches, Mourning Cloak, Diana, Milbert's Tortoiseshell, Malachite, and others	645–660, 667, 673–675, 679, 681–687
	hackberries, painted ladies, and Red Admiral	661–666, 668–672
	longwings, Elf, Pixie, and Crimson-banded Black	643, 644, 677, 678, 680
	Golden-banded Skipper	676
	browns, pearly eyes, satyrs, arctics, alpines, and ringlets	694–757
	buckeyes, calicoes, and White Peacock	688–693, 758, 759

The color plates on the following pages are numbered to correspond with the numbers preceding the text descriptions. The caption under each photograph gives the plate number, common name, measurement, and page number of the text description. The measurement in inches indicates wingspan for butterflies and maximum body length for caterpillars (*ct*) and chrysalises (*ch*). Egg measurements, given in millimeters, are maximum size. In some cases male (♂), female (♀), wings above, or dorsal, (*d*), and wings below, or ventral, (*v*), are also indicated. The color plate number is repeated at the beginning of each text description.

Eggs, Caterpillars, and Chrysalises

Butterfly eggs vary greatly in size, color, and shape. Some eggs are laid singly; others are deposited in clusters of several hundred. Caterpillars are usually long and cylindrical. The coloring of many species blends with the flowers and leaves on which they feed, while others have bold patterns and bright shades. Most chrysalises are irregularly shaped and resemble leaves, thorns, or bits of wood.

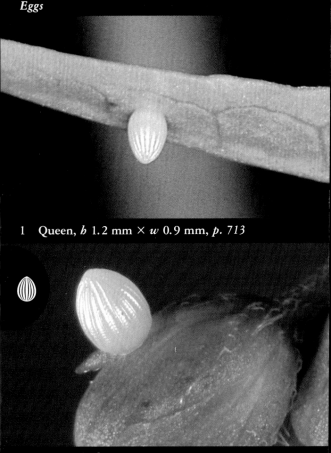

1   Queen, *h* 1.2 mm × *w* 0.9 mm, *p. 713*

2   Monarch, *h* 1.2 mm × *w* 0.9 mm, *p. 711*

3   Silver-bordered Fritillary, *h* 0.9 mm, *p. 559*

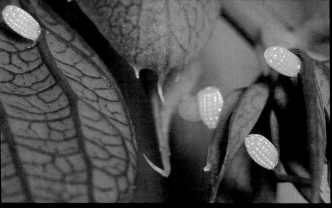

4 Zebra Longwing, *h* 1.2 mm × *w* 0.7 mm, *p. 541*

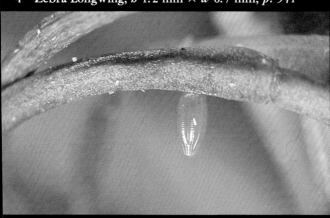

5 Falcate Orangetip, *h* 1.0 mm × *w* 0.4 mm, *p. 369*

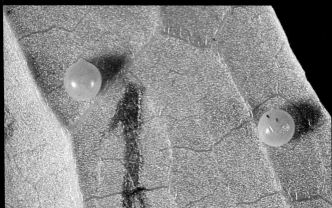

6 Tiger Swallowtail, *h* 0.8 mm × *w* 1.2 mm, *p. 340*

7   Hessel's Hairstreak, *h* 0.3 mm × *w* 0.6 mm, *p. 439*

8   Long-tailed Skipper, *w* 0.1 mm, *p. 725*

9   Question Mark, *h* 1.0 mm × *w* 0.7 mm, *p. 609*

10 Mourning Cloak, *h* 0.9 mm × *w* 0.7 mm, *p. 621*

11 Baltimore, *h* 0.8 mm × *w* 0.6 mm, *p. 602*

12 Tawny Emperor, *h* 0.6 mm × *w* 0.5 mm, *p. 657*

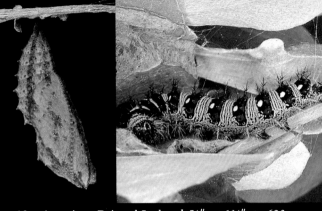

13   American Painted Lady, *ch* ⅞″, *ct* 1⅜″, *p. 623*

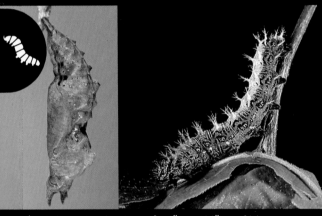

14   Sylvan Anglewing, *ch* ⅞″, *ct* 1¼″, *p. 614*

15   Question Mark, *ch* ⅞″, *ct* 1⅝″, *p. 609*

16   Silver-bordered Fritillary, *ch* ½″, *ct* ⅝″, *p. 559*

17   Mylitta Crescentspot, *ch* ⅜″, *ct* ⅞″, *p. 579*

18   Painted Lady, *ch* ⅞″, *ct* 1¼″, *p. 625*

19 Gulf Fritillary, *ch* 1⅛″, *ct* 1½″, *p. 538*

20 Variegated Fritillary, *ch* ¾″, *ct* 1¼″, *p. 542*

21 Baltimore, *ch* ¾″, *ct* 1″, *p. 602*

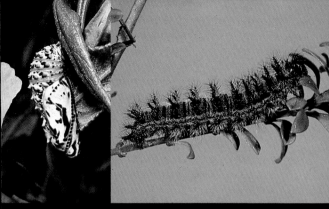

22 Chalcedon Checkerspot, *ch* ¾″, *ct* 1″, *p. 603*

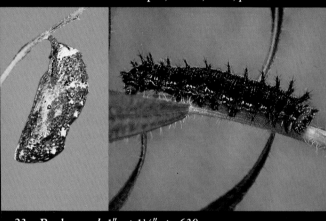

23 Buckeye, *ch* 1″, *ct* 1¼″, *p. 630*

24 Mourning Cloak, *ch* ⅞″, *ct* 2″, *p. 621*

25   Pipevine Swallowtail, *ch* 1⅛″, *ct* 2⅛″, *p.* 325

26   Viceroy, *ch* ⅞″, *ct* 1¼″, *p.* 637

27   Giant Swallowtail, *ch* 1⅝″, *ct* 2⅜″, *p.* 337

28 **Phoebus Parnassian,** *ct 1″, p. 324*

29 **Atala,** *ct 1″, p. 419*

30 **Moss Elfin,** *ct ½″, p. 426*

31 California Sister, *ct* 1¼", *p. 642*

32 Cloudless Giant Sulphur, *ct* 1¾", *p. 387*

33 Queen, *ct* 2", *p. 713*

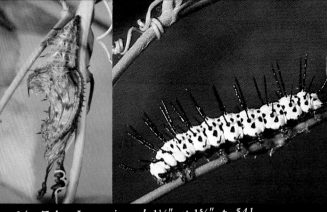

34   Zebra Longwing, *ch* 1⅛″, *ct* 1⅝″, *p. 541*

35   Monarch, *ch* ⅞″, *ct* 2″, *p. 711*

36   Anise Swallowtail, *ch* 1¼″, *ct* 2″, *p. 334*

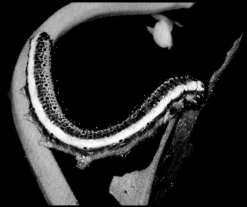

37 Falcate Orangetip, *ct* ⅞", *p. 369*

38 Eastern Pine Elfin, *ct* ⅝", *p. 430*

39 Wright's Metalmark, *ct* ⅞", *p. 522*

40 Creamy Marblewing, *ch* ¾″, *ct* ¾″, *p.* 362

41 Hessel's Hairstreak, *ch* ⅜″, *ct* ⅝″, *p.* 439

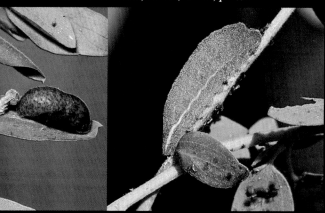

42 Hedgerow Hairstreak, *ch* ⅜″, *ct* ½″, *p.* 459

43   Blue Copper, *ch* ⅜″, *ct* ⅝″, *p. 406*

44   Common Blue, *ch* ⁵⁄₁₆″, *ct* ⅜″, *p. 510*

45   Hackberry Butterfly, *ch* ⅞″, *ct* 1¼″, *p. 653*

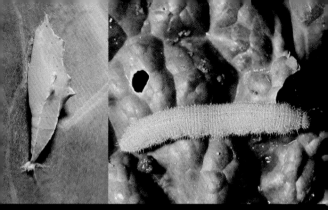

46   Cabbage White, *ch* ¾″, *ct* ¾″, *p.* 360

47   Long-tailed Skipper, *ch* ⅞″, *ct* 1¼″, *p.* 725

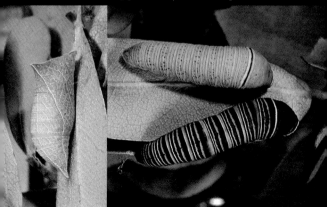

48   Zebra Swallowtail, *ch* 1″, *ct* 2⅛″, *p.* 347

49 Tiger Swallowtail, *ch* 1¼", *ct* 2", *p. 340*

50 Palamedes Swallowtail, *ch* 1⅝", *ct* 2", *p. 346*

51 Spicebush Swallowtail, *ch* 1¼", *ct* 1⅝", *p. 345*

White Butterflies

These familiar butterflies have white wings, often with black borders or black and white checkered fore wing tips. Most of them are members of the white and sulphur family, except for the parnassians, whose white or yellowish wings have black checkering as well as red spots. The orangetips have orange fore wing tips; the marblewings have greenish-yellow marbling below. Three white skippers and the Amymone also appear in this group.

52   Great Southern White ♂, 1¾–2¼″, *p. 361*

53   Great Southern White ♀, 1¾–2¼″, *p. 361*

54   Veined White, 1½–1⅝″, *p. 358*

55   Giant White ♂, 2½–2⅞″, *p. 362*

56   Cabbage White *d* ♀, 1¼–1⅞″, *p. 360*

57   Falcate Orangetip ♀, 1⅜–1½″, *p. 369*

58 Pearly Marblewing, 1¼–1⅜″, *p. 365*

59 Large White Skipper ♂, 1⅛–1⅝″, *p. 766*

60 Checkered White ♀, 1¼–1¾″, *p. 355*

61 Pine White *v* ♀ *d* ♂, 1¾–2″, *p. 351*

62 Western White, 1¼–1¾″, *p. 357*

63 Checkered White *v* ♂ *d* ♀, 1¼–1¾″, *p. 355*

64 Spring White, 1¼–1½″, *p. 355*

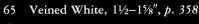

65 Veined White, 1½–1⅝″, *p. 358*

66 Chiricahua Pine White ♂, 1⅞–2¼″, *p. 352*

67   West Virginia White, 1⅜–1⅝″, *p. 359*

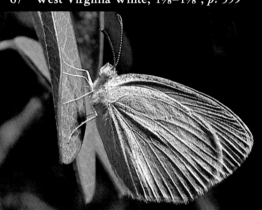

68   Fairy Yellow ♂ (*summer*), 1–1⅜″, *p. 395*

69   Florida White, 1⅝–2⅜″, *p. 352*

70 Clodius Parnassian ♀, 2⅜–3″, *p. 323*

71 Phoebus Parnassian ♂, 2⅛–3″, *p. 324*

72 Arsalte Skipper *d* ♀, 1⅜–1⅝″, *p. 767*

73   Amymone, 1⅜–1⅝″, *p. 645*

74   Falcate Orangetip ♂, 1⅜–1½″, *p. 369*

75   Sara Orangetip, 1¼–1⅝″, *p. 368*

76 Desert Orangetip ♂, 1⅛–1½″, *p. 366*

77 Creamy Marblewing, 1½–1¾″, *p. 362*

78 Olympia Marblewing, 1½–1¾″, *p. 364*

79 Becker's White, 1⅜–1⅞″, *p. 353*

80 Olympia Marblewing, 1½–1¾″, *p. 364*

81 Checkered White, 1¼–1¾″, *p. 355*

82 Large White Skipper, 1⅛–1⅝″, *p. 766*

83 Gray Marble, 1⅝–1¾″, *p. 370*

84 Great Southern White, 1¾–2¼″, *p. 361*

Sulphurs

Sulphurs range in hue from pale
yellowish-white to bright yellow,
orange, or yellowish-green. Several have
black borders above. Others bear
delicate pink fringes and pink-rimmed
spots below. Many sulphurs are
common in meadows and parks, but a
few species fly only in cold, high
mountains.

85 Common Sulphur ♀ (*albino*), 1⅜–2″, *p. 371*

86 Orange Sulphur ♀ (*albino*), 1⅝–2⅜″, *p. 372*

87 Blueberry Sulphur ♀, 1¼–1½″, *p. 381*

88 Thula Sulphur, 1⅛–1⅝″, *p. 378*

89 Greenland Sulphur ♂, 1⅜–1⅞″, *p. 376*

90 Palaeno Sulphur ♀, 1⅜–1¾″, *p. 379*

91   Behr's Sulphur ♀, 1¼–1⅝″, *p. 379*

92   Behr's Sulphur ♂, 1¼–1⅝″, *p. 379*

93   Scudder's Willow Sulphur, 1½–1¾″, *p. 383*

94   Labrador Sulphur, 1⅛–1⅝″, *p. 377*

95   Pink-edged Sulphur, 1⅜–1¾″, *p. 380*

96   Mead's Sulphur , 1¼–1½″, *p. 375*

97   Orange Sulphur , 1⅝–2⅜″, *p. 372*

98   Dogface Butterfly, 1⅞–2½″, *p. 384*

99   California Dogface, 1⅝–2½″, *p. 385*

100 Orange Sulphur, 1⅛–2⅜″, *p. 372*

101 Dogface Butterfly, 1⅞–2½″, *p. 384*

102 California Dogface ♀, 1⅝–2½″, *p. 385*

103　　Queen Alexandra's Sulphur ♀, 1½–1⅞″, *p. 373*

104　　Lyside, 1½–2″, *p. 391*

105　　Yellow Angled Sulphur, 2¾–3½″, *p. 386*

106　Mexican Yellow ♂, 1⅜–1⅞″, *p. 397*

107　Little Yellow, 1–1½″, *p. 392*

108　Argante Giant Sulphur, 2¼–2½″, *p. 389*

109 Dwarf Yellow ♀, ¾–1⅛″, *p. 398*

110 Queen Alexandra's Sulphur ♂, 1½–1⅞″, *p. 373*

111 Common Sulphur ♂, 1⅜–2″, *p. 371*

112 Dwarf Yellow, ¾–1⅛″, *p. 398*

113 Queen Alexandra's Sulphur ♂, 1½–1⅞″, *p. 373*

114 Common Sulphur, 1¾–2″, *p. 371*

115   Cloudless Giant Sulphur, 2⅛–2¾″, *p. 387*

116   Orange-barred Giant Sulphur, 2¾–3¼″, *p. 388*

117   Sleepy Orange, 1⅜–1⅞″, *p. 394*

118 Cloudless Giant Sulphur, 2⅛–2¾″, *p. 387*

119 Orange-barred Giant Sulphur, 2¾–3¼″, *p. 388*

120 Sleepy Orange, 1⅜–1⅞″, *p. 394*

121   Orange Giant Sulphur ♂, 2¼–2½″, *p. 390*

122   Mimosa Yellow ♀ ♂, 1–1¼″, *p. 393*

123   Boisduval's Yellow ♀ ♂, 1⅛–1⅝″, *p. 396*

124   Fairy Yellow, 1–1⅜″, *p. 395*

125   Tailed Orange ♂ *(winter)*, 1⅜–1¾″, *p. 392*

126   Tailed Orange ♀ *(winter)*, 1⅜–1¾″, *p. 392*

127 Statira, 2¼–2½″, *p. 390*

128 Tailed Orange ♂ (*summer*), 1⅜–1¾″, *p. 392*

129 Little Yellow ♂, 1–1½″, *p. 392*

130 Orange Sulphur ♂, 1⅝–2⅜″, *p. 372*

131 Western Sulphur, 1½–2″, *p. 374*

132 Pink-edged Sulphur ♂, 1¼–1¼″, *p. 391*

Folded-winged Skippers

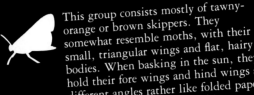

This group consists mostly of tawny-orange or brown skippers. They somewhat resemble moths, with their small, triangular wings and flat, hairy bodies. When basking in the sun, they hold their fore wings and hind wings at different angles rather like folded paper airplanes. The larger giant skippers, which are usually black or brown with yellow markings and frosty scaling, are also included here.

133　Ottoe Skipper, 1¼–1⅝″, *p.* 790

134　Rural Skipper ♂, ¾–1″, *p.* 814

135　Delaware Skipper, 1–1⅜″, *p.* 811

136 Southern Skipperling *d* ♂, ½–¾″, *p. 782*

137 Orange Skipperling, ¾–⅞″, *p. 781*

138 Delaware Skipper ♂, 1–1⅜″, *p. 811*

139   Sachem ♂, 1–1⅜″, *p. 799*

140   Sandhill Skipper ♂, ¾–1⅛″, *p. 801*

141   Draco Skipper ♂, ⅞–1″, *p. 803*

142 Least Skipperling *d* ♂, ¾–1″, *p. 778*

143 European Skipper *d* ♂, ¾–1″, *p. 783*

144 Rural Skipper *v* ♂, ¾–1″, *p. 814*

145 Sunrise Skipper, ⅞–1⅛″, *p. 782*

146 Florida Swamp Skipper ♀, 1⅛–1½″, *p. 826*

147 Crossline Skipper ♂, 1–1⅛″, *p. 805*

148   Yehl Skipper ♂, 1⅛–1½″, *p. 820*

149   Ottoe Skipper ♀, 1¼–1⅝″, *p. 790*

150   Broken Dash ♂, 1–1¼″, *p. 808*

151   Whirlabout ♀, 1–1¼″, *p.* 807

152   Columbian Skipper ♂, ⅞–1⅜″, *p.* 793

153   Leonardus Skipper, ⅞–1⅜″, *p.* 791

154 Fiery Skipper ♂, 1–1¼″, *p. 784*

155 Whirlabout ♂, 1–1¼″, *p. 807*

156 Common Branded Skipper ♂, ⅞–1″, *p. 788*

157   Zabulon Skipper ♂, 1–1⅜″, *p. 818*

158   Mulberry Wing *v* ♂, 1–1⅛″, *p. 817*

159   Hobomok Skipper ♂, 1–1⅜″, *p. 817*

160 Golden Skipper ♂, 1¼″–1⅜″, *p. 819*

161 Yellowpatch Skipper *d* ♂, ¾–1″, *p. 800*

162 Long Dash *d* ♀ *v* ♂, 1–1¼″, *p. 806*

163 Woodland Skipper *d* ♂, ¾–1⅛″, *p. 813*

164 Umber Skipper *d* ♂, 1⅛–1⅜″, *p. 822*

165 Sonora Skipper *v* ♀, 1–1¼″, *p. 807*

166 Indian Skipper ♂, 1–1⅜″, *p. 797*

167 Bunchgrass Skipper ♂, 1¼–1½″, *p. 812*

168 Sachem, 1–1⅜″, *p. 799*

169 Broad-winged Skipper, 1¼–1¾″, *p. 821*

170 Scarce Swamp Skipper, 1¼–1½″, *p. 825*

171 Saffron Skipper, 1–1½″, *p. 819*

172 **Black Dash** *v* ♂, 1–1⅜″, *p. 825*

173 **Northern Broken Dash** *d* ♂, 1–1¼″, *p. 809*

174 **Tawny-edged Skipper**, ¾–1″, *p. 804*

175 Bunchgrass Skipper ♀, 1¼–1½″, *p. 812*

·176 Sonora Skipper *d* ♂ *v* ♀, 1–1¼″, *p. 807*

177 Dakota Skipper *d* ♀ *v* ♂♀, 1–1⅜″, *p. 796*

178 Dixie Skipper ♂, 1⅛–1⅜", *p. 795*

179 Obscure Skipper ♀, 1–1¼", *p. 842*

180 Fawn-spotted Skipper, 1–1¼", *p. 776*

181   Sandhill Skipper ♀, ¾–1⅛″, *p. 801*

182   Nevada Skipper *d* ♂ *v* ♀, 1–1⅜″, *p. 798*

183   Common Branded Skipper, ⅞–1″, *p. 788*

184 Green Skipper, 1–1⅜″, *p. 794*

185 Common Branded Skipper, ⅞–1″, *p. 788*

186 Pahaska Skipper, 1⅛–1½″, *p. 792*

187   Lace-winged Roadside Skipper, 1–1¼″, *p.* 835

188   Desert Gray Skipper, ⅞–1¼″, *p.* 785

189   Uncas Skipper, 1–1⅝″, *p.* 787

190   **Pepper-and-salt Skipper**, ⅞–1″, *p. 834*

191   **Bronze Roadside Skipper**, ⅞–1⅛″, *p. 833*

192   **Cobweb Skipper** ♂, 1–1⅜″, *p. 793*

193   Sierra Skipper ♂, 1⅛–1⅜″, *p.* 797

194   California Giant Skipper, 2–2⅛″, *p.* 850

195   Orange Giant Skipper ♀, 2–2⅜″, *p.* 847

196 Aryxna Giant Skipper, 2–2⅜″, *p. 847*

197 Brigadier, 2″, *p. 849*

198 Violet-banded Skipper, 1⅜–1⅝″, *p. 845*

199   Bell's Roadside Skipper, ⅞–1¼″, *p. 837*

200   Spotted Skipperling, ¾–1″, *p. 773*

201   Apache Skipper, 1⅛–1⅝″, *p. 789*

202 Twin-spot Skipper, 1¼–1⅜″, *p. 840*

203 Snow's Skipper, 1⅛–1⅜″, *p. 815*

204 Brazilian Skipper ♂ ♂, 1¾–2¼″, *p. 841*

205 Twin-spot Skipper, 1¼–1⅜″, *p. 840*

206 Little Glassywing, 1–1¼″, *p. 809*

207 Palmetto Skipper ♂, 1⅜–1¾″, *p. 822*

208 Russet Skipperling *d* ♂, ⅞–1″, *p. 772*

209 Wandering Skipper ♂, 1–1¼″, *p. 843*

210 Salt Marsh Skipper, 1¼″, *p. 841*

211　Baracoa Skipper ♂, ¾–1″, *p. 803*

212　Golden Skipper ♀, 1¼–1⅜″, *p. 819*

213　Clouded Skipper ♂, 1–1⅜″, *p. 777*

214 Sedge Skipper ♀, 1¼–1⅝″, *p. 824*

215 Three-spotted Skipper *d* ♀ *v* ♂, ⅞–1⅛″, *p. 775*

216 Dun Skipper ♀, 1–1¼″, *p. 827*

217 Long-winged Skipper, 1¼–1⅜″, *p. 843*

218 Monk, 1⅝–2″, *p. 828*

219 Saw-grass Skipper ♂, 1½–1¾″, *p. 823*

220　Swarthy Skipper ♂, ⅞–1″, *p. 774*

221　Julia Skipper ♂♀, 1–1⅛″, *p. 775*

223 **Beard-grass Skipper** *d* ♂, ⅞–1¼″, *p. 810*

224 **Two-spotted Skipper** *d* ♀, 1⅛–1¼″, *p. 827*

225 **Eufala Skipper,** ⅞–1¼″, *p. 839*

226 Poweshiek Skipperling, 1–1¼″, *p. 779*

227 Garita Skipperling, ¾–1″, *p. 780*

228 European Skipper *d* ♂, ¾–1″, *p. 783*

229   Blue-dusted Roadside Skipper, ¾–1″, *p. 838*

230   Malicious Skipper, 1–1¼″, *p. 773*

231   Purple-washed Skipper, 1⅜″, *p. 844*

232   Roadside Skipper, ⅞–1″, *p. 836*

233   Hobomok Skipper *v* ♀, 1–1⅜″, *p. 817*

234   Dusted Skipper, 1¼–1½″, *p. 829*

235   Zabulon Skipper ♀, 1–1⅜″, *p. 818*

236   Green-backed Skipper, 1¾–2⅛″, *p. 777*

237   Yucca Giant Skipper, 2–2⅞″, *p. 851*

238 Violet-patch Skipper, ¾–1″, *p.* 774

239 Clouded Skipper ♂, 1–1⅜″, *p.* 777

240 Dun Skipper ♂, 1–1¼″, *p.* 827

Spread-winged Skippers

 These skippers tend to perch with their wings held open or partly open in the same plane. They include duskywings, cloudywings, sootywings, and checkered skippers, as well as several long-tailed skippers.

241   Arsalte Skipper *d* ♀, 1⅜–1⅝″, *p. 767*

242   Large White Skipper *d* ♂, 1⅛–1⅝″, *p. 766*

243   Tropical Checkered Skipper, 1–1⅜″, *p. 764*

244 White Patch, 1¼–1½″, *p. 746*

245 White Checkered Skipper ♂, ⅞–1⅛″, *p. 764*

246 Desert Checkered Skipper, ⅞–1⅛″, *p. 765*

247 Two-banded Checkered Skipper, ¾–1⅛″, *p. 760*

248 Small Checkered Skipper, ⅝–1″, *p. 762*

249 Common Streaky Skipper, ¾–1″, *p. 768*

250 Alpine Checkered Skipper, ⅞–1¼″, *p. 759*

251 Common Checkered Skipper, ¾–1¼″, *p. 762*

252 Arctic Skipper, ¾–1¼″, *p. 771*

253 Golden-banded Skipper, 1⅝–2″, *p.* 730

254 Horace's Duskywing ♀, 1¼–1¾″, *p.* 752

255 Hermit, 1¾–2¼″, *p.* 745

256 Hoary Skipper, 1⅛–1⅜″, *p. 742*

257 Window-winged Skipper, 1⅛–1⅜″, *p. 742*

258 Sickle-winged Skipper, 1⅝–1⅞″, *p. 744*

259 Arizona Powdered Skipper, 1–1½″, *p. 744*

260 Texas Powdered Skipper, 1–1¼″, *p. 743*

261 Variegated Skipper, ⅞–1⅛″, *p. 741*

262 Scalloped Sootywing ♂, 1–1¼″, *p. 741*

263 Southern Scalloped Sootywing ♂, 1–1¼″, *p. 740*

264 Golden-headed Sootywing ♂, 1–1⅛″, *p. 740*

265 Florida Duskywing ♀, 1¼–1⅝", *p. 747*

266 Rocky Mountain Duskywing ♀, 1¼–1⅝", *p. 750*

267 Propertius Duskywing, 1¼–1¾", *p. 751*

268 Sleepy Duskywing ♀, 1⅛–1⅝″, *p. 748*

269 Juvenal's Duskywing ♂, 1¼–1¾″, *p. 749*

270 Horace's Duskywing ♀, 1¼–1¾″, *p. 752*

271 Sleepy Duskywing ♂, 1⅛–1⅝″, *p. 748*

272 Wild Indigo Duskywing ♂, 1⅛–1⅝″, *p. 757*

273 Mottled Duskywing ♀, 1–1⅜″, *p. 753*

274 Dreamy Duskywing, 1–1⅜″, *p.* 748

275 Pacuvius Duskywing, 1⅛–1½″, *p.* 754

276 Zarucco Duskywing ♂, 1⅛–1¾″, *p.* 755

277　Funereal Duskywing, 1⅛–1¾″, *p. 756*

278　Mournful Duskywing, 1¼–1⅝″, *p. 753*

279　Blue-banded Skipper, 1⅛–1⅜″, *p. 746*

280   Florida Duskywing ♂, 1¼–1⅝″, *p. 747*

281   Purplish Black Skipper, 1¼–1½″, *p. 739*

282   Common Sootywing, ⅞–1¼″, *p. 769*

283   Northern Cloudywing, 1¼–1¾″, *p. 734*

284   Eastern Cloudywing *v ♂*, 1¼–1⅝″, *p. 735*

285   Southern Cloudywing, 1¼–1⅝″, *p. 733*

286 Northern Cloudywing, 1¼–1¾″, *p. 734*

287 Potrillo Skipper, 1⅛–1½″, *p. 737*

288 Mexican Cloudywing, 1¼–1½″, *p. 735*

289 Caicus Skipper, 1⅜–1⅝″, *p. 739*

290 Coyote Skipper, 1⅝–2″, *p. 732*

291 Acacia Skipper, 1½–1⅞″, *p. 738*

292   Mercurial Skipper, 2¼–2⅝″, *p. 719*

293   Laviana Skipper, 1¼–1⅝″, *p. 767*

294   Hoary Edge, 1½–1¾″, *p. 731*

295   Great Basin Sootywing, 1–1⅜″, *p. 769*

296   Araxes Skipper, 1¾–2½″, *p. 717*

297   Short-tailed Arizona Skipper, 1½–1⅝″, *p. 724*

298 Hammock Skipper, 1¾–2″, *p. 721*

299 Guava Skipper, 2–2½″, *p. 718*

300 Mangrove Skipper, 2–2½″, *p. 717*

301 Mercurial Skipper, 2¼–2⅝″, *p. 719*

302 Gold-spot Aguna, 2–2¼″, *p. 723*

303 Falcate Skipper, 1⅝–1¾″, *p. 737*

304 Mimosa Skipper, 1¼–1¾", *p. 738*

305 Arizona Skipper, 1⅝–2¼", *p. 725*

306 White-tailed Skipper, 1⅛–1¾", *p. 728*

307 Dull Astraptes, 2–2¼", *p. 729*

308 Silver-spotted Skipper, 1¾–2⅜", *p. 719*

309 Flashing Astraptes, 1⅞–2⅜", *p. 729*

310 Long-tailed Skipper, 1½–2″, *p. 725*

311 Teleus Longtail, 1⅝–1⅞″, *p. 727*

312 Brown Longtail, 1⅝–1⅞″, *p. 728*

313 Zilpa Longtail, 1½–1⅞″, *p. 723*

314 White-striped Longtail, 1¾–2″, *p. 722*

315 Lilac-banded Longtail, 1½–2″, *p. 726*

Swallowtails

Swallowtails are easily recognized by their large wings and long tails. All have patterns of contrasting light and dark colors—yellow and black, white and black, or blue and black. Often their wings also bear conspicuous orange and blue spots. The Red-spotted Purple and the female Diana are included in this group because they mimic the Pipevine Swallowtail.

316   Spicebush Swallowtail ♂, 3½–4½", *p. 345*

317   Red-spotted Purple ♂, 3–3⅜", *p. 636*

318   Pipevine Swallowtail ♂, 2¾–3⅜", *p. 325*

319   Spicebush Swallowtail ♂, 3½–4½″, p. 345

320   Red-spotted Purple ♂, 3–3⅜″, p. 636

321   Pipevine Swallowtail ♀, 2¾–3⅜″, p. 325

322 Tiger Swallowtail ♀ (*dark*), 3⅛–5½″, *p. 340*

323 Diana ♀, 3–3⅞″, *p. 544*

324 Eastern Black Swallowtail ♀, 2⅝–3½″, *p. 327*

325   Desert Swallowtail ♀ (*dark*), 2⅝–2¾″, *p. 329*

326   Short-tailed Swallowtail, 2¾–3⅜″, *p. 330*

327   Thoas Swallowtail, 4–5½″, *p. 336*

328   Giant Swallowtail, 3⅜–5½″, *p. 337*

329   Polydamas Swallowtail ♂, 3–4″, *p. 326*

330   Short-tailed Black Swallowtail, 2⅛–3⅜″, *p. 335*

331 Giant Swallowtail, 3⅜–5½", p. 337

332 Polydamas Swallowtail ♂, 3–4", p. 326

333 Short-tailed Black Swallowtail, 2⅛–3⅜", p. 335

334 Western Black Swallowtail ♀, 3–3½″, *p. 331*

335 Eastern Black Swallowtail ♂, 2⅝–3½″, *p. 327*

336 Palamedes Swallowtail ♂, 3⅛–5½″, *p. 346*

337 Western Black Swallowtail ♂, 3–3½", *p. 331*

338 Eastern Black Swallowtail ♂, 2⅝–3½", *p. 327*

339 Palamedes Swallowtail ♂, 3⅛–5½", *p. 346*

340 Zebra Swallowtail ♀, 2⅜–3½", *p. 347*

341 Dark Zebra Swallowtail ♂, 2½–3½", *p. 349*

342 Pale Tiger Swallowtail ♂, 3–3¾", *p. 344*

343 Zebra Swallowtail ♀, 2⅜–3½″, *p. 347*

344 Dark Zebra Swallowtail ♂, 2½–3½″, *p. 349*

345 Pale Tiger Swallowtail ♂, 3–3¾″, *p. 344*

346 Two-tailed Tiger Swallowtail ♂, 3⅜–5⅛″, *p. 343*

347 Western Tiger Swallowtail ♂, 2¾–3⅞″, *p. 342*

348 Tiger Swallowtail ♂, 3⅛–5½″, *p. 340*

349 Two-tailed Tiger Swallowtail ♂, 3⅜–5⅛", *p. 343*

350 Western Tiger Swallowtail ♂, 2¾–3⅞", *p. 342*

351 Tiger Swallowtail ♂, 3⅛–5⅛", *p. 340*

352 Oregon Swallowtail ♂, 2⅝–3⅜″, *p. 332*

353 Anise Swallowtail ♀, 2⅝–3″, *p. 334*

354 Old World Swallowtail ♀, 2⅝–2¾″, *p. 333*

355 Oregon Swallowtail ♀, 2⅝–3⅜″, p. 332

356 Anise Swallowtail, 2⅝–3″, p. 334

357 Old World Swallowtail, 2⅝–2¾″, p. 333

358 Western Black (*"brucei"*), 3–3½", *p. 331*

359 Desert Swallowtail ♂ *(light)*, 2⅝– 2¾", *p. 329*

360 Two-tailed Tiger Swallowtail ♂, 3⅜–5⅛", *p. 343*

Angled-winged Butterflies

These butterflies have sharply angled, somewhat irregular wings. When the wings are folded at rest, the drab undersides resemble leaves or pieces of bark, making them difficult to see. The uppersides tend to be somewhat brighter shades of orange and rust-brown. This group includes anglewings, tortoiseshells, commas, and question marks, as well as leafwings, snouts, and daggerwings.

361   Florida Leafwing ♂, 2¾–3″, *p. 651*

362   Goatweed Butterfly, 2⅜–3″, *p. 652*

363   Ruddy Daggerwing, 2⅝–2⅞″, *p. 650*

364   Banded Daggerwing, 2–2⅜″, *p. 650*

365   Southern Snout, 1⅝–1⅞″, *p. 535*

366   Snout Butterfly, 1⅝–1⅞″, *p. 534*

367   California Tortoiseshell, 1⅞–2⅜″, *p. 620*

368   Hoary Comma, 1⅜–1⅝″, *p. 616*

369   Compton Tortoiseshell, 2½–2⅞″, *p. 619*

370   Zephyr Anglewing *v*♂, 1¾–2″, *p. 615*

371   Comma (*fall*), 1¾–2″, *p. 610*

372   Satyr Anglewing, 1¾–2″, *p. 611*

373   Question Mark (*fall*), 2⅜–2⅝″, *p. 609*

374   Question Mark (*summer*), 2⅜–2⅝″, *p. 609*

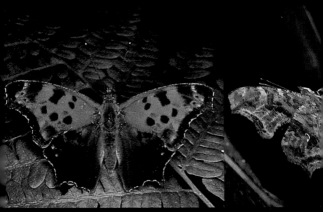

375   Comma (*summer*), 1¾–2″, *p. 610*

376 Gray Comma (*summer*), 1⅝–1⅞″, *p. 618*

377 Oreas Anglewing, 1⅝–2″, *p. 617*

378 Faunus Anglewing, 1⅞–2″, *p. 612*

379 Sylvan Anglewing, 2–2⅜″, *p. 614*

380 Laure, 2–2½″, *p. 660*

381 Waiter, 2–2⅝″, *p. 649*

Hairstreaks and Elfins

Both hairstreaks and elfins perch with the wings pulled back over the body, exposing the underside surface. Most hairstreaks have thin, hairlike tails projecting from the hind wings and delicate streaks below. Each hind wing often bears a burst of bright red, orange, or blue below. However, some hairstreaks lack tails and have faint markings on their undersides. Most elfins are mottled brown or purplish-gray. Some have tail stumps on the hind wing. Three butterflies that might be mistaken for hairstreaks—the Eastern Tailed Blue, the Tailed Copper, and the Hermes Copper—are also included here.

382   Colorado Hairstreak, 1⅜–1½″, *p. 417*

383   Martial Hairstreak ♂, 1″, *p. 481*

384   Acis Hairstreak, 1″, *p. 480*

385 Zebina Hairstreak ♂, 1⅛–1¼″, *p. 421*

386 Gray Hairstreak, 1–1¼″, *p. 472*

387 Northern Hairstreak, 1–1¼″, *p. 460*

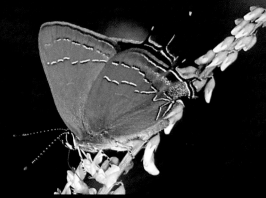

388 Southern Hairstreak, 1–1¼", *p. 460*

389 White M Hairstreak ♂, 1⅛–1½", *p. 422*

390 Mexican Gray Hairstreak, ⅞–1⅛", *p. 475*

391   Endymion Hairstreak, ¾–⅞″, *p. 462*

392   Dusky Blue Hairstreak, ¾–⅞″, *p. 468*

393   Red-banded Hairstreak, ¾–1″, *p. 467*

394 Banded Hairstreak, 1–1¼″, *p. 453*

395 Striped Hairstreak, 1–1⅜″, *p. 456*

396 Hickory Hairstreak, 1–1¼″, *p. 454*

397 King's Hairstreak, 1–1¼″, *p. 455*

398 Edwards' Hairstreak, 1–1¼″, *p. 452*

399 California Hairstreak, 1–1¼″, *p. 450*

400   Behr's Hairstreak, 7/8–1 1/8″, *p. 448*

401   Eastern Tailed Blue ♂, 3/4–1″, *p. 491*

402   Eastern Tailed Blue ♀, 3/4–1″, *p. 491*

403 Coral Hairstreak ♂, 1–1¼″, *p. 482*

404 Acadian Hairstreak ♂, 1⅛–1¼″, *p. 449*

405 Sylvan Hairstreak ♂, ⅞–1¼″, *p. 451*

406 Columella Hairstreak, 7/8–1", *p. 477*

407 Leda Hairstreak, 3/4–7/8", *p. 470*

408 Larger Lantana Butterfly, 7/8–1", *p. 471*

409   Early Hairstreak ♀, ¾–1″, *p. 466*

410   Arizona Hairstreak, ¾–1″, *p. 467*

411   Sooty Hairstreak, 1–1¼″, *p. 449*

412   **Aquamarine Hairstreak,** 1⅛″, *p. 423*

413   **Marsyas Hairstreak,** 1¾–2″, *p. 420*

414   **Avalon Hairstreak,** ¾–1″, *p. 474*

415 Mountain Mahogany ♂, 1⅛–1¼″, *p. 458*

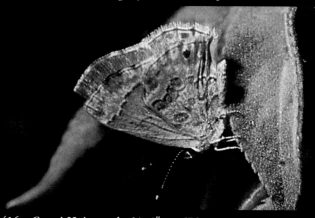

416 Cestri Hairstreak, ¾–1″, *p. 479*

417 Smaller Lantana Butterfly ¾–1″, *p. 479*

418   White Hairstreak, ⅞–1⅛″, *p. 482*

419   Yojoa Hairstreak, ⅞–1⅛″, *p. 476*

420   Reddish Hairstreak, ⅞″, *p. 478*

421 Angelic Hairstreak, ¾–⅞″, *p. 463*

422 Tailed Copper ♀, ⅞–1¼″, *p. 402*

423 Jade-blue Hairstreak, 1″, *p. 424*

424   Hedgerow Hairstreak ♂, 1–1⅛″, *p. 459*

425   Hermes Copper, 1–1⅛″, *p. 407*

426   Golden Hairstreak ♀, 1–1¼″, *p. 418*

427　Golden Hairstreak ♂, 1–1¼″, *p. 418*

428　Gold Hunter's Hairstreak ♂, 1–1¼″, *p. 457*

429　Goodson's Hairstreak ♀, ¾″, *p. 447*

430   **Immaculate Green Hairstreak,** 7/8–1 1/8″, *p. 444*

431   **Bramble Green Hairstreak,** 7/8–1 1/8″, *p. 440*

432   **Canyon Green Hairstreak,** 7/8–1 1/8″, *p. 443*

433   **Alpine Green Hairstreak,** ⅞–1¼″, *p. 442*

434   **Miserabilis Hairstreak,** ¾–1″, *p. 446*

435   Bluish Green Hairstreak, ⅞–1⅛″, *p. 444*

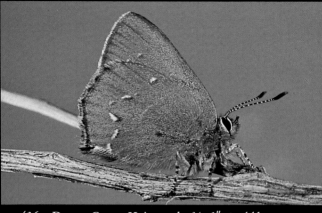

436   Desert Green Hairstreak, ¾–1″, *p. 441*

437   White-lined Green Hairstreak, ¾–⅞″, *p. 445*

438   Telea Hairstreak, ¾″, *p. 464*

439   Silver-banded Hairstreak ♂, ¾–⅞″, *p. 465*

440   Xami Hairstreak, ⅞–1″, *p. 433*

441   Juniper Hairstreak, ⅞–1⅛″, *p. 437*

442 Hessel's Hairstreak, ⅞–1″, *p. 439*

443 Olive Hairstreak, ⅞–1″, *p. 438*

444 Loki Hairstreak, ♀, ⅞–1″, *p. 438*

445 Loki Hairstreak ♂, ⅞–1", *p. 438*

446 Thicket Hairstreak, 1–1¼", *p. 434*

448   Nelson's Hairstreak, ⅞–1″, *p. 436*

449   Brown Elfin, ¾–1⅛″, *p. 424*

450   Moss Elfin, ⅞–1″, *p. 426*

451 Early Elfin, ⅞–1″, *p. 425*

452 Brown Elfin, ¾–1⅛″, *p. 424*

453 Hoary Elfin, ¾–1″, *p. 427*

454   Frosted Elfin, ⅞–1¼″, *p. 428*

455   Henry's Elfin ♀, ⅞–1⅛″, *p. 429*

456   Henry's Elfin ♂, ⅞–1⅛″, *p. 429*

457　**Bog Elfin,** ⅝–1″, *p. 429*

458　**Western Pine Elfin,** ¾–1¼″, *p. 431*

459　**Eastern Pine Elfin,** ¾–1¼″, *p. 430*

460 Great Purple Hairstreak, 1¼–1½″, *p. 421*

461 Atala ♀, 1¾″, *p. 419*

462 Cycad Butterfly, 1¾–2″, *p. 419*

Blue Butterflies

Several of these small- to medium-sized butterflies are familiar in gardens and woods. Most are blues, but several blue hairstreaks, a bluish copper, and a blue metalmark are also included here.

463 Greenish Blue ♂, ⅞–1¼″, *p. 507*

464 Sooty Azure ♂, ¾–1¼″, *p. 495*

465 Small Blue ♂, ½–⅝″, *p. 501*

466 Blue Copper ♂, 1⅛–1¼″, *p. 406*

467 Common Blue ♂, 1–1⅜″, *p. 510*

468 Silvery Blue *d* ♂, 1–1¼″, *p. 503*

469 Pale Blue ♀, ¾–⅞″, *p. 499*

470 Dotted Blue, ¾–1″, *p. 498*

471 Western Tailed Blue ♂, ⅞–1⅛″, *p. 492*

472 Orange-bordered Blue ♂, ⅞–1¼″, *p. 506*

473 Northern Blue ♂, ⅞–1¼″, *p. 505*

474 Reakirt's Blue ♂, ¾–1⅛″, *p. 490*

475   Xerces Blue ♂, 1⅛–1¼″, *p. 504*

476   Arrowhead Blue ♂, 1–1¼″, *p. 502*

477   Spring Azure, ¾–1¼″, *p. 493*

478 **Greenish Blue** *v* ♀ *d* ♂, ⅞–1¼″, **p. 507**

479 **Yukon Blue** ♂, ¾–⅞″, **p. 516**

480 **Miami Blue** *d* ♂, ¾–1⅛″, **p. 488**

481 Early Hairstreak ♀, ¾–1″, *p. 466*

482 Spring Azure ♀, ¾–1¼″, *p. 493*

483 High Mountain Blue ♂, ¾–1″, *p. 516*

484   Arizona Hairstreak *v* ♂ *d* ♀, ¾–1″, *p. 467*

485   Blackburn's Bluet *d* ♀, ⅞–1″, *p. 487*

486   Square-spotted Blue ♂, ¾–1″, *p. 497*

487 Marine Blue ♂, ⅝–1″, *p. 486*

488 Antillean Blue *d* ♂, ¾–1″, *p. 489*

489 Sonoran Blue *v* ♂ *d* ♀, ¾–⅞″, *p. 496*

490 **Eastern Tailed Blue** ♂, ¾–1″, *p. 491*

491 **Blue Metalmark**, ¾–1¼″, *p. 526*

492 **Blue Copper** ♀, 1⅛–1¼″, *p. 406*

493   San Emigdio Blue *d* ♂, ⅞–1⅛″, *p. 508*

494   Acmon Blue ♂, ¾–1″, *p. 512*

495   Lupine Blue *v* ♀ *d* ♂, ¾–1⅛″, *p. 514*

496 Northern Blue ♀, ⅞–1¼″, *p. 505*

497 Greenish Blue ♀, ⅞–1¼″, *p. 507*

498 Colorado Hairstreak ♂, 1⅜–1½″, *p. 417*

499   Shasta Blue ♀, ⅞–1″, *p. 511*

500   Orange-bordered Blue ♀, ⅞–1¼″, *p. 506*

501   Orange-bordered Blue ♀, ⅞–1¼″, *p. 506*

502　High Mountain Blue ♀, ¾–1″, *p. 516*

503　Reakirt's Blue ♀, ¾–1⅛″, *p. 490*

504　Antillean Blue ♀, ¾–1″, *p. 489*

505   Western Pygmy Blue *v ♂ d ♀*, ⅜–¾″, *p. 483*

506   Eastern Pygmy Blue ♂, ½–¾″, *p. 484*

507   High Mountain Blue *d ♀*, ¾–1″, *p. 516*

508 Cassius Blue ♀, ½–¾″, *p. 485*

509 Sooty Azure ♂, ¾–1¼″, *p. 495*

510 Cestri Hairstreak, ¾–1″, *p. 479*

Copper Butterflies

These butterflies usually have bright orange markings or bands. Many bear black spots on all wing surfaces. Most butterflies in this well-known group are called coppers; copper-colored blues and hairstreaks also appear here, along with the distinctive Harvester.

511   Ruddy Copper ♂, 1⅛–1¼″, *p. 405*

512   Lustrous Copper ♂, 1–1¼″, *p. 404*

513   American Copper, ⅞–1⅛″, *p. 403*

514 Bronze Copper ♂, 1¼–1⅜″, *p. 411*

515 Purplish Copper ♀, 1–1¼″, *p. 414*

516 Blue Copper ♀, 1⅛–1¼″, *p. 406*

517 Tailed Copper ♀, ⅞–1¼″, *p. 402*

518 Harvester ♂, 1⅛–1¼″, *p. 401*

519 Mariposa Copper *d* ♀ *v* ♂, 1–1⅛″, *p. 412*

520 Gorgon Copper ♀, 1⅛–1¼″, *p. 410*

521 Edith's Copper ♂ ♀, 1⅛–1¼″, *p. 409*

522 Ruddy Copper ♀, 1⅛–1¼″, *p. 405*

523   Dorcas Copper *v* ♂ *d* ♀, 1–1¼″, *p. 416*

524   Gorgon Copper ♂, 1⅛–1¼″, *p. 410*

525   Bog Copper ♂, ⅞–1″, *p. 414*

526　Purplish Copper ♂, 1–1¼″, *p. 414*

527　Bronze Copper *v* ♀ *d* ♂ , 1¼–1⅜″, *p. 411*

528　Great Gray Copper ♀, 1¼–1¾″, *p. 408*

529 Edith's Copper ♂, 1⅛–1¼″, *p. 409*

530 Bog Copper ♀, ⅞–1″, *p. 414*

531 Acmon Blue *v* ♂ *d* ♀, ¾–1″, *p. 512*

532 Square-spotted Blue ♀, ¾–1″, *p. 497*

533 Orange-veined Blue ♂, ¾–⅞″, *p. 515*

534 Behr's Hairstreak ♂, ⅞–1⅛″, *p. 448*

535   Sylvan Hairstreak ♂, ⅞–1¼″, *p. 451*

536   Great Gray Copper ♂, 1¼–1¾″, *p. 408*

537   Great Gray Copper ♂, 1¼–1¾″, *p. 408*

538 Northern Blue ♀, ⅞–1¼″, *p. 505*

539 Tailed Copper ♂, ⅞–1¼″, *p. 402*

540 Nivalis Copper ♂, 1⅛–1¼″, *p. 412*

Metalmarks

 Metalmarks get their names from the shiny metallic markings of many species. Most North American metalmarks are brown, gray, or rust, with black spots and checkered patterns.

541   Red-bordered Metalmark, ¾–1″, *p. 525*

542   Northern Metalmark, 1–1¼″, *p. 519*

543   Rawson's Metalmark, ¾–1″, *p. 523*

544 Lost Metalmark, ⅝–⅞″, *p. 521*

545 Fatal Metalmark, ¾–1″, *p. 520*

546 Lost Metalmark, ⅝–⅞″, *p. 521*

547　Wright's Metalmark ♂, ¾–1″, *p. 522*

548　Red-bordered Metalmark ♀, ¾–1″, *p. 525*

549　Little Metalmark ♂, ⅝–¾″, *p. 519*

550 Arizona Metalmark ♂, 1⅛″, *p. 524*

551 Swamp Metalmark, ⅞–1⅛″, *p. 522*

552 Little Metalmark ♀, ⅝–¾″, *p. 519*

553　**Ares Metalmark** ♀, 1¼–1½″, *p. 528*

554　**Ares Metalmark** ♂, 1¼–1½″, *p. 528*

555　**Zela Metalmark** ♂, 1⅛–1½″, *p. 527*

Checkered Butterflies

These butterflies have checkered patterns. Most are known as checkerspots, but this group also includes some crescentspots and metalmarks.

556　Anicia Checkerspot, 1⅛–1⅞″, *p. 606*

557　Chalcedon Checkerspot ♂, 1⅜–2″, *p. 603*

558　Fulvia Checkerspot ♀, 1⅛–1½″, *p. 594*

559   Leanira *d* ♀ *(desert)*, 1¼–1¾″, *p. 593*

560   Northern Checkerspot *d* ♂, 1¼–1⅝″, *p. 584*

561   Edith's Checkerspot, 1⅛–1⅞″, *p. 607*

562   Anicia Checkerspot, 1⅛–1⅞″, *p. 606*

563   Baltimore ♂, 1⅝–2½″, *p. 602*

564   Gillette's Checkerspot, 1⅜–1¾″, *p. 608*

565 Edith's Checkerspot, 1⅛–1⅞″, *p. 607*

566 Northern Checkerspot ♀, 1¼–1⅝″, *p. 584*

567 Cyneas Checkerspot, 1⅛–1⅜″, *p. 595*

568   Colon Checkerspot *v* ♂, 1⅜–1⅞″, *p. 605*

569   Mormon Metalmark *v* ♀, ¾–1¼″, *p. 528*

570   Mormon Metalmark *v* ♂, ¾–1¼″, *p. 528*

571   Chalcedon Checkerspot, 1⅜–2″, *p. 603*

572   Leanira Checkerspot ♂, 1¼–1¾″, *p. 593*

573   Texan Crescentspot, 1–1½″, *p. 570*

574   Phaon Crescentspot *d ♀*, ⅞–1¼″, *p. 572*

575   Tawny Crescentspot, 1¼–1½″, *p. 574*

576   Field Crescentspot *d ♀*, 1⅛–1⅜″, *p. 575*

577 Gray Metalmark *d* ♂, ¾–⅞″, *p. 530*

578 Definite Patch, 1–1⅜″, *p. 591*

579 Elada Checkerspot, ⅞–1⅛″, *p. 600*

580   Arachne Checkerspot ♂, 1¼–1⅝″, *p. 601*

581   Sagebrush Checkerspot *d* ♂, 1¼–1¾″, *p. 585*

582   Orseis Crescentspot ♂, 1¼–1½″, *p. 578*

583   Theona Checkerspot ♂, 1–1⅝″, *p. 596*

584   Vesta Crescentspot ♀, ¾–1⅛″, *p. 577*

585   Gabb's Checkerspot ♂, 1¼–1½″, *p. 587*

586   Aster Checkerspot, 1¼–1⅝″, *p. 589*

587   Rockslide Checkerspot ♂, 1⅛–1⅝″, *p. 587*

588   Harris' Checkerspot ♂, 1¼–1¾″, *p. 583*

589   Sagebrush Checkerspot ♀, 1¼–1¾″, *p. 585*

590   Field Crescentspot ♂, 1⅛–1⅜″, *p. 575*

591   Nais Metalmark, 1⅛–1½″, *p. 532*

Fritillaries and Orange Patterned Butterflies

Most butterflies in this group are fritillaries and crescentspots. Their tawny-orange wings bear many black zigzags, dots, crescents, and bars. A few checkerspots that might be confused with fritillaries also appear here. Other orange-patterned butterflies include the well-known Monarch, Viceroy, and Queen.

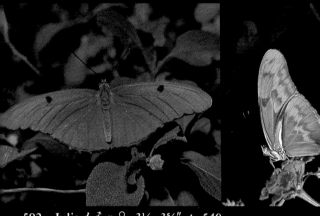

592   Julia *d* ♂ *v* ♀, 3⅛–3⅝″, *p. 540*

593   Gulf Fritillary *d* ♂, 2½–2⅞″, *p. 538*

594   Queen *d* ♂, 3–3⅜″, *p. 713*

595 Tropic Queen *d* ♂, 3⅛–3¼″, *p. 714*

596 Monarch ♂, 3½–4″, *p. 711*

597 Viceroy, 2⅝–3″, *p. 637*

598   Chiricahua Pine White ♀, 1⅞–2¼″, *p. 352*

599   Mexican Fritillary ♂, 1¾–2¼″, *p. 543*

600   Pearly Crescentspot ♂ *(type A)*, 1–1½″, *p. 573*

601 **Mexican Silverspot,** 2⅝–2⅞″, *p. 539*

602 **Desert Checkerspot** *d* ♂, 1¼–1¾″, *p. 586*

603 **Chara Checkerspot,** ¾–1⅛″, *p. 599*

604 Nokomis Fritillary *v* ♂ ♀ *d* ♂, 2¾–3″, *p. 548*

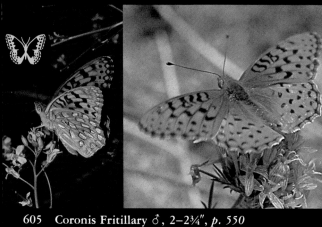

605 Coronis Fritillary ♂, 2–2¾″, *p. 550*

606 Mormon Fritillary *d* ♂, 1⅝–2″, *p. 557*

607 Zerene Fritillary *v* ♀, 1⅞–2½″, *p. 551*

608 Aphrodite ♂, 2–2⅞″, *p. 546*

609 Great Spangled Fritillary *d* ♂, 2⅛–3″, *p. 545*

Fritillaries and Orange Patterned Butterflies

610   Callippe Fritillary *d* ♂, 1⅞–2⅜″, *p. 552*

611   Hydaspe Fritillary ♂, 1¾–2⅜″, *p. 555*

612   Egleis Fritillary ♂, 1½–2⅜″, *p. 553*

613 Titania's Fritillary ♂ ♀, 1¼–1¾", *p. 568*

614 Meadow Fritillary ♂, 1¼–1⅞", *p. 560*

615 Western Meadow Fritillary ♀, 1⅜–1⅝", *p. 564*

616 Kriemhild Fritillary ♂, 1⅜–1¾″, *p. 564*

617 Aphrodite ♀, 2–2⅞″, *p. 546*

618 Atlantis Fritillary ♂, 1¾–2⅛″, *p. 554*

619　Titania's Fritillary ♂, 1¼–1¾″, *p. 568*

620　Mormon Fritillary ♀, 1⅝–2″, *p. 557*

621　Callippe Fritillary ♀, 1⅞–2⅜″, *p. 552*

622　Edwards' Fritillary *d* ♀, 2⅜–2¾″, *p. 549*

623　Aphrodite *d* ♀, 2–2⅞″, *p. 546*

624　Atlantis Fritillary, 1¾–2⅝″, *p. 554*

625 Silver-bordered Fritillary, 1⅜–2″, *p. 559*

626 Variegated Fritillary *d ♂*, 1¾–2¼″, *p. 542*

627 Mylitta Crescentspot, 1⅛–1⅜″, *p. 579*

628   Pearly Crescentspot ♀, 1–1½″, *p. 573*

629   Cuban Crescentspot ♂ ♀, 1⅛–1⅜″, *p. 571*

630   Pearly Crescentspot ♂ *(type B)*, 1–1½″, *p. 573*

631 Silvery Crescentspot, 1⅜–1¾″, p. 581

632 Gorgone Crescentspot ♂, 1⅛–1⅜″, p. 580

633 Bog Fritillary ♀, 1¼–1½″, p. 558

634   Polaris Fritillary ♂, 1¼–1½″, *p. 565*

635   Astarte Fritillary, 1⅜–1⅞″, *p. 568*

636   Dingy Arctic Fritillary, 1⅛–1⅝″, *p. 562*

637 Frigga's Fritillary, 1¼–1⅝″, *p. 561*

638 Napaea Fritillary *v* ♀ *d* ♂, 1⅛–1½″, *p. 557*

639 Arctic Fritillary *v* ♂ *d* ♀, 1¼–1⅜″, *p. 569*

640   Great Spangled Fritillary ♀, 2⅛–3″, *p. 545*

641   Great Spangled Fritillary ♀, 2⅛–3″, *p. 545*

642   Nokomis Fritillary ♀, 2¾–3″, *p. 548*

Boldly Patterned Butterflies

Butterflies in this group have bold stripes, spots, or patches across the upperside. Most admirals and sisters can be recognized by the large white "V" design. Hackberry butterflies, painted ladies, and the Red Admiral have boldly marked fore wing tips. This group also includes the Mourning Cloak, the Malachite, and the Diana, as well as a few tropical longwings and the Golden-banded Skipper.

643   Crimson-patched Longwing, 3–3⅜″, *p. 541*

644   Zebra Longwing, 3–3⅜″, *p. 541*

645   Blue Wing, 2⅞–3¼″, *p. 643*

646 Atala, 1¾″, *p. 419*

647 Mylitta Greenwing *d* ♂ *v* ♂♀, 1½–1⅞″, *p. 644*

648 Malachite, 2½–3″, *p. 634*

649 Weidemeyer's Admiral, 2¾–3⅜″, *p. 638*

650 White Admiral, 2⅞–3⅛″, *p. 635*

651 Lorquin's Admiral, 2¼–2¾″, *p. 639*

652   California Sister, 2⅞–3⅜″, *p. 642*

653   Laure ♀, 2–2½″, *p. 660*

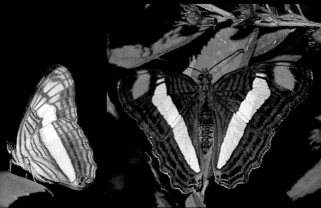

654   Mexican Sister, 2–2½″, *p. 641*

655 | Pavon ♂, 1¾–2⅜″, *p. 659*

656   Diana ♀, 3–3⅞″, *p. 544*

657   Diana ♂, 3–3⅞″, *p. 544*

658   Milbert's Tortoiseshell, 1¾–2″, *p. 622*

659   Bordered Patch *d* ♀, 1⅝–1⅞″, *p. 590*

660   California Patch, 1¼–1⅝″, *p. 589*

661   Empress Louisa ♀, 1⅞–2¼″, *p. 659*

662   Empress Leilia ♂, 1¾–2″, *p. 656*

663   Empress Antonia *v* ♂ *d* ♀, 1¾–2⅛″, *p. 655*

664 Hackberry Butterfly ♂, 1¾–2¼″, *p. 653*

665 Empress Flora ♂, 2–2¾″, *p. 658*

666 Tawny Emperor *v* ♂ *d* ♀, 1⅞–2⅜″, *p. 657*

667   Regal Fritillary, 2⅝–3⅝″, *p. 547*

668   West Coast Lady, 1¾–2″, *p. 626*

669   American Painted Lady, 1¾–2⅛″, *p. 623*

670 Painted Lady, 2–2¼″, *p. 625*

671 Kamehameha ♀, 2⅜–2¾″, *p. 628*

672 Red Admiral, 1¾–2¼″, *p. 627*

673 Mimic ♀, 2–2⅞", *p. 629*

674 Mimic ♂, 2–2⅞", *p. 629*

675 Bordered Patch ♂, 1⅝–1⅞", *p. 590*

676 Golden-banded Skipper, 1⅝–2″, *p. 730*

677 Elf, ¾–1⅜″, *p. 598*

678 Pixie, 1⅝–2″, *p. 526*

679   Amymone, 1⅜–1⅝″, *p. 645*

680   Crimson-banded Black, 2–2⅛″, *p. 646*

681   Mourning Cloak, 2⅞–3⅜″, *p. 621*

682   Fatima, 2–2⅛″, *p. 633*

683   Janais Patch, 1¾–2″, *p. 592*

684   Florida Purplewing, 1⅝–2″, *p. 643*

685   Great Purple Hairstreak, 1¼–1½″, *p. 421*

686   Blomfild's Beauty, 3⅜–3½″, *p. 648*

687   Eighty-eight Butterfly, 1⅝–1⅞″, *p. 645*

Eyespot Patterned Butterflies

These butterflies have conspicuous
eyespots on their somber brown or gray
wings. Most are satyrs, arctics, or
alpines, denizens of woodlands and
high mountains. A few more brightly
hued species, such as the White
Peacock and the Buckeye, are most
common in the South.

688   Buckeye, 2–2½″, *p. 630*

689   West Indian Buckeye, 2–2½″, *p. 632*

690   White Peacock, 2–2⅜″, *p. 633*

691 Buckeye, 2–2½″, *p. 630*

692 West Indian Buckeye, 2–2½″, *p. 632*

693 White Peacock, 2–2⅜″, *p. 633*

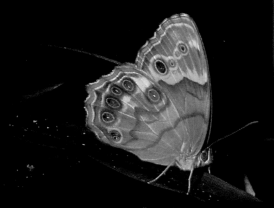

694 **Pearly Eye**, 1¾–2″, *p. 663*

695 **Pearly Eye** ♀, 1¾–2″, *p. 663*

696 **Creole Pearly Eye** ♂, 2–2¼″, *p. 665*

697   Pine Satyr, 1⅜–1¾″, *p. 677*

698   Large Wood Nymph, 2–2⅞″, *p. 683*

699   Dark Wood Nymph ♂, 1½–1⅞″, *p. 686*

700   Northern Pearly Eye, 1⅝–2″, *p. 664*

701   Eyed Brown, 1⅝–2″, *p. 666*

702   Appalachian Brown, 1⅝–2″, *p. 667*

703   Northern Pearly Eye, 1⅝–2″, *p. 664*

704   Eyed Brown, 1⅝–2″, *p. 666*

705   Appalachian Brown, 1⅝–2″, *p. 667*

706 Little Wood Satyr, 1¾–1⅞″, *p. 675*

707 Red Satyr, 1¾–1⅞″, *p. 676*

708 Great Basin Wood Nymph, 1⅜–2″, *p. 685*

709 Little Wood Satyr, 1¾–1⅞″, *p. 675*

710 Red Satyr, 1¾–1⅞″, *p. 676*

711 Great Basin Wood Nymph, 1⅜–2″, *p. 685*

712 Carolina Satyr, 1⅛–1⅝″, *p. 672*

713 Mitchell's Marsh Satyr, 1½–1¾″, *p. 674*

714 Georgia Satyr, 1½–1¾″, *p. 673*

715   **Hayden's Ringlet,** 1½–1⅞″, *p. 677*

716   **Gemmed Satyr,** 1¼–1⅜″, *p. 670*

717   **Canyonland Satyr,** 1½–1¾″, *p. 669*

718 Banded Alpine, 2–2¼″, *p. 691*

719 Ocher Ringlet, 1–1⅞″, *p. 680*

720 California Ringlet, 1–1¾″, *p. 682*

721   Prairie Ringlet, 1–1⅞″, *p. 679*

722   Kodiak Ringlet, 1–1⅜″, *p. 678*

723   Northwest Ringlet, 1–1⅞″, *p. 681*

724   Ocher Ringlet (*Alaska*), 1–1⅞″, p. 680

725   Large Wood Nymph ♀, 2–2⅞″, p. 683

726   Red-eyed Wood Nymph, 1⅝–2″, p. 684

727   Great Arctic, 2–2½″, p. 700

728   Uhler's Arctic, 1½–1⅞″, p. 703

729   Polixenes Arctic (*Alaska*), 1½–1¾″, p. 709

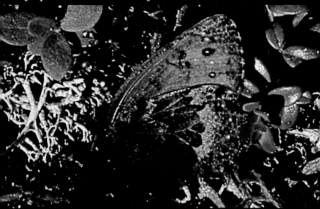

730   Polixenes Arctic ♀ (*Maine*), 1½–1¾″, *p. 709*

731   White-veined Arctic, 1¾–1⅞″, *p. 705*

732   Arctic Grayling, 1¾–1⅞″, *p. 706*

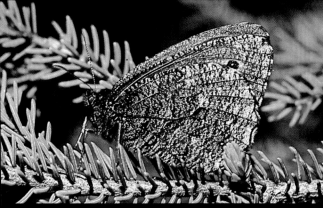

733 Jutta Arctic, 1⅞–2⅛", *p. 707*

734 California Arctic, 2–2⅛", *p. 699*

735 Melissa Arctic, 1⅝–1⅞", *p. 708*

736 Hermes Satyr, 1⅛–1½″, *p. 671*

737 Alberta Arctic ♂, ⅜–1⅞″, *p. 704*

738 California Arctic, 2–2⅛″, *p. 699*

739 Hermes Satyr, 1⅛–1½″, *p. 671*

740 Alberta Arctic ♂, ⅜–1⅞″, *p. 704*

741 California Arctic ♂, 2–2⅛″, *p. 699*

742 Riding's Satyr ♂, 1½–1⅞″, *p. 698*

743 Chryxus Arctic, 1¾–2″, *p. 702*

744 Canada Arctic, 2–2¼″, *p. 701*

745　Riding's Satyr, 1½–1⅞″, *p. 698*

746　Chryxus Arctic, 1¾–2″, *p. 702*

747　Canada Arctic, 2–2¼″, *p. 701*

748 Red-disked Alpine, 1¾–2″, *p. 692*

749 Spruce Bog Alpine ♀, 1¾–2″, *p. 689*

750 Common Alpine ♂, 1¾–2″, *p. 695*

751 Red-disked Alpine, 1¾–2″, *p. 692*

752 Spruce Bog Alpine ♀, 1¾–2″, *p. 689*

753 Common Alpine, 1¾–2″, *p. 695*

754   Theano Alpine, 1¼–1½″, *p. 693*

755   Red-bordered Brown, 2¼–2⅞″, *p. 697*

756   Magdalena Alpine, 2–2½″, *p. 690*

757 Northwest Alpine, 1¾–2″, *p. 688*

758 Ferentina Calico, 2–2⅝″, *p. 647*

759 Guatemalan Calico, 3⅛–3¼″, *p. 647*

Part II
Text

The numbers preceding the species descriptions in the following pages correspond to the plate numbers in the color section. If the description has no plate number, it is illustrated by a drawing that accompanies the text.